NERIKOMI

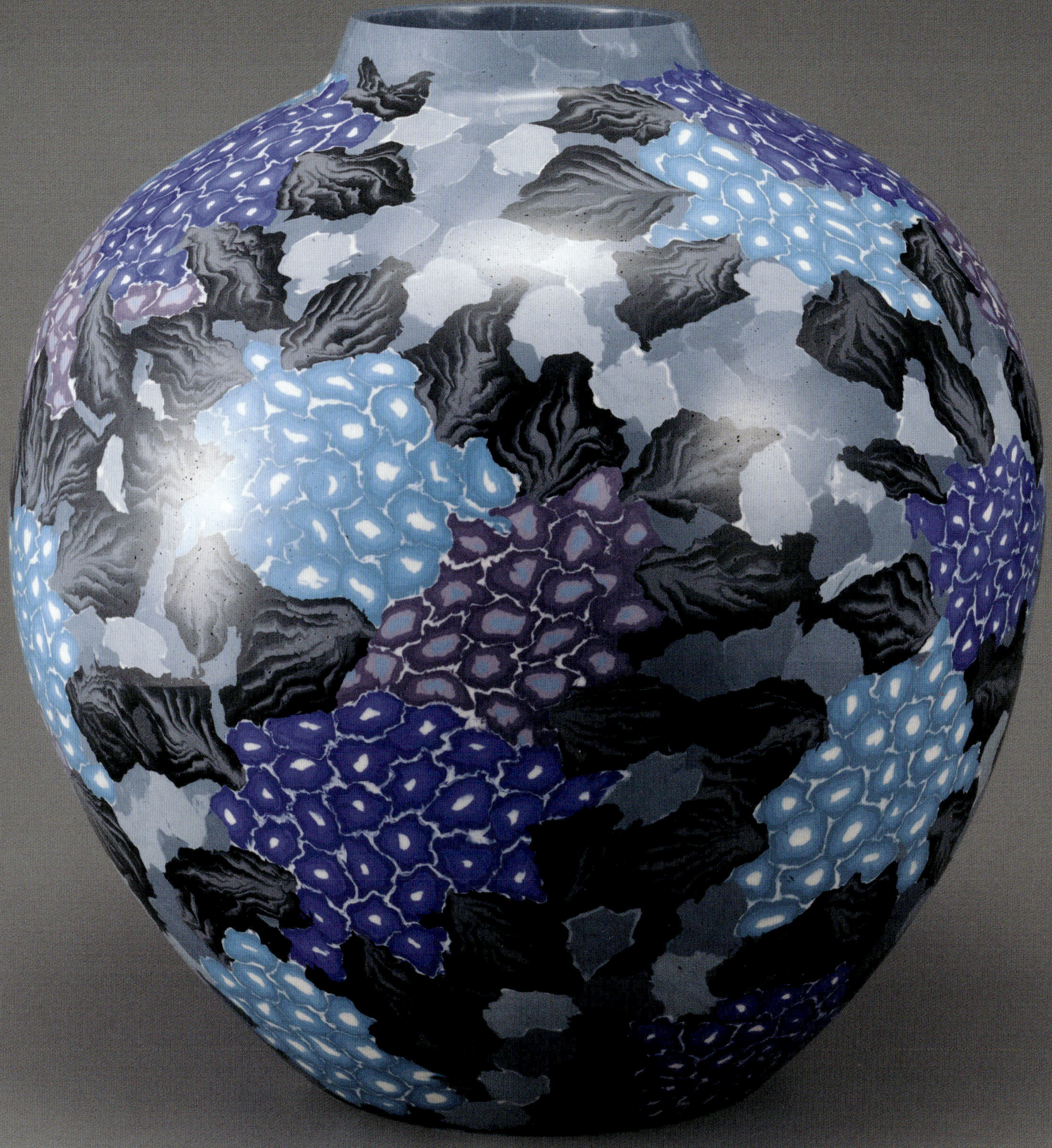

HERBERT PRESS
Bloomsbury Publishing Plc
50 Bedford Square, London, WC1B 3DP, UK
29 Earlsfort Terrace, Dublin 2, Ireland

Bloomsbury, Herbert Press and the Herbert Press logo are trademarks of Bloomsbury Publishing Plc

First published in Great Britain in 2024

A catalog record for this book is available from the British Library
Library of Congress Cataloguing-in-Publication data has been applied for

ISBN: 978-1-7899-4169-2; ePub: 978-1-7899-4170-8; epdf 987-1-7899-4171-5

2 4 6 8 10 9 7 5 3

Design by Laura Shaw Design, Inc.
Printed and bound in Italy by L.E.G.O. S.p.A.

FSC www.fsc.org MIX Paper | Supporting responsible forestry FSC® C023419

To find out more about our authors and books visit www.bloomsbury.com and sign up for our newsletters

Front cover: Large bowl with gold leaf by Thomas Hoadley (© Patina Gallery, Inc./2023). Back cover, upper to lower: Vase by Nichi Koichi (© The Stratford Gallery Ltd/2022) – see pages 46–7; *Caterpillar is dreaming series 6* (© Aya Murata) – see page 140; Landscape vessel (© Cody Hoyt) – see page 115

Page 1: *Nerijo glass, light hydrangea bundai, tsubo*, by Matsui Kōsei (see pages 26–30). Height: $14\frac{1}{3}$ inches, diameter: $14\frac{1}{4}$ inches (37.2 x 36.2 cm) © Ibaraki Museum

Opposite: *Nerikomi Vase #210*, by Robert Hessler (see pages 108–12), 2022. Height: 11 inches, diameter: $4\frac{1}{2}$ inches (28 x 11.4 cm)

Page 4: A vase by the author from 1988 (see details in caption on page 54)

Page 6: *Blue Geometric Charger*, by Dorothy Feibleman (see pages 96–102), 2017. Height $1\frac{3}{10}$ inches, diameter: $12\frac{1}{5}$ inches (3.3 x 31 cm). This charger is part of a tableware collection made for Nikko company, Japan. Cast bone china with "in glaze" nerikomi. The original imaging was constructed by hand and then made into an "in glaze" decal that melts a structure into the glaze. When you use cutlery, you do not damage the design because it is a structure in the glaze. This was a new way of using "in glaze" and nerikomi together. © Dorothy Feibleman; photo by Ken Yanoviak

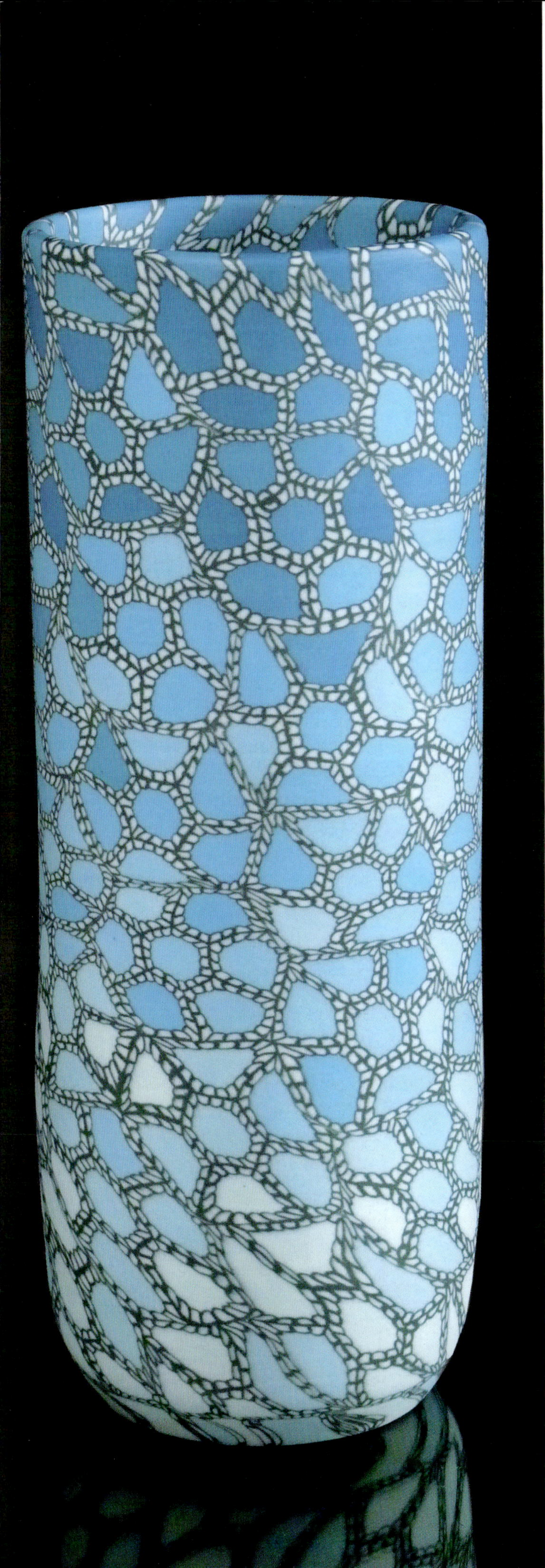

NERIKOMI

The Art of Colored Clay

THOMAS HOADLEY

HERBERT PRESS
LONDON • OXFORD • NEW YORK • NEW DELHI • SYDNEY

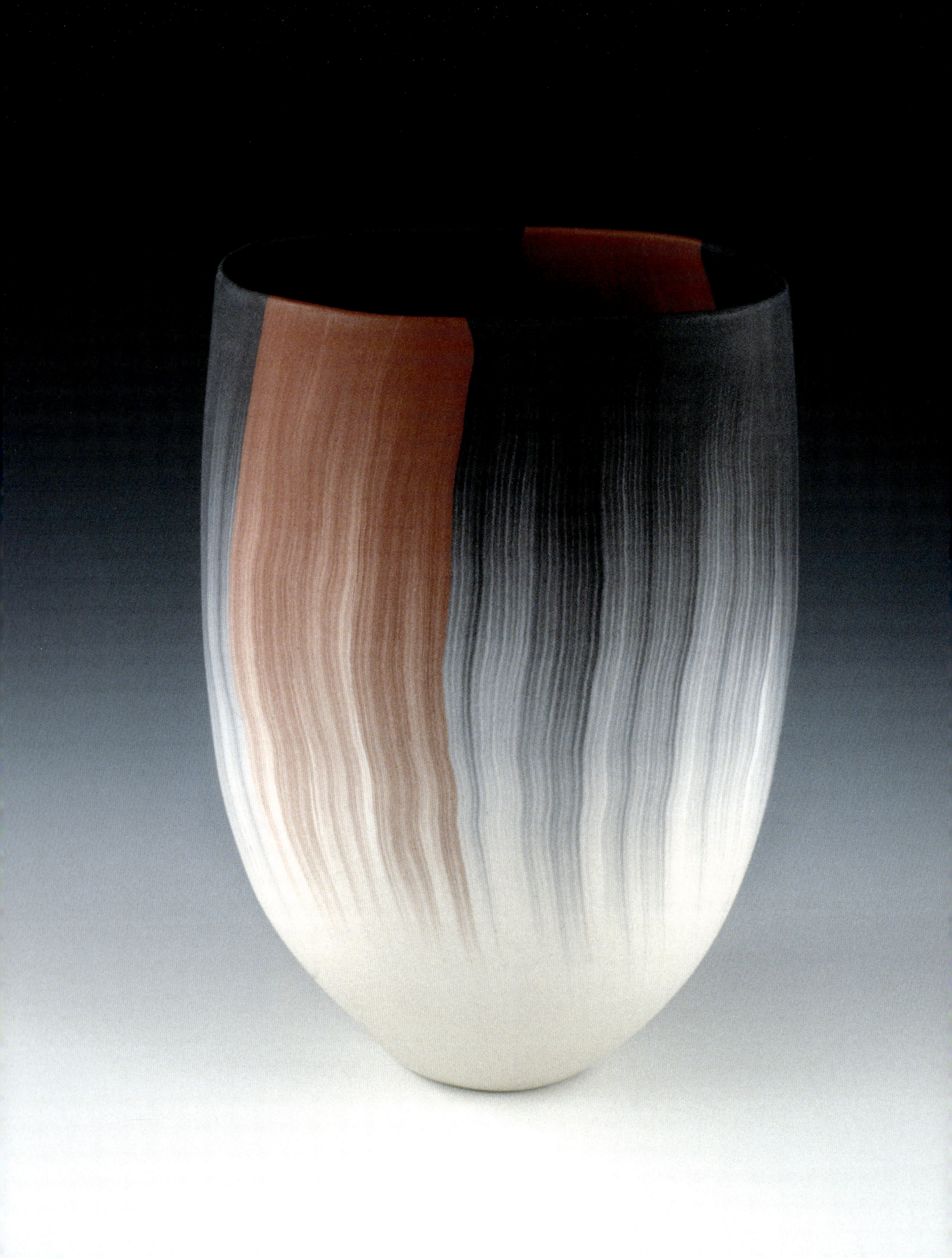

CONTENTS

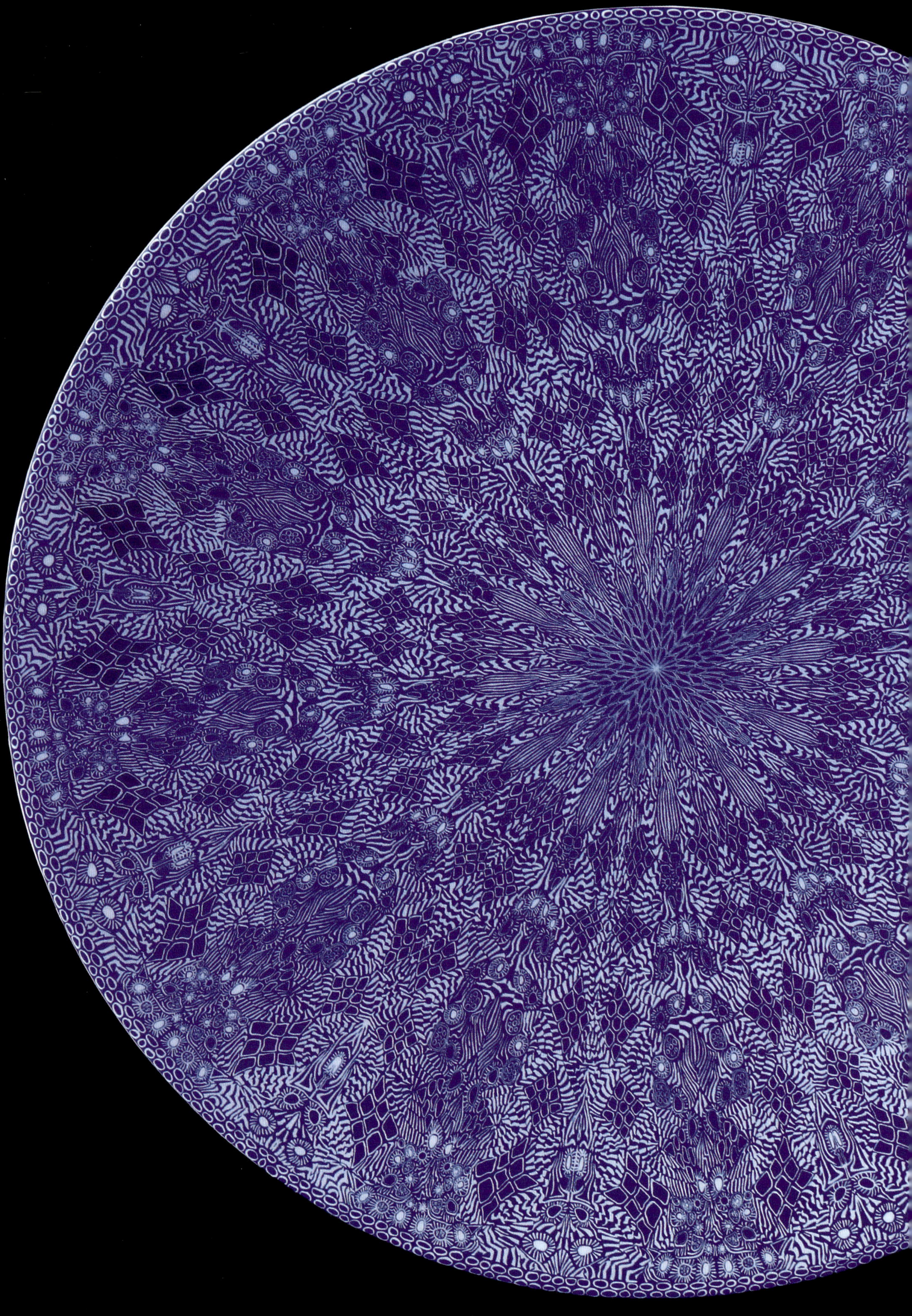

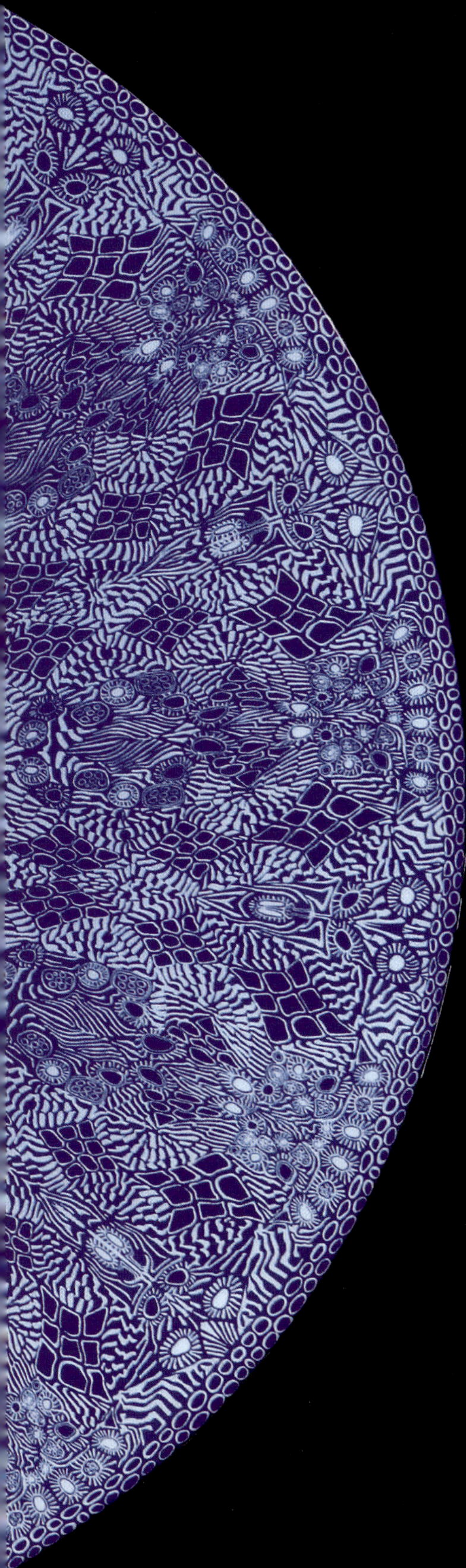

DEFINITIONS AND SCOPE

To be invited to write a book on a subject that has been one's lifelong work would, for most, be quite an honor. This is exactly what happened to me many months ago and since I had never written much more than a magazine article, I hesitated briefly before saying yes and taking the plunge. The following many months were both exciting and life changing, and the experience has broadened my perspective on ceramics in general, and colored clay work in particular.

When I first began using the Japanese nerikomi technique in a dedicated way as my life's work, I was faced with a linguistic dilemma. I had heard both the terms "nerikomi" and "neriage" seemingly being used interchangeably and felt I needed to understand the difference. I asked a Japanese acquaintance of mine if she could help. She told me that the term "neri" is a root word meaning "to mix," and the term "komi" means "to press into." Nerikomi, therefore, refers to mixed clays which are pressed into, presumably, molds, to make pottery forms. The term "age" means "to pull up," which indicates the word "neriage" refers to pottery made from mixed clays that are "thrown" on the wheel, the principal action being to "pull" the mass of clay upwards to create the form.

While researching this project I realized early on that I wanted to expand the scope of my search to include not only examples of the nerikomi technique but of neriage as well. Delving even deeper I discovered beautiful and exciting work being done with colored clay, but outside either of these two Japanese traditions. Artists were inventing remarkable methods with colored clay that, for me at least, were completely new. A few of the artists in this book fall into that exciting category.

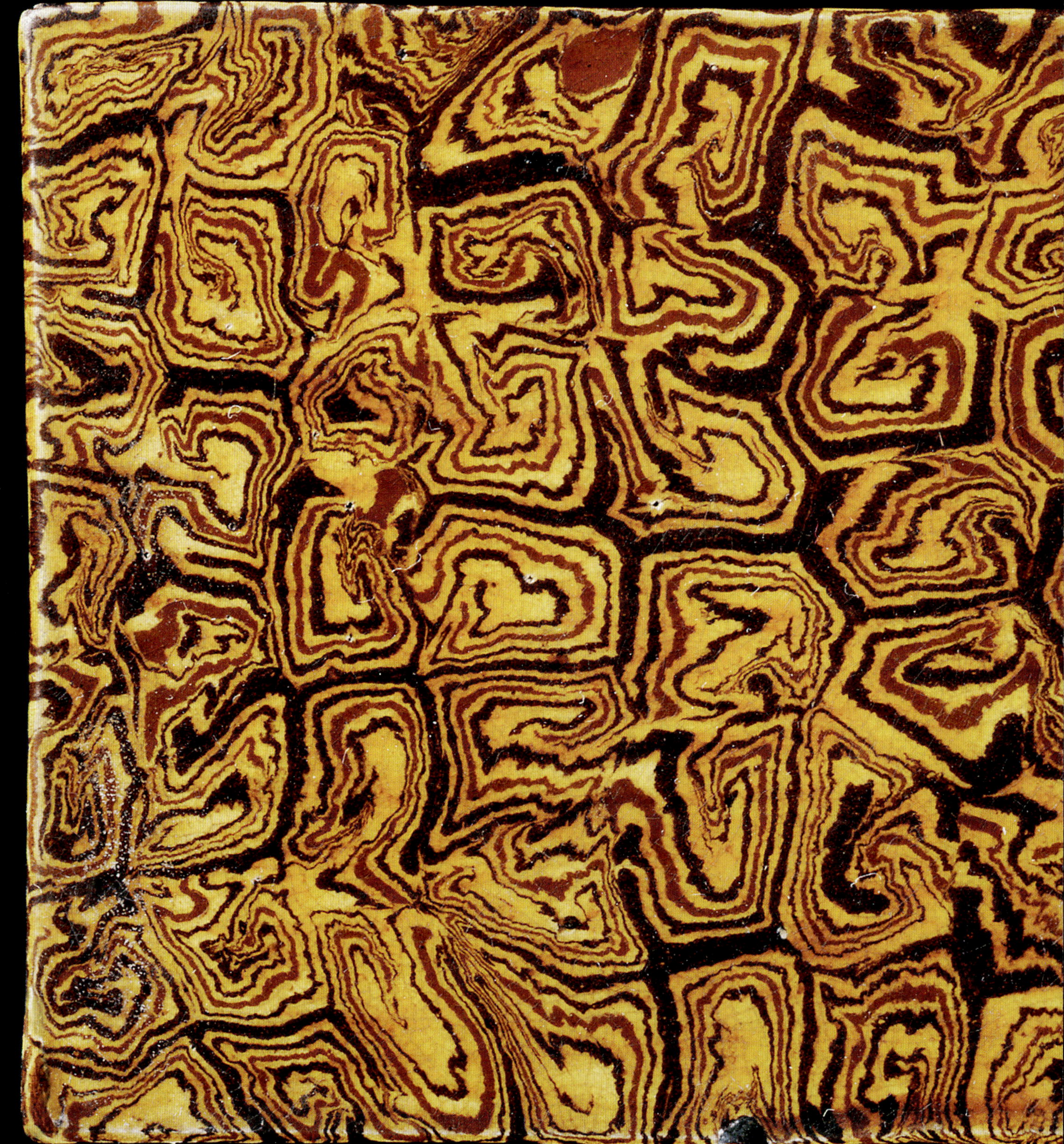

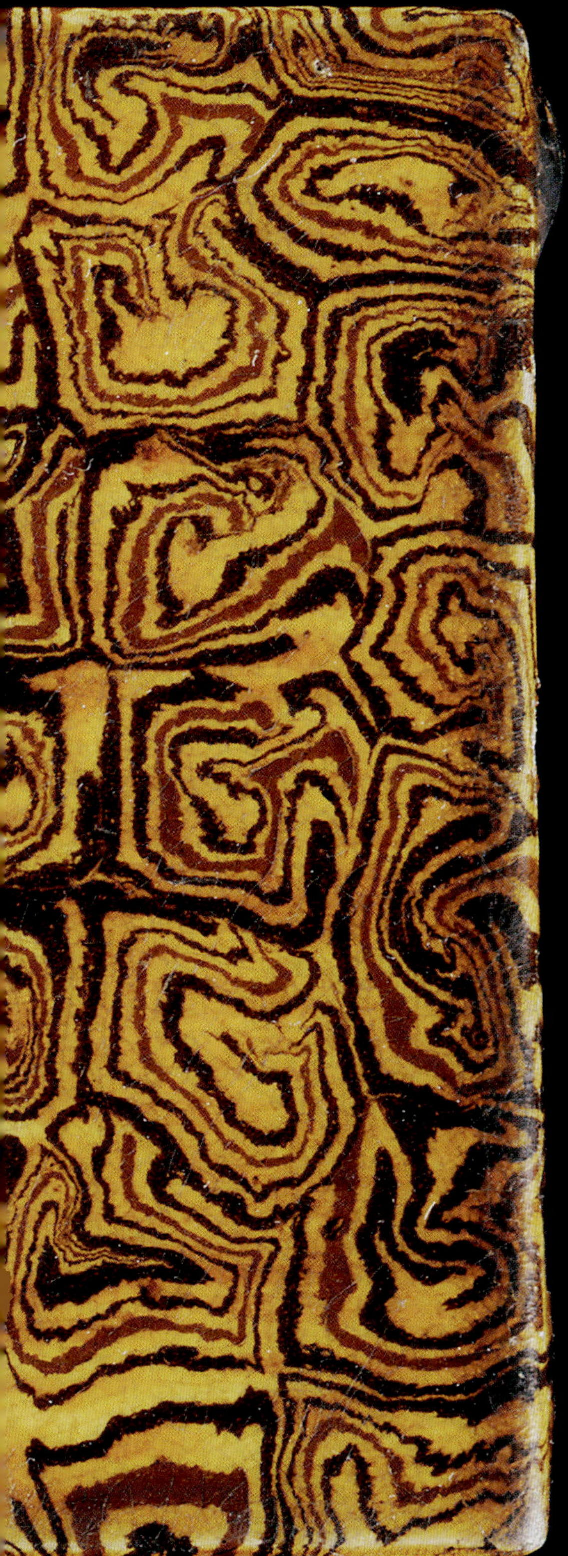

A COLORFUL HISTORY

Early Chinese and Korean Colored-Clay Ceramics

It can be said that the history of civilization was written in clay – both literally, in the form of cuneiform tablets, and figuratively, through the many beautifully preserved ceramic objects and vessels created in cultures around the world since the discovery of pottery, the transformation of naturally occurring clay into permanent objects through the method of firing. The focus of this chapter is to trace the development of colored clay techniques as they were used in ancient and more modern times, delving into influences from one culture to another, as well as similarities among varied crafts media.

Some of the earliest examples of the use of colored clay to create designs in pottery come from ancient China, particularly during the Tang Dynasty (618–907). An intriguing speculation suggests a link between this work and much earlier vessel production in glass, particularly from

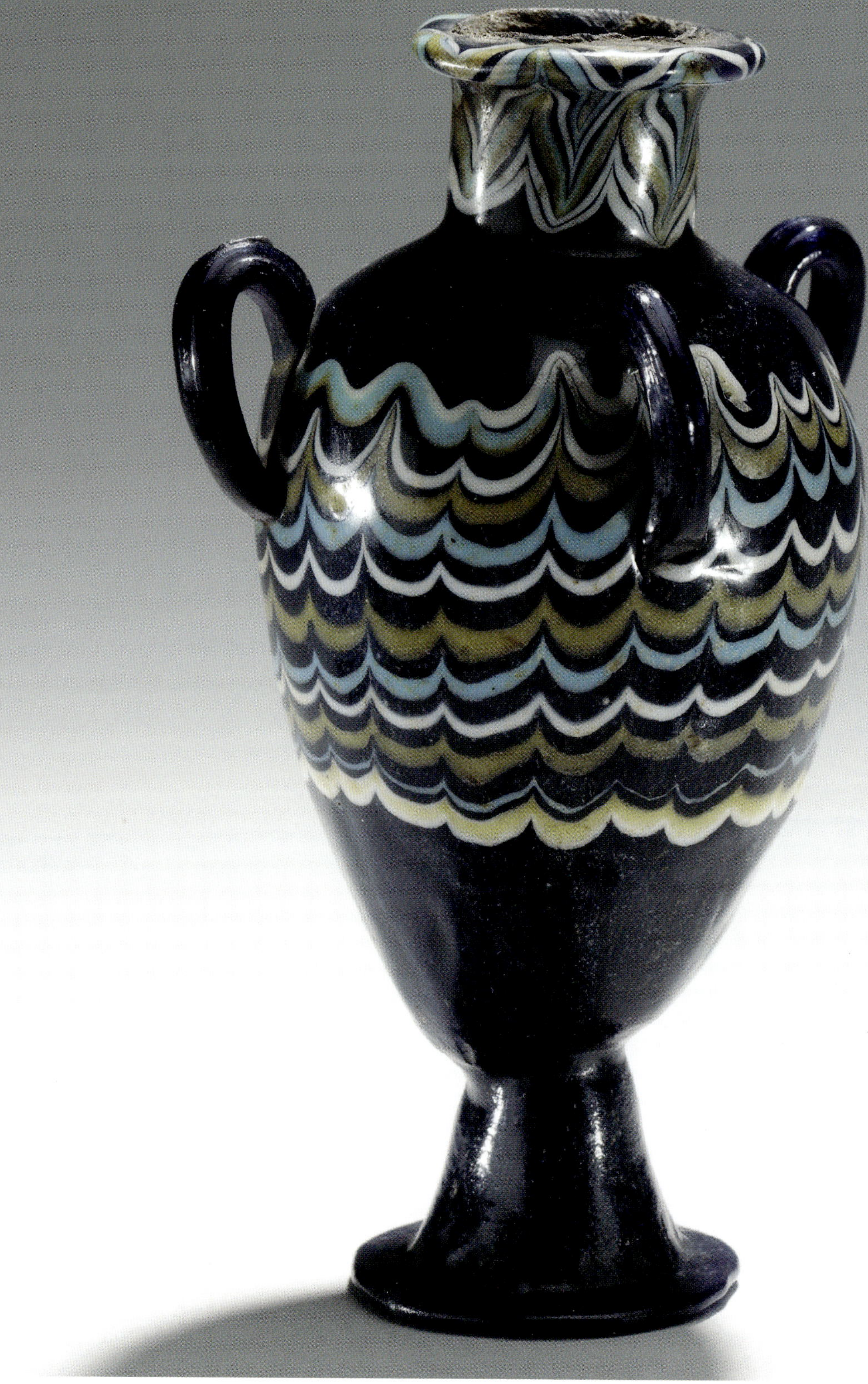

Egyptian cosmetic jar, Memphis, 18th Dynasty (1549–1292 BC). Height: 4 inches (10 cm)

Four-handled ovoid cosmetic jar, with a cylindrical neck and a hollow trumpet-shaped base with flat rim. The body is of opaque dark-blue glass, a color repeated in the four vertical monochrome handles. A chevron pattern in opaque yellow, turquoise-blue, and white decorates the neck and the flat rim. These colors are repeated in the simple festoons of the body. *Image: © Trustees of the British Museum*

◀ previous Detail of the top of the "Doctor's Pillow" shown on page 15. *Image: © 2024 Museum of Fine Arts, Boston*

Mesopotamia and Egypt, as well as Rome. As early as 2500 BCE glass technology had developed in the making of beads and amulets, and by the 1300s BCE ingots of colored glass were being traded in the Mediterranean. About that same time the first glass bottles were created by Egyptian craftsmen using the "core" technique. Predating actual blown glass, the core method utilized clay mixed with dung and plant fibers, shaped into a form and mounted on a rod. The core was then covered with glass (opinions vary as to the methods) and rolled on a stone slab to even out the shape. Softened rods of colored glass, or trails of more molten glass, were then wound around the form and further manipulated with a blade to create chevron designs in contrasting colors. After a prolonged cooling, the core was carefully extracted through abrasion, leaving traces where it was in contact with the glass. These tiny bottles were often used for oils and fragrances by the elite classes and found their way to China via the Silk Road.

A vast network of trade routes, more properly called the "Silk Roads," developed over time, reaching its height between the years 600 and 1000, to connect much of Eurasia and Africa. Caravans of camels and horses carried exotic goods – and ideas – in both directions, influencing cultures and traditions often in profound ways. Some of the most prominent goods being traded were silk, horses, paper, spices, jade, and glassware. Purple and multicolored Roman glass bowls have been discovered in Chinese tombs in Nanjing from the Eastern Jin Period (317–420). These were probably traded for silk, that magical fabric invented by the Chinese more than 1,000 years earlier.

In examining the patterns created with colored clay in the early Chinese examples, one can easily imagine how a potter, in seeing a beautiful multicolored glass bottle from the West, might want to create something similar in clay. The deep blue Egyptian cosmetic jar (shown opposite) is an interesting example of this idea. While the glass was still semi-soft and pliable, and before the handles were added, the glassmaker dragged a tool across the fine lines of white, pale yellow, and light blue, separated from each other to reveal the deep blue beneath them, to create the scalloped design.

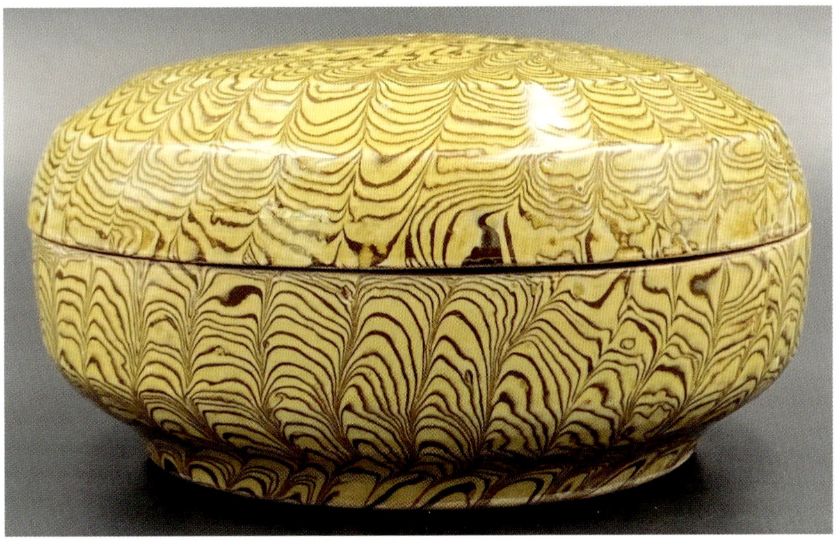

Box with cover, China, Tang Dynasty (618–907). Height: 2⅛ inches, diameter: 4⅓ inches (5.4 x 11 cm)

Press-molded round box with cover made of yellow and brown marble clay, glazed with a clear glaze inside and outside. This covered box shows a beautiful and skilled use of a curved chevron pattern which may originally have been inspired by ancient Egyptian glassware, an example of which is illustrated opposite, showing a similar design, which was traded on the Silk Roads and later found in China. *Image: Varshavsky Collection, San Francisco, CA*

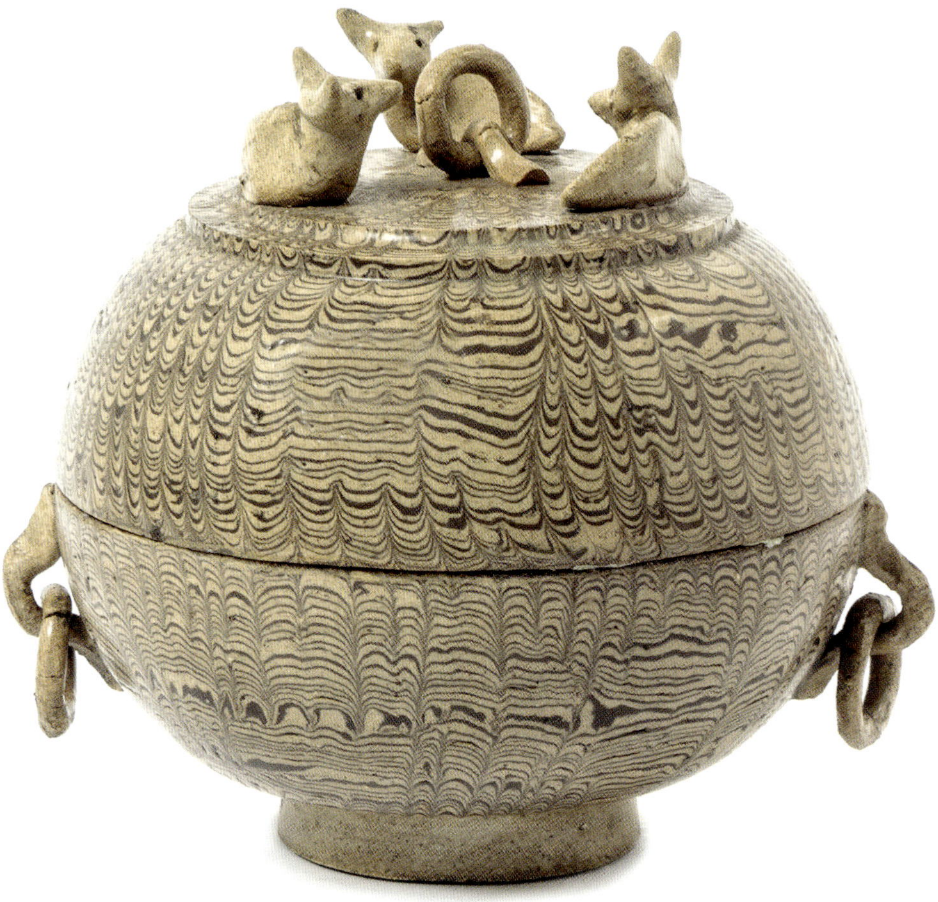

Covered Jiaotai bowl, China, Tang Dynasty (618–907) or later. Diameter: 5⅜ inches (13.6 cm)

An intriguing example of Chinese Jiaotai (mixed clay) pottery, this bowl with an elaborate domed cover is embellished with loose-ringed handles on the sides and lid as well as three recumbent animals on top. All parts were covered with a clear glaze and a raised foot ring was added to give elegance to the form. Since the overall shape of this bowl and added decorations were prominent, the artist was sensitive to create a Jiaotai fine-line design that is consistent around the shape, thus complementing rather than distracting from the more sculptural elements. The particular design of curved chevron shapes harkens back to the early Egyptian jar on page 10. *Image: Hindman Auctions*

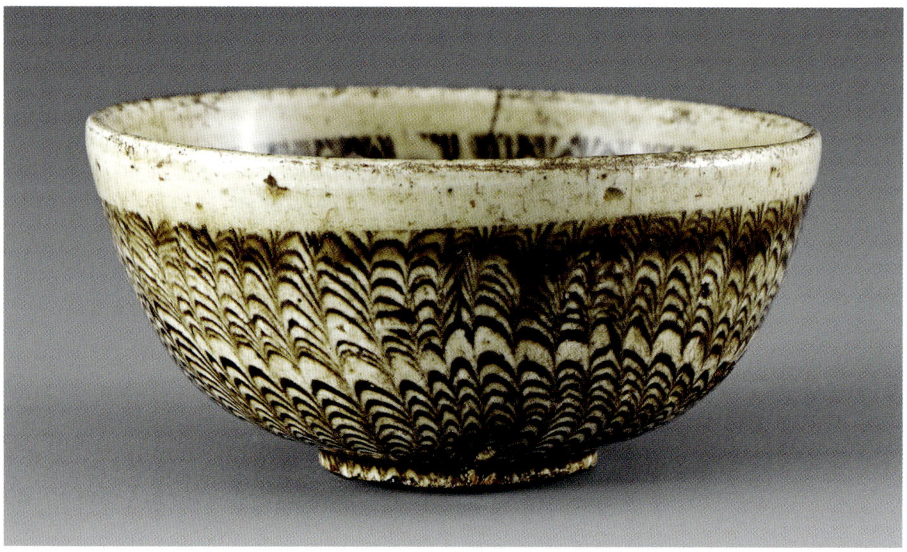

◂ **Cup, Korea, Goryeo Dynasty (918–1392). Height: 1¾ inches, diameter: 3⅜ inches (4.4 x 8.6 cm)**

This marbleized Korean cup is an example of the possible influence of ancient Egyptian glassware, such as the Egyptian cosmetic jar depicted on page 10, on later ceramics in China and Korea. The curved chevron pattern of this cup may also have been influenced indirectly by the much earlier glass jar. *Image: In the collection of the National Museum of Korea*

A similar design is seen in three works pictured on pages 11–12. In all three examples, tools have been used to force the parallel stripes of soft clay into the curved repeating shapes that demonstrate the potter's skill and are intriguingly beautiful. The first two were probably made in the general time frame of the Tang Dynasty (618–907), and the last is from Korea during the Goryeo Dynasty (918–1392).

While Chinese marbled ware was made from two colors of clay, white and reddish brown, covered with a low-fired amber or green glaze, Korean marbled ware was made from three colors of clay and fired to a high temperature, becoming white, black, and gray. Both the Tang Dynasty in China and the Goryeo Dynasty in Korea were periods of profound artistic achievements that were encouraged and supported by the ruling class. In addition, a new wealthy middle class had developed as a result of the Silk Road trade. Ceramic production also expanded, partly due to an increased demand for funerary art in China, as well as the need for ceramic containers for many of the goods being traded. One effect of this trade and monetary expansion was a stabilization of governments which provided a fertile atmosphere for artistic development and experimentation.

Chinese pottery made from naturally occurring colored clays was called Jiaotai, (mixed clay). Kilns in Gongxian, Henan province, were important centers

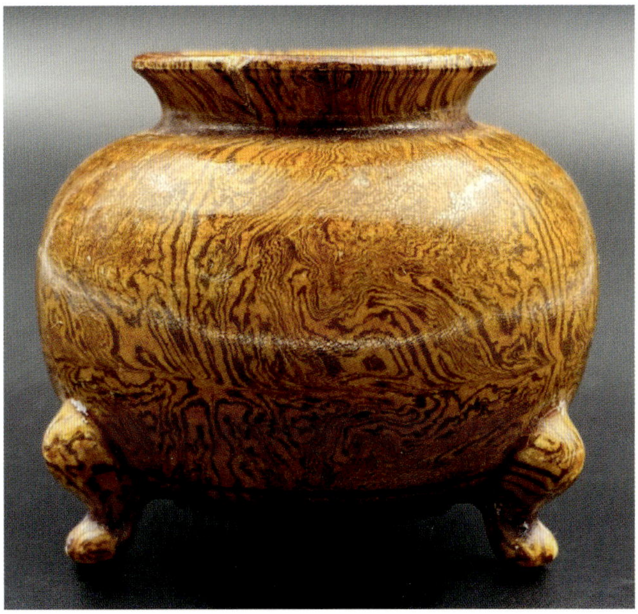

Jiaotai censer, Henan province, China, Tang Dynasty (618–907). Height: 2¼ inches, diameter: 2½ inches (5.7 x 6.4 cm)

This small amber-glazed Jiaotai (mixed clay) pottery censer was used to hold burning incense, usually for religious purposes. The three ceramic feet held the vessel away from the table on which it stood, protecting it from the heat. This censer has a monochrome interior, evidence that the marbleized clay was added as a thin veneer to the surface of a vessel made from a single type of clay. *Image: Varshavsky Collection, San Francisco, CA*

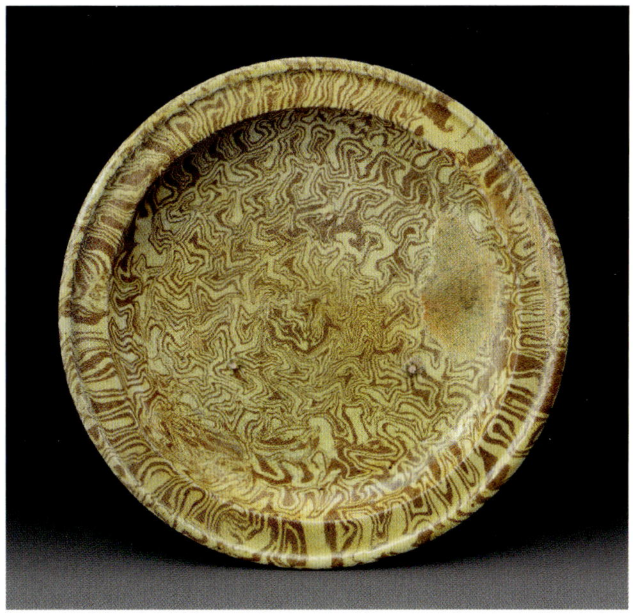

Tripod dish, China, Tang Dynasty (618–907). Diameter: 5¼ inches (13.3 cm)

A rare, yellow-glazed marbled pottery tripod dish, of brown and white clays with a pale-yellow glaze, supported by three feet. A stunning example of the potter's proficiency in Jiaotai (mixed clay) technique, this small dish is also remarkable for the skilful use of a wave pattern which organically flows from the center out to the wide rim. The inside edge of the rim reveals the fact that the colored clay slices were pressed into a mold, as the pattern flows seamlessly over the break between rim and interior curved space. *Image: © 2015 Christie's Images Limited*

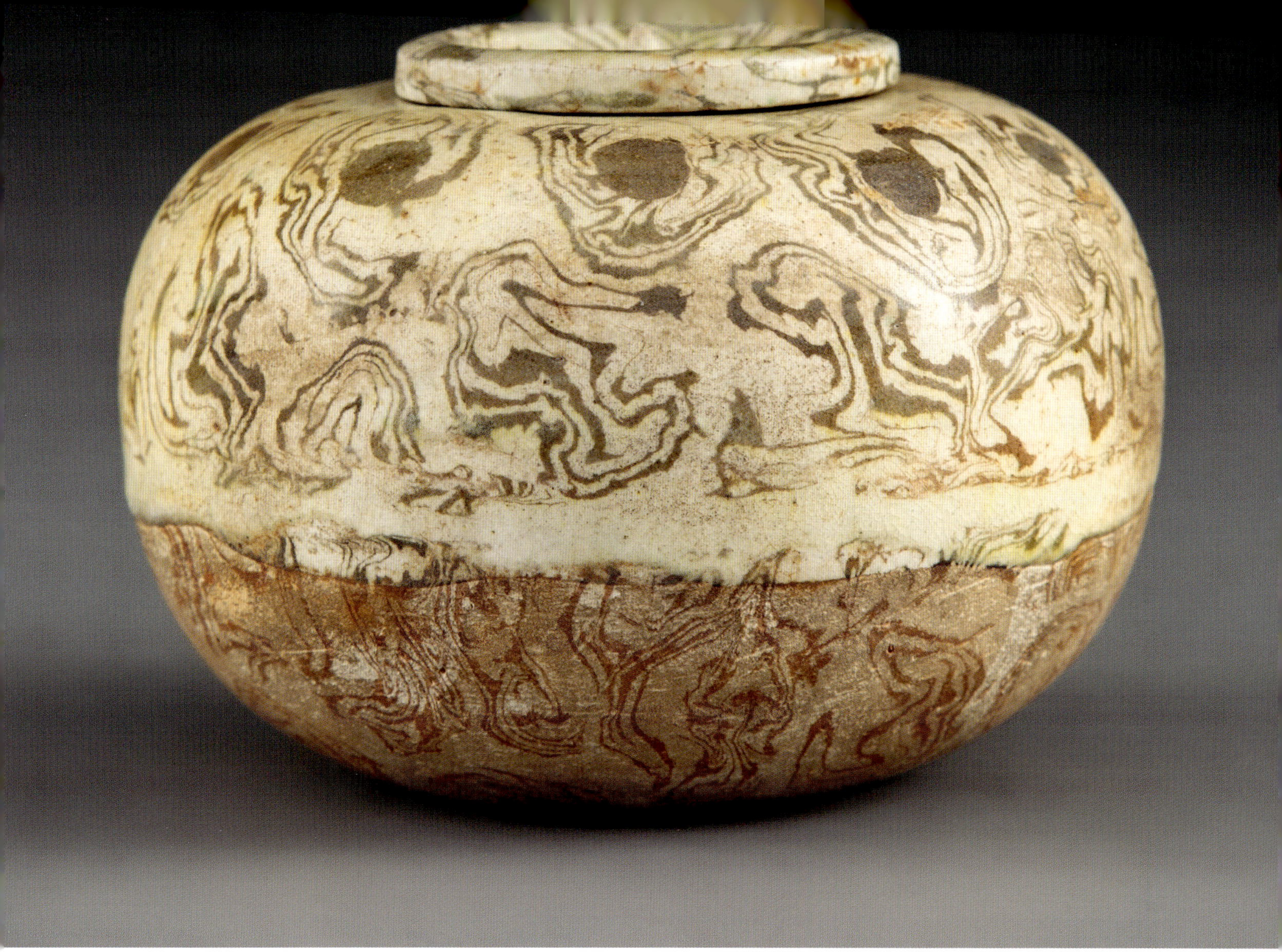

Covered jar, China, Tang Dynasty (618–907). Height: 2⅞ inches, width: 3⅛ inches (7.3 x 7.9 cm)

This Chinese Jiaotai (marbled) earthenware covered jar is an interesting example for a few reasons. First, the clear glaze covers just the top three-fifths, leaving the bottom area unglazed, so we can see how the clear glaze interacts with the colored clay. What looks red-brown on the bottom becomes a deep gray-brown when covered with a clear glaze. Also, since earthenware is low fired (as opposed to high-fired porcelain and stoneware), it is semi-porous, resulting in significant darkening of the raw clay surface over the centuries that it may have been buried or in storage.

The jar was assembled from two hemispherical forms which may have been pressed using the same mold. A clear area around the widest part, just above the bottom glaze edge, is probably where the two parts were joined together, with a band of white clay. The overall pattern, composed of folded and twisted dark and light clay slices, was enhanced by the potter with the addition of dark disks spaced regularly around the shoulder of the jar, an indication that the artist was sensitive to the interaction between decoration and form, and wanted to draw attention to the top of the jar as opposed to the unglazed bottom. The cover has a recessed center and rests over the jar's opening. *Image: Heritage Auctions, HA.com/2022*

▼ **Doctor's Pillow, China, Tang Dynasty (618–907). Height: 2³⁄₁₆ inches, width: 3⁹⁄₁₆ inches (5.6 x 9.1 cm)**

Brick-shaped "Doctor's Pillow" marbleized stoneware with brownish lead glaze. Despite the title Doctor's Pillow, this Jiaotai (mixed clay) form is unlikely to have served that function, both due to its miniature size and the strict geometrical form and flat surfaces. It remains a puzzle what its use was, although one theory suggests it may have been used as an arm rest for a patient during a doctor's examination.

Nearly all other ceramic pillows were created with curved surfaces and rounded edges for comfort. Ceramic pillows were often used as burial wares but also to keep the body cool while sleeping during the warmer months. While not in use, the pillows served as decorative objects to be admired, both for their formal qualities and their imagery, often symbolic. They were frequently given as wedding gifts with suggestions, through the often-intricate designs, of good fortune and happy times. It was thought that certain animals possessed magical powers, and carvings or paintings of those animals were also popular. *Image: © 2024 Museum of Fine Arts, Boston*

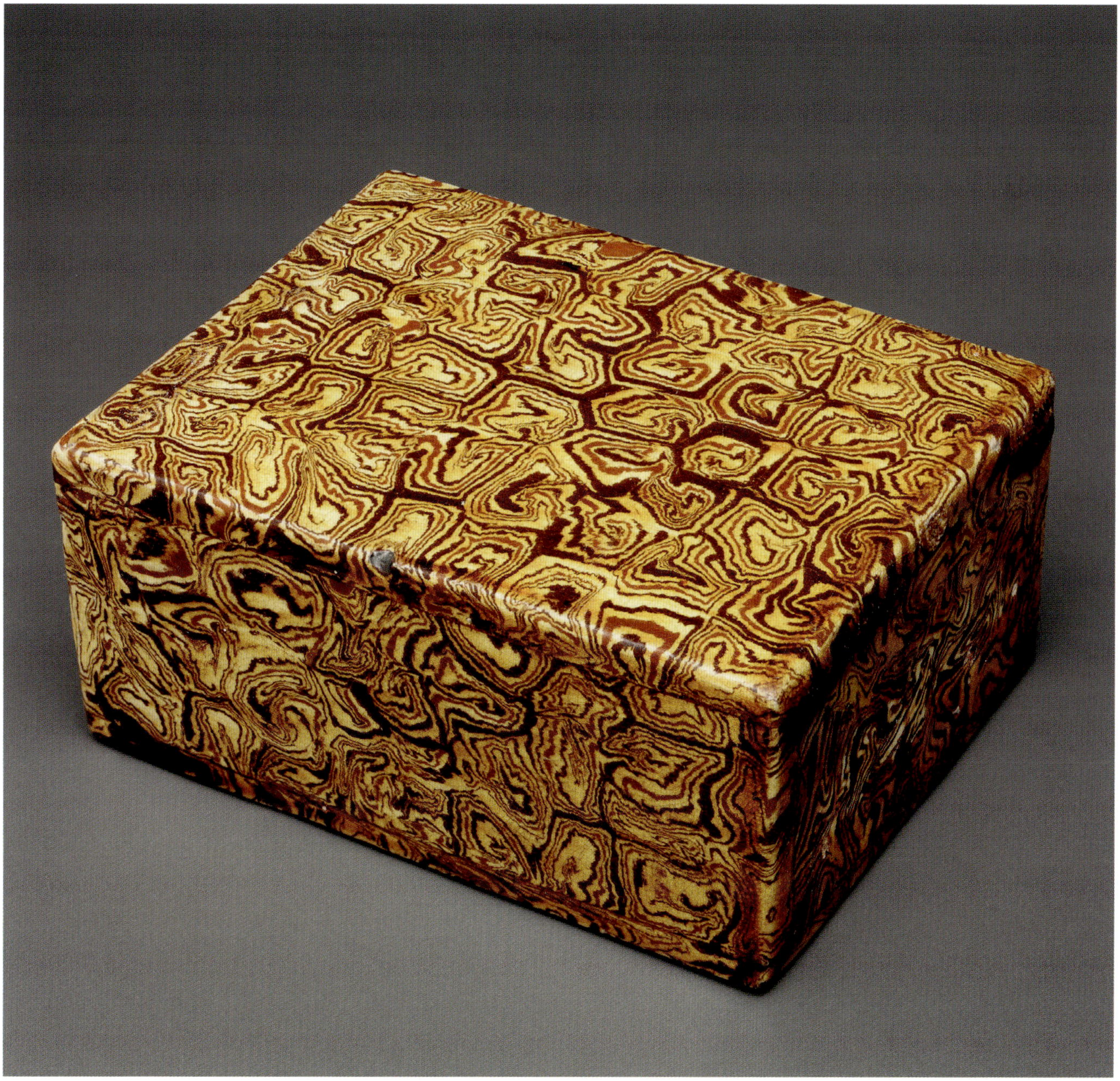

of production, but kilns in many other parts of China have also been known to produce this work. Its sophisticated and appealing detailed designs, reminiscent of agate or marble patterns, made it an appropriate style to be used for both ritual items and household vessels for the wealthy. This unique method of pottery-making was extremely time consuming as well as technically difficult. A large percentage of pots were lost due to cracking, which made this work, in general, more rare and expensive. The clay used for Jiaotai was usually white and red-brown earthenware during the Tang Dynasty, and white and brown or black stoneware or porcelain during the Song Dynasty (960–1279). Jiaotai pottery continued to be produced through the Yuan Dynasty (1271–1368) but seems to have died out at that point as highly refined porcelain attracted most of the attention.

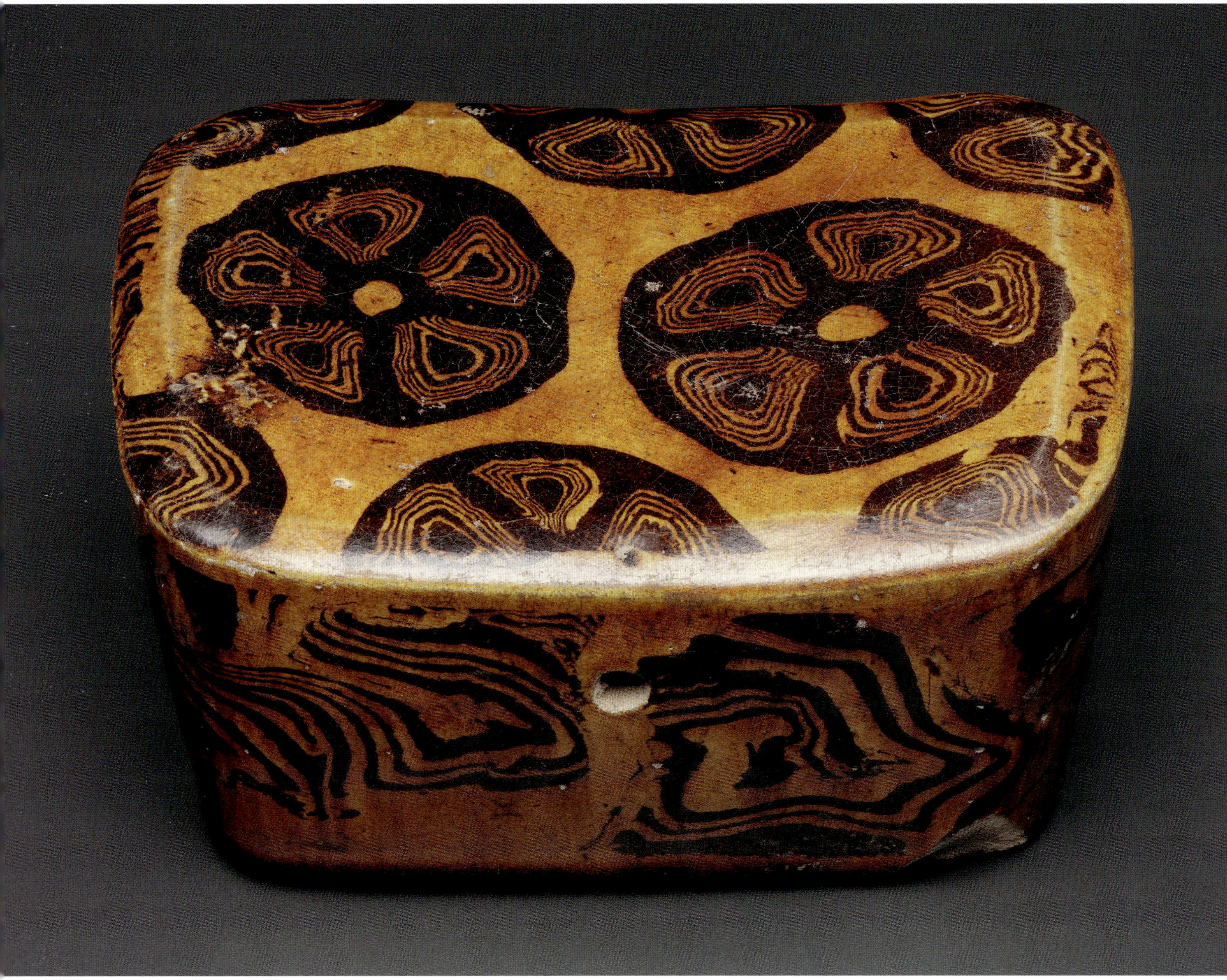

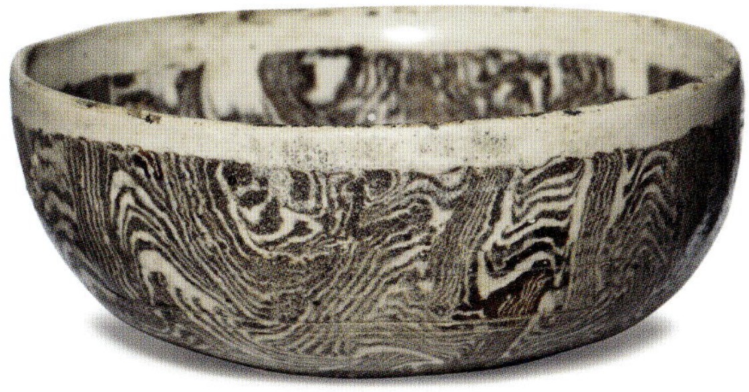

Bowl, China, Northern Song Dynasty (960–1279). Diameter: 5 inches (12.7 cm)

Cizhou ware is a term used to describe a wide range of ceramic styles produced in China during the Northern Song Dynasty. They were stoneware, or high-fired ceramics which were marketed to the middle classes, as opposed to the "relentlessly monochrome" porcelains of the court and upper class. This humble bowl with a very utilitarian shape would have been prized by its owner due to the intricate and intriguing designs made with colored clays. The white rim provides a visual contrast to the highly patterned body while also adding stability, from a technical point of view, to the bowl's structure. The use of marbling techniques in pottery unfortunately yields an often-fragile object, before and during the firing, and cracks frequently appear. The addition of a solid clay band for the rim adds a level of stability which helps prevent cracking. *Image: © 2018 Christie's Images Limited*

◀ **Trapezoidal pillow, China, Tang Dynasty (618–907). Width: 5 11/16 inches, depth: 3 1/8 inches (14.4 x 7.9 cm)**

Aside from the potter's demonstrated skill in creating highly controlled and random designs on this glazed earthenware pillow, we can see clues as to its construction. The hole in the side and broken corner at the bottom right both show a solid color of clay rather than the two colors that make up the designs. This suggests that the original form was constructed of a single type of earthenware while the designs were applied as a veneer of thin slices that were added on top of the basic form. The very evident hole also suggests the possibility that the entire pillow may have been submerged in cool water, then drained, to provide a long-lasting coolness for the sleeper – earthenware being permeable and able to soak up the water. *Image: © 2024 Museum of Fine Arts, Boston*

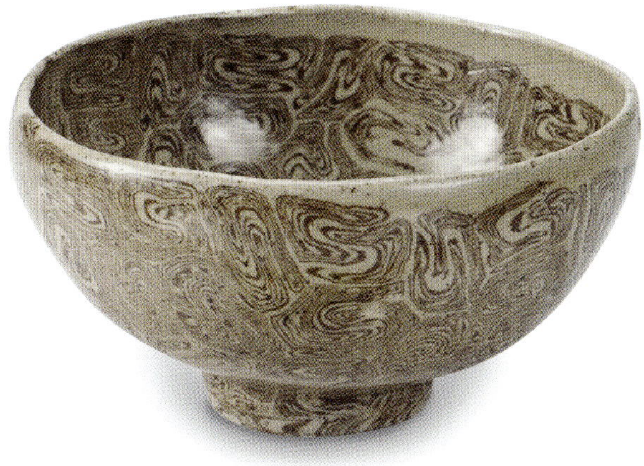

Bowl, Jiaozou, Henan province, China, Northern Song Dynasty (960–1279). Height: 2 3/8 inches, width: 4 3/8 inches (6.2 x 11.1 cm)

A beautifully sensitive example of Chinese Jiaotai (mixed clay) artistry, this small bowl demonstrates a high level of craftsmanship and design. With the Northern Song Dynasty which came after the Tang Dynasty, we see a more careful manipulation of the colored clay designs. Looking carefully, we can see that layers of white and dark clays were rolled together, then flattened slightly. Two such semi-flat rolls were formed into interlocking "C" shapes, which formed a larger roll. Slices were then taken from this roll and placed side by side into a mold, where pressure on the soft clay from the inside joined the pieces together. Some took on a rectangular shape while others can be seen as hexagonal. Since the pattern was not regular, the design ended close to the rim on one side and lower down on the other. This irregularity was corrected by adding a coil of white clay up to the top of the mold, creating a pleasing upper rim and, at the same time, revealing the potter's hand and enabling us all to appreciate the human being who created this timeless object. *Image: © Arthur M. Sackler Gallery, Smithsonian Institution, Washington, DC: The Dr. Paul Singer Collection of Chinese Art of the Arthur M. Sackler Gallery, Smithsonian Institution; a joint gift of the Arthur M. Sackler Foundation, Paul Singer, and the AMS Foundation for the Arts, Sciences, and Humanities, and the Children of Arthur M. Sackler, S2012.9.3944*

Agateware

By the 1600s ancient Chinese marbled ceramics may have found their way to England and, perhaps, inspired local potters to experiment with this fascinating technique. Historically, agateware is classified as two types, "thrown agateware" and "laid agateware," sometimes also called "solid agateware" to distinguish it from marbleized surface decorations applied to solid color pottery using various colored slips. Thrown agateware was created by carefully wedging together two or three differently colored clays and then "throwing" the result on a potter's wheel. The surface of the resulting form would be quite blurry but when the piece was partially dried, a metal tool would be used to trim off the outer layer to reveal the resulting marbleized diagonal swirling pattern.

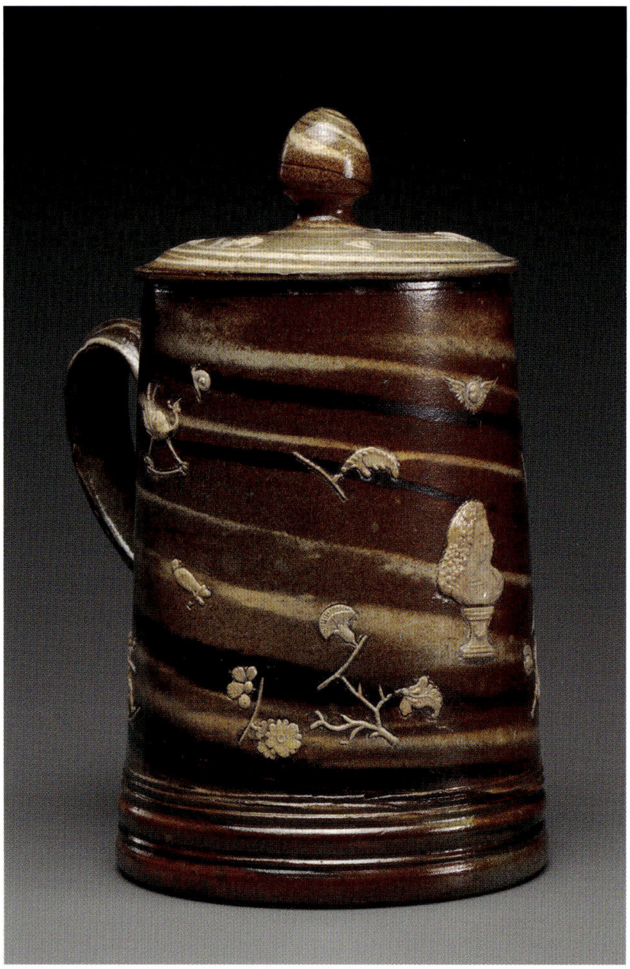

Laid agateware, on the other hand, was the result of a very complicated technique starting with thin slices of two or more colored clays. These were laid, one on top of the other, to form a thicker slab or an even larger loaf. A variety of steps were then taken, including folding, twisting, and stretching, to create more elaborate designs. Cross-sectional slices were then taken, resulting in a layer of soft clay which showed the pattern or design on both sides, all the colors penetrating through the thickness of the slice. This description would also characterize the techniques used by the ancient Chinese and Korean potters to create their colored clay patterns. The slices of patterned clay were then pressed into plaster molds which had been taken from hand-built or wheel-thrown originals. For more complicated forms, such a teapots, separate two-part molds were made for each part, that is, spout, handle, lid, and main body.

In England, the earliest known work of thrown agateware was made by John Dwight, considered by many as the father of English pottery, in the 1670s. He established Fulham Pottery in Wigan, England, and is known to have experimented widely, searching for the secrets of porcelain and working with colored clays. Dwight developed the technique of mixing different colored clays and later patented

◄ **Covered tankard, Fulham, England, c.1685–90. Height: 10½ inches (26.7 cm)**

This covered tankard made by the factory of John Dwight of Fulham (1670–1859) incorporates thrown agateware as well as three-dimensional applied-clay images of snails, shells, flowers, birds, and profile busts of King William III and Queen Mary II of England (r.1689–1702). These were made by pressing the soft clay into brass stamps and later applying them to the surface of the tankard. Dwight used both stoneware – a non-porous clay body which is fired at a high temperature – and salt, which was added to the kiln during the firing to create a permanent glaze. These processes had both originated in Germany. *Image: ©The Nelson-Atkins Museum of Art, Kansas City, Missouri. Gift of Frank P. Burnap, 55–77*

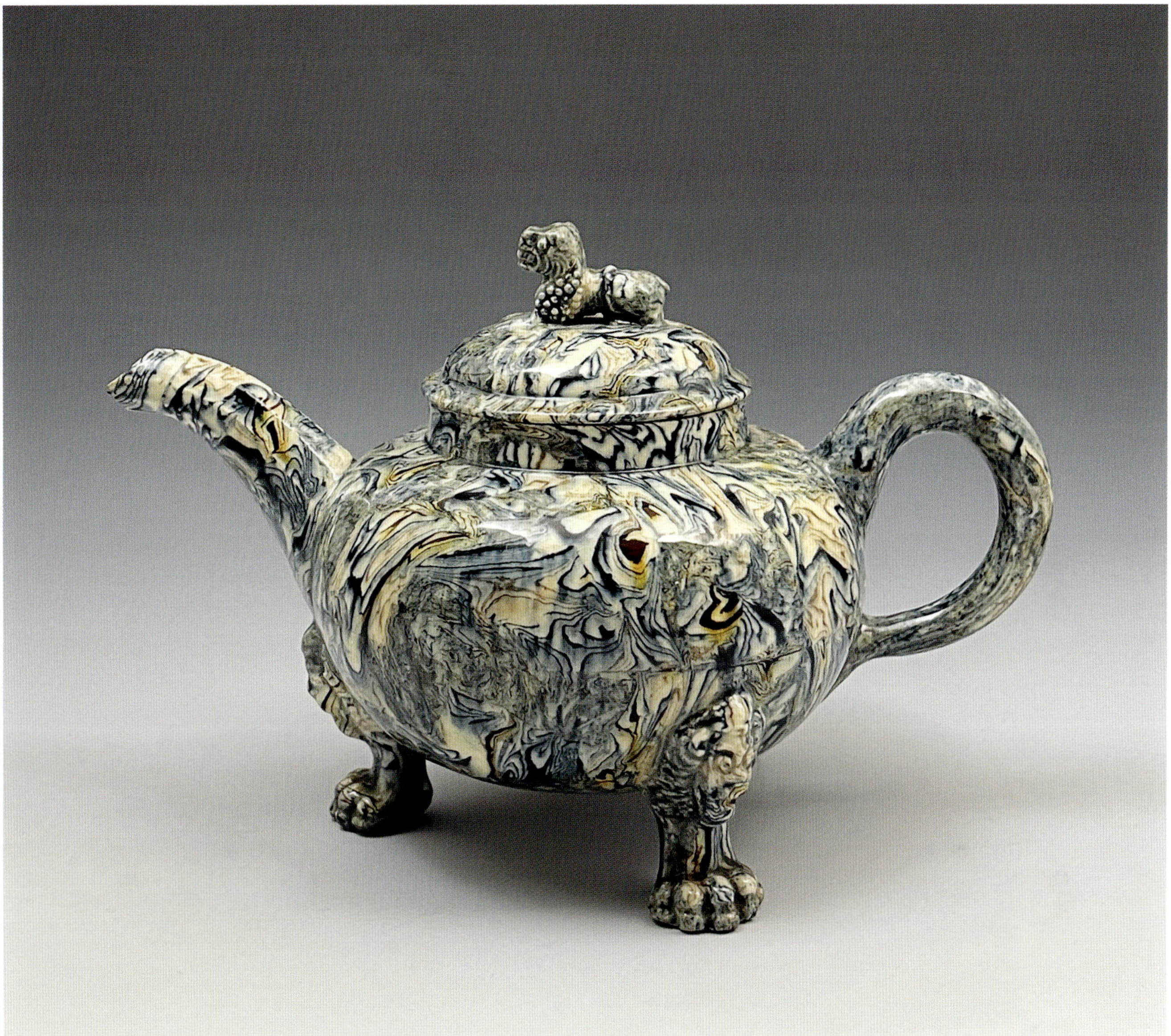

Teapot, Staffordshire, England, 1750. Height: 4¼ inches (10.8 cm)

This teapot is a fine example of the elaborately embellished formal style that was paired with the agateware technique. Teapots such as this were assembled from – in this case up to seven – one- or two-piece plaster molds to create the individual parts. The body was assembled from a top mold and a bottom mold. The separation line where the two parts were joined together can be seen running horizontally around the widest part. Slices taken from variegated pattern blocks, made with white and either dark-brown or blue accent clay, were pressed into the open mold and the two halves were then joined to form the teapot body. While the spout is obviously hollow, the handle might be solid and looks to have been created by hand rather than in a mold. Solid semi-hard clay was carefully carved to create the foot shapes as well as the *kylin* (Chinese mythical creature) decoration that serves as the handle for the lid. Plaster molds were taken from these carvings and later used for pressing the agateware slices to create the final version of these parts. The cobalt which was used to create the blue clay can be seen bleeding into the lead glaze as it melted, running down the form. A few spots of a high concentration of iron in the dark clay can also be seen melting with the glaze. *Image: John Howard, The Antique English Pottery Specialist*

his techniques in 1684. For the next 100 years thrown agateware table and tea ware were popular.

Dr. Thomas Wedgwood of Burslem, Staffordshire, (1695–1737) was probably the first British potter to work with laid agateware, according to ceramic historian Simeon Shaw, who writes that he produced "Coffee and Tea Pots, of clays compounded and mixed to resemble agate, marble, and other natural bodies, ... in various shapes; some glazed with lead ore; and the white, all salt glaze."

Another innovator with this technique was Thomas Whieldon of Fenton Low (1719–95). He is known for his use of metallic oxides to stain white clay to create predictable colors, ranging from black to blue and green clays, rather than relying on the few colors that are found naturally. An additional benefit of this strategy is that the shrinkage rate of these colored clays would be the same since they were all based on the same white clay. The use of naturally found colored clays would often present technical problems if their shrinkage rates differed,

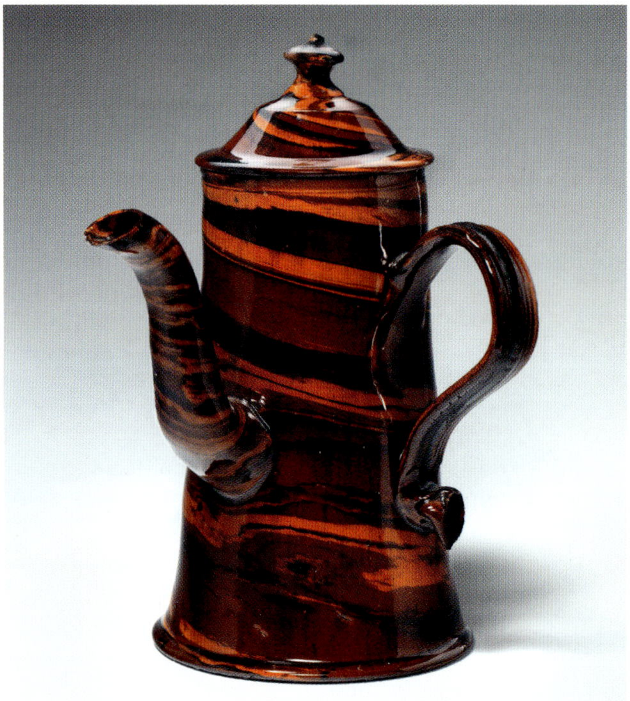

Coffeepot, Staffordshire, England. c.1745.
Height: 7¾ inches (19.7 cm)

The complex form of this lead-glazed earthenware coffeepot by John Astbury demonstrates the skilful use of the thrown agateware technique. The body, spout, and lid were each thrown separately from the same three colored clays, the spout being especially difficult to throw due to its very narrow diameter. The handle was hand-formed from the same three clays. The simple tapered cylinder of the main body shows off the decorative striping of the clays to optimum effect. *Image: Metropolitan Museum, OA Public domain*

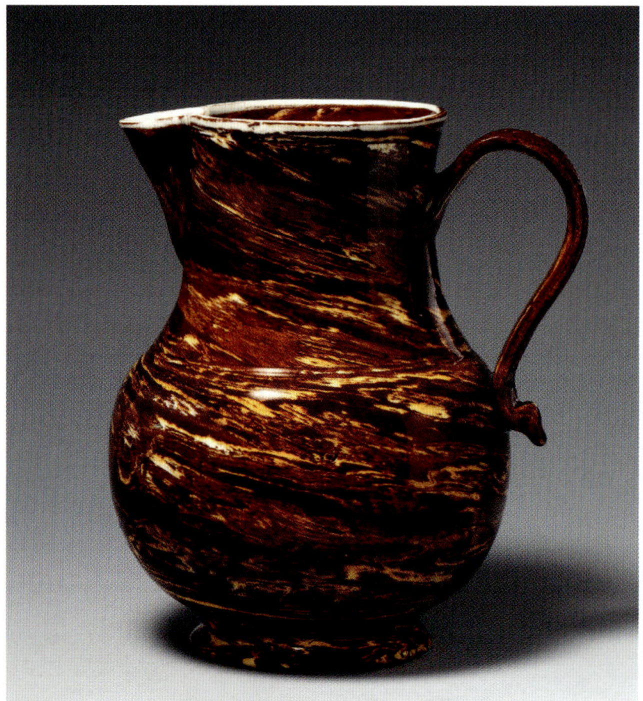

Agateware jug, Staffordshire, England, c.1745.
Height: 7 inches (17.8 cm)

This marbleized jug in the style of John Astbury mimics the natural designs found in real marble or agate, the visual inspiration for agateware. The colored clays used – two shades of brown with white – were partially mixed before being thrown on the potter's wheel, resulting in the more complex intermingling of clays to create this effect. The foot ring appears to be part of the main body and most likely was trimmed into shape when the pot was leather-hard. The rim is highlighted by a brushed-on white slip before final glazing. *Image: Metropolitan Museum, OA Public domain*

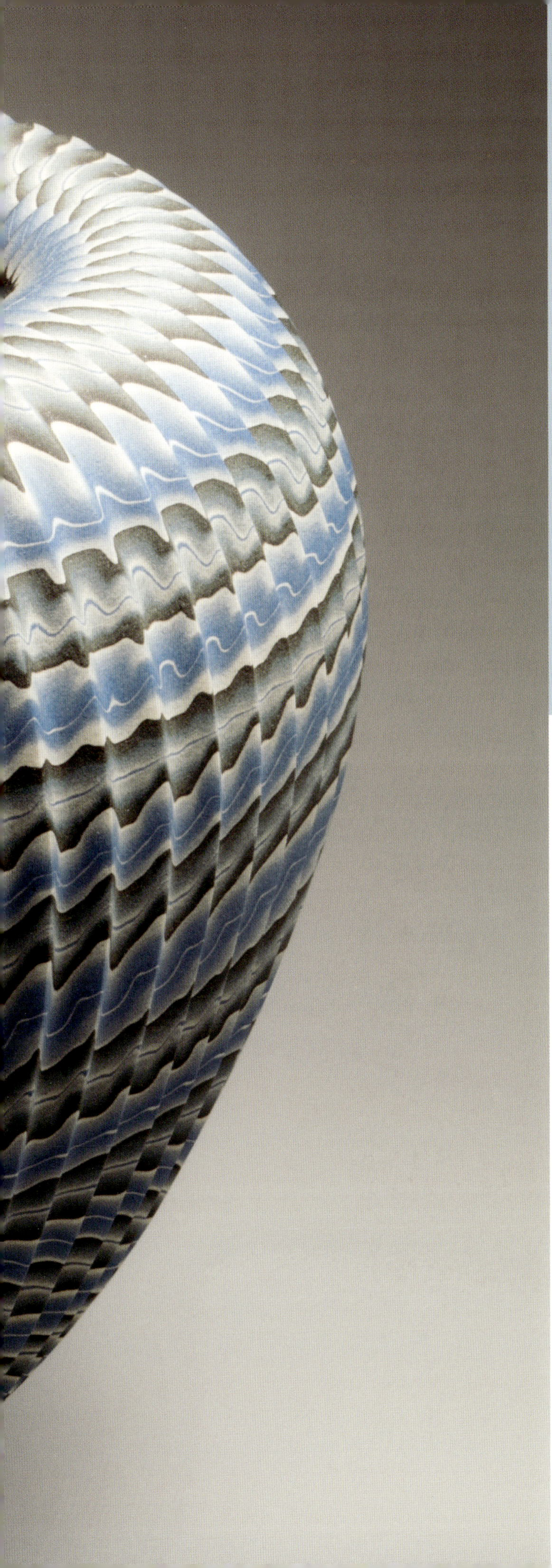

JAPANESE MASTERS

IN RECENT DECADES the arts of nerikomi and neriage have seen tremendous growth and development, starting in Japan and ultimately spreading worldwide. This chapter serves to honor five artists who have chosen these techniques as their life's work and who have provided beautiful and inspiring examples of what can be done with colored clay. Matsui Kōsei and Itō Sekisui V were both designated "Living National Treasure" for contributions to their Japanese heritage through their remarkable work. Ogata Kamio chose wheel-thrown neriage as his primary direction, exploring added texture and a sense of emotion to his pieces. Kondō Takahiro uses colored clay as just one of many techniques as he explores traditional ceramic forms as well as sculpture, often influenced by cultural and social issues of the present day. Nishi Koichi focuses on exploring highly detailed and controlled pattern-making, often with the addition of surface textures that add a beautiful level of mystery to his work.

The names of these artists are presented here in the traditional form, family name followed by given name.

Matsui Kōsei (1927–2003), Living National Treasure 1993

Matsui Kōsei was perhaps the most famous 20th-century potter working with colored clay. After graduating from Meiji University with a degree in literature and the intention of being a novelist, he instead pursued the role of becoming a Buddhist priest, ultimately being inaugurated as the 24th chief priest of the Gesso-ji Temple in Kasama, Ibaraki Prefecture, in 1957. Previously he had begun to study ceramics in 1946, apprenticing at the Okuda Pottery Factory in Ibaraki province. Often visiting the Tokyo National Museum, he became fascinated with ancient Jiaotai (mixed clay) ceramic works from China (Tang Dynasty, 618–907), Japan, and Korea, studying them carefully to understand their secrets, as well as attempting to duplicate their designs. Matsui may also have been aware of the work of Kawai Kanjirō who was one of the first modern ceramic artists to work with nerikomi (then referred to as neriage).

Starting in 1959 he began restoring the old kiln at the temple's gate and continued his research and experimentation. In 1966 his teacher and mentor, ceramic artist and professor Tamura Kōichi (who later was himself designated Living National Treasure), recognized Matsui's combination of interests and abilities and encouraged him to pursue this direction as his specialty. Matsui then embarked on a fascinating decades-long discovery of colored clay techniques while creating unrivaled masterpieces.

While "neriage" refers to wheel-thrown colored clay pottery and "nerikomi" refers to hand-built, often press-molded colored clay pottery, Matsui referred to all his work as "neriage." This may be

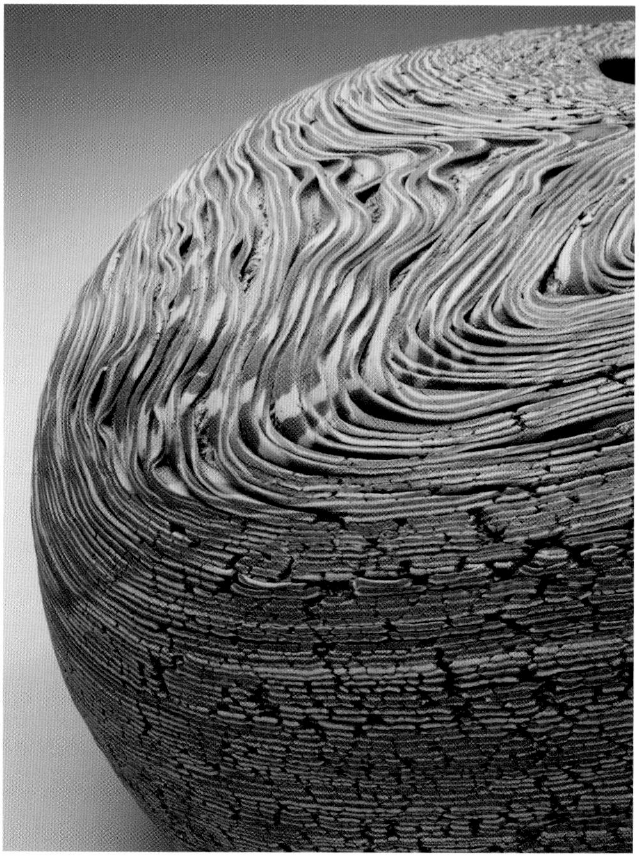

Detail of *Tsubo*, 1978. Height: 12 inches, diameter: 17 inches (30.5 x 43 cm)

Globular large *tsubo* (vessel) of marbleized stoneware with rough layers of dark and light-gray colored clays, using the "Shoretsu" technique. *Image: © Richard Goodbody. Courtesy of Joan B Mirviss Ltd*

▶ Nenjo vase *Fossil Forest*, 1973. Height: 16 7/10 inches, diameter: 10 7/10 inches (42.4 x 27 cm)
Image: © Ibaraki Museum

The designation of "Living National Treasure" (more officially called "Preserver of Important Intangible Cultural Properties") is an honor given by the Japanese government to individuals who have attained high levels of mastery in their specific fields. To date Living National Treasures have been certified for 16 categories of Intangible Cultural Properties in the performing arts and Japanese crafts. By December 1, 2020 there had been 371 persons designated as Living National Treasures. Currently there are 111 who are still living.

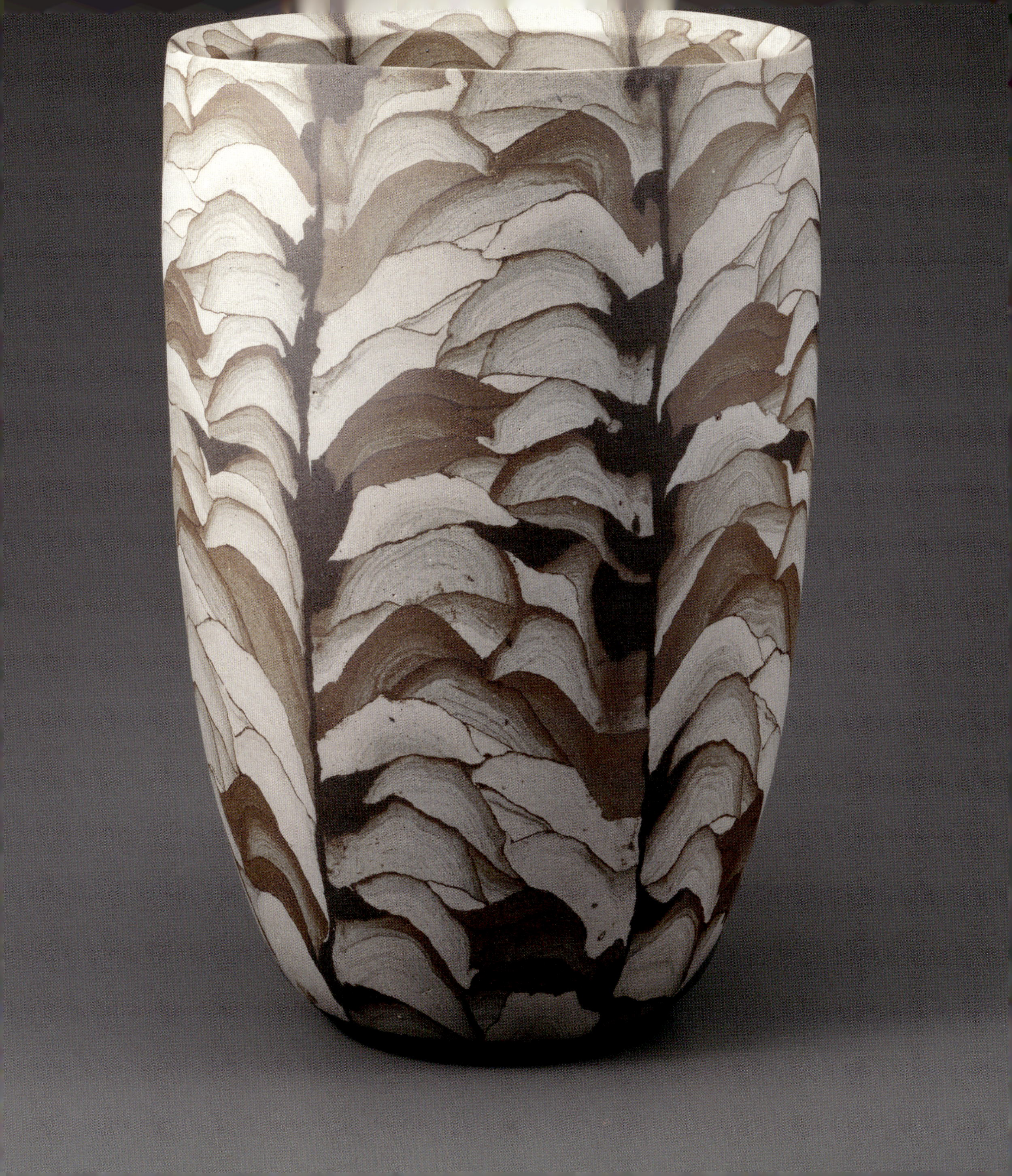

due to his discovery of an innovative method of using a large slab of already highly patterned colored clay (made using the nerikomi technique) and wrapping it around a form on the potter's wheel. Upon the removal of the form, the resulting cylinder is shaped into a vessel – essentially "throwing" it on the wheel – the definition of neriage. Matsui was the first to do this.

Between 1976 and 2001 Matsui developed eight new techniques within the general neriage category ultimately leading to him being designated Living National Treasure in 1993, as an artisan of the neriage technique. Stylistic groups of work were associated with these eight techniques. At times Matsui would combine techniques to create even more creative variety. They include:

Kezurimono, translated as "shavings," scraped and trimmed (before 1975).
Shoretsu, translated as "various lines" or "rows," scored with a variety of combing tools (1975).
Shoretsu Fine Porcelain, a cracked surface layer through which one can see the colored body (1976).
Akanede, with characteristic red color (1981).
Laminated, including laminated inlay and under inlay techniques, not technically nerikomi or neriage but still made with colored clays (1983).
Capricious, irregular and indefinite patterns with several colors, often with surface-combing textures similar to those of the earlier Shoretsu work (1985).
Fuhakuji, translated as "no night," (sandblasted), described by Matsui as resembling "layers of earth revealed after a sandstorm," earth tones predominate with brighter-colored accents. This technique specifically refers to the surface texture (1987).
Seihaku, "clear white," made with a bright white New Zealand clay with a wide variety of colorful pigments and metal oxides, resulting in bright clear colors. These pieces were usually polished to a very smooth, almost shiny surface (1989).

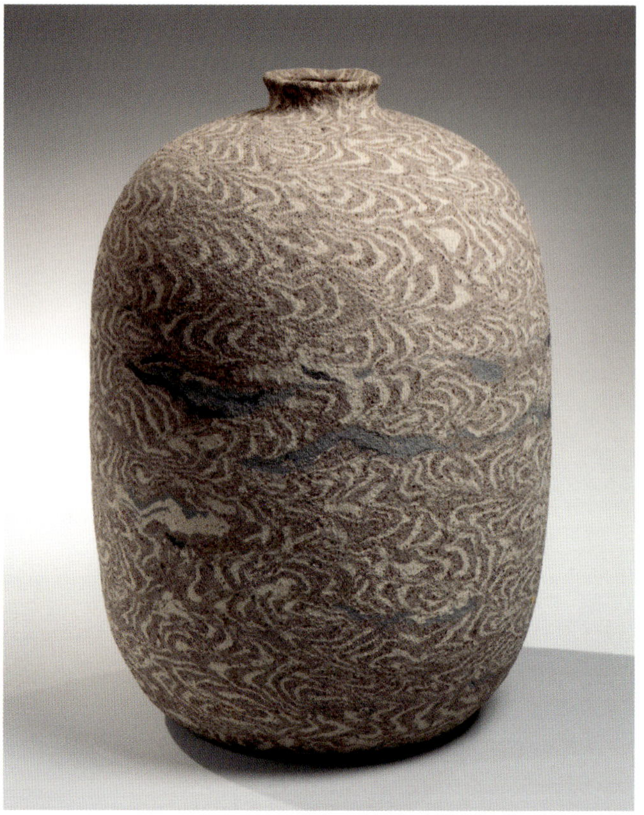

Columnar vessel, 1973. Height: 16½ inches, width: 11¼ inches, depth: 10¾ inches (42 x 28.5 x 27.3 cm)
Large, light-brown, beige, and pastel blue neriage columnar vessel of marbleized stoneware, with small raised mouth and "Fuhakuji" texture. *Image: © Richard Goodbody. Courtesy of Joan B Mirviss Ltd*

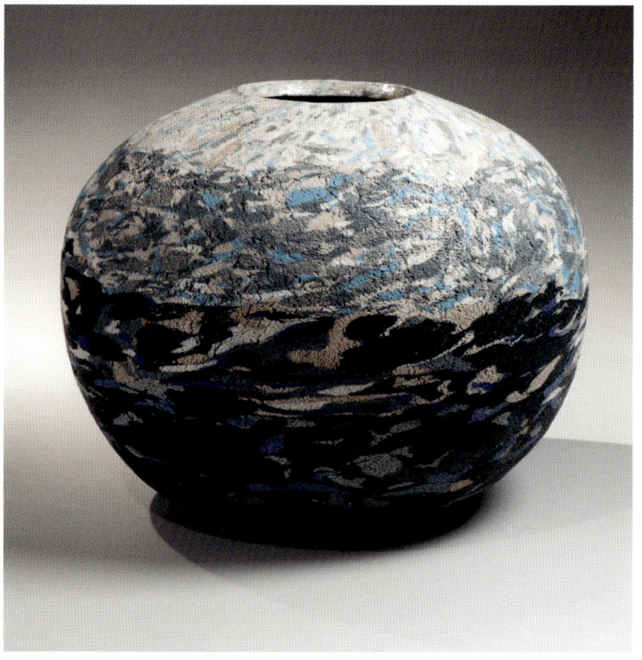

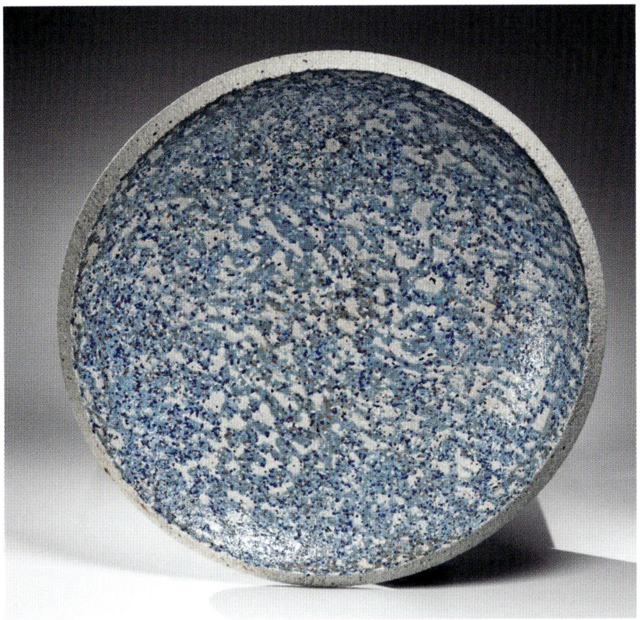

◂ ***Streaming River and Floating Clouds*, c.1986. Height: 3⅛ inches, diameter: 20¾ inches (8.6 x 52.7 cm)**

Large neriage laminated stoneware platter with dotted meandering pattern in deep blue, light blue, brown and creamy white colored clays. *Image: © Richard Goodbody. Courtesy of Joan B Mirviss Ltd*

Some of Matsui's Shoretsu works (late 1970s to early 1980s) are the result of various experiments with techniques involving the use of different clays, with varying shrinkage rates, in a single piece. If the outside layer of clay, perhaps a thick brushed-on layer of slip, had a higher shrinkage rate than the underlying layer, often stained with different colors, it would tend to shrink during the drying phase as well as the firing cycles. This shrinkage would cause that layer to pull away from itself creating a delicate texture of cracks over the entire surface. The contrasting underlying layer would then reveal both texture and color components of the surface decoration. Variations of this technique included scoring the outside layer with a fine comb which would give direction to the beautiful pattern of delicate cracks as they developed over the surface during the shrinkage period.

The use of a larger combing tool became the dominant decorating device in other neriage jars whereby the comb was inserted more deeply into the thick wall of the vessel, creating expressively flowing, and often multicolored parallel ribbons of clay protruding from the surface. These ribbons would crack as the work dried and shrank, creating a complex combination of subtle swirling colored clay overlaid by an astoundingly unique texture (see detail of *Tsubo* on page 26).

Matsui's use of color in creating his compositions has varied greatly over his oeuvre, from quiet and understated to more vibrant, with many stages in between. The nerikomi artist Dorothy Feibleman, who knew Matsui and interviewed him, spoke of his use of color, stating: "The interesting thing about Matsui is his father was a kimono maker and I think he was very much influenced by fabric in kimonos. The way he colors things especially is related to fabrics of different types used in Obi, Kimono and Obiage."

Matsui Kōsei began his exhibition career in 1969 at the Dento Kogei Shinsakuten (9th Traditional Crafts New Work Exhibition) where he was awarded an honorable mention. His entry the following year won him the Japan Crafts Society Prize and his work was included every year thereafter, winning him several more prizes (he later became the director of the Japan Crafts Society). Following several other important awards he was granted the prestigious Japan Ceramics Society Award in 1974. In 1988 he was awarded the Order of Cultural Merit and in 1990 the Japan Ceramics Society Gold Prize. He was given the title Ningen Kokuho in 1993 and the Order of the Rising Sun from his majesty three years before his death in 2003.

Matsui was given numerous solo exhibitions throughout his career, including many at the Mitsukoshi Department Store in Tokyo, and participated in many national and international exhibitions. He traveled extensively worldwide,

◂ ***Himalaya*, 1985. Height: 14 inches, diameter: 16½ inches (35.5 x 42 cm)**

Large globular neriage vase of marbleized stoneware with rough surface of navy, blue, light-blue, gray, ochre, and white colored-clay inlays, using the "Capricious" technique. *Image: © Richard Goodbody. Courtesy of Joan B Mirviss Ltd*

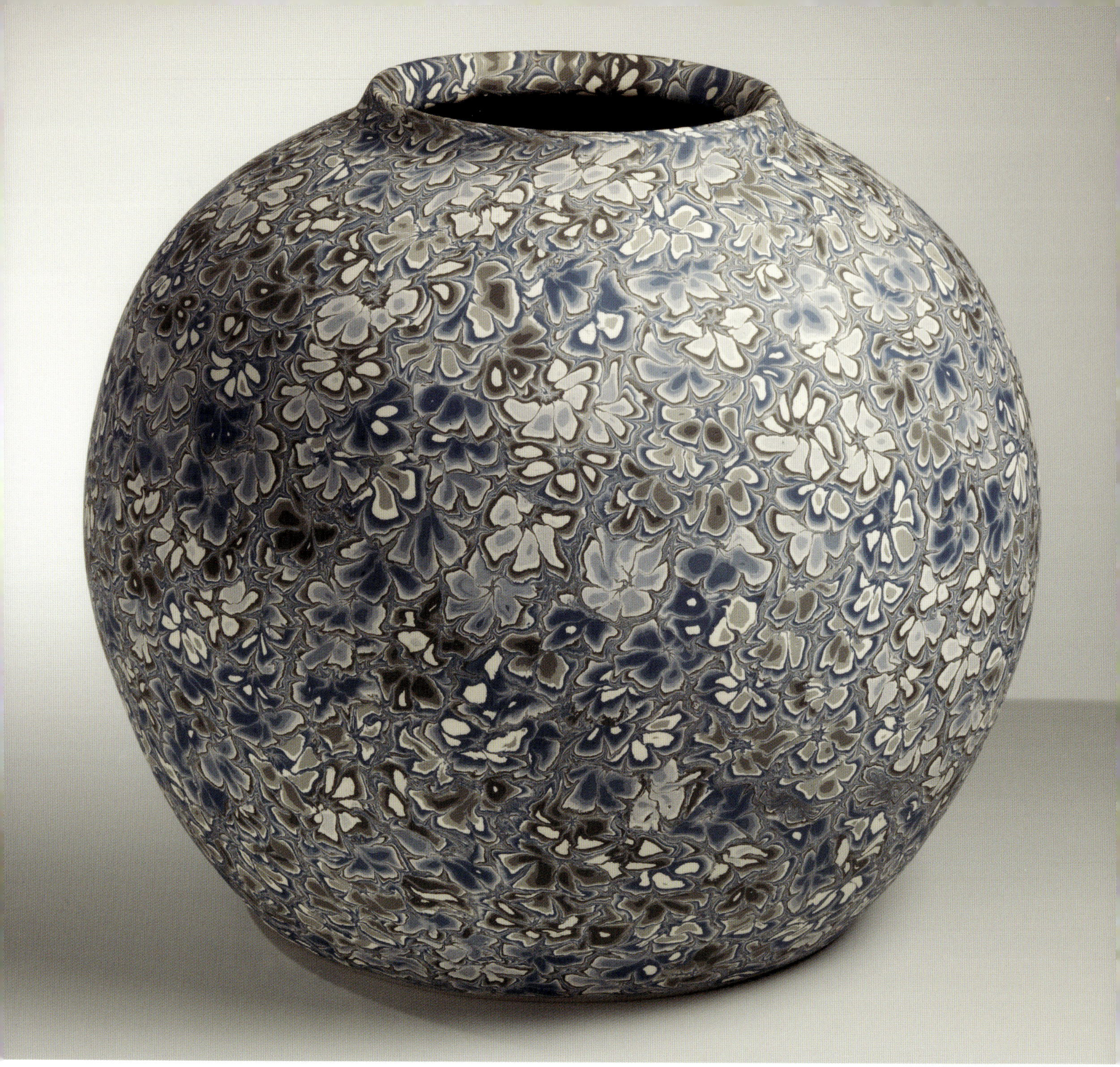

visiting many cultural areas and researching a wide variety of art and craft traditions.

Matsui's work can be found in numerous Japanese and international museums including the Ibaraki Ceramic Art Museum, Japan; both the Tokyo and Kyoto National Museums of Modern Art; the Hiroshima Prefectural Art Museum, Japan; the Hokkaido Prefectural Museum of Modern Art, Sapporo, Japan; the Seto City Museum, Japan; the Shiseido Museum, Kakegawa, Japan; the New York Metropolitan Museum of Art; the Victoria and Albert Museum, London; the Minneapolis Institute of Arts; the Saint Louis Art Museum; the Art Institute of Chicago; and the Brooklyn Museum, New York City.

▲ **Blue Thoughts, 1988. Height: 12 inches, diameter: 13½ inches (30.5 x 34.2 cm)**

Large neriage vase of marbleized stoneware composed of blue abstract floral patterning. *Image: © Richard Goodbody. Courtesy of Joan B Mirviss Ltd*

Itō Sekisui V (b. 1941), Living National Treasure 2003

An ancestor of Itō Sekisui V, Itō Ihei moved from the mainland to the island of Sado, off the west coast of Japan near the city of Niigata, around 1640 when a gold rush brought many to the island. His son Jinbei started a pottery business which supplied ceramic tuyere pipes used in gold smelting along with functional ceramic items for everyday use. For generations the Itō family carried on the pottery business which later produced sophisticated tea bowls and plates. Years later, in 1844 a red clay called *mumyoi* was discovered in a local gold mine. It was distinctive for its vibrant red color due to its high iron content and its extremely fine clay particles which gave it a smooth texture and remarkable hardness when high fired. Itō Tomitaro who was the ninth-generation descendant of Itō Ihei was able to fire this clay to a high temperature, initiating a new ceramic style known as *mumyoi-yaki*. At this high temperature the clay shrinks a remarkable 30 percent, unlike the more common 10–15 percent of other clays, which contributes to its unusual hardness. In 1868 Tomitaro took on the name Sekisui which now has been passed down through five generations to Itō Sekisui V. The Sekisui Itō Pottery Museum in Sado exhibits ceramic art by all five generations.

Itō Sekisui V was born with the name Itō Yoichi, the first character "yo" meaning "kiln" in Japanese. In 1966, several years after Itō Yoichi's father had died, he graduated from the Kyoto Institute of Technology, Ceramics Industry Department, the first in his family to receive a formal education. He then returned to his family on Sado Island where his elderly grandfather, Itō Sekisui III, taught him the traditions of *mumyoi-yaki*. One important technique is called the *namamigaki* process, which involves burnishing the partly dried vessel with a smooth metal rod or polished stone. This technique creates a very smooth and hard surface which then will be left unglazed. The rich and vibrant red color is primarily

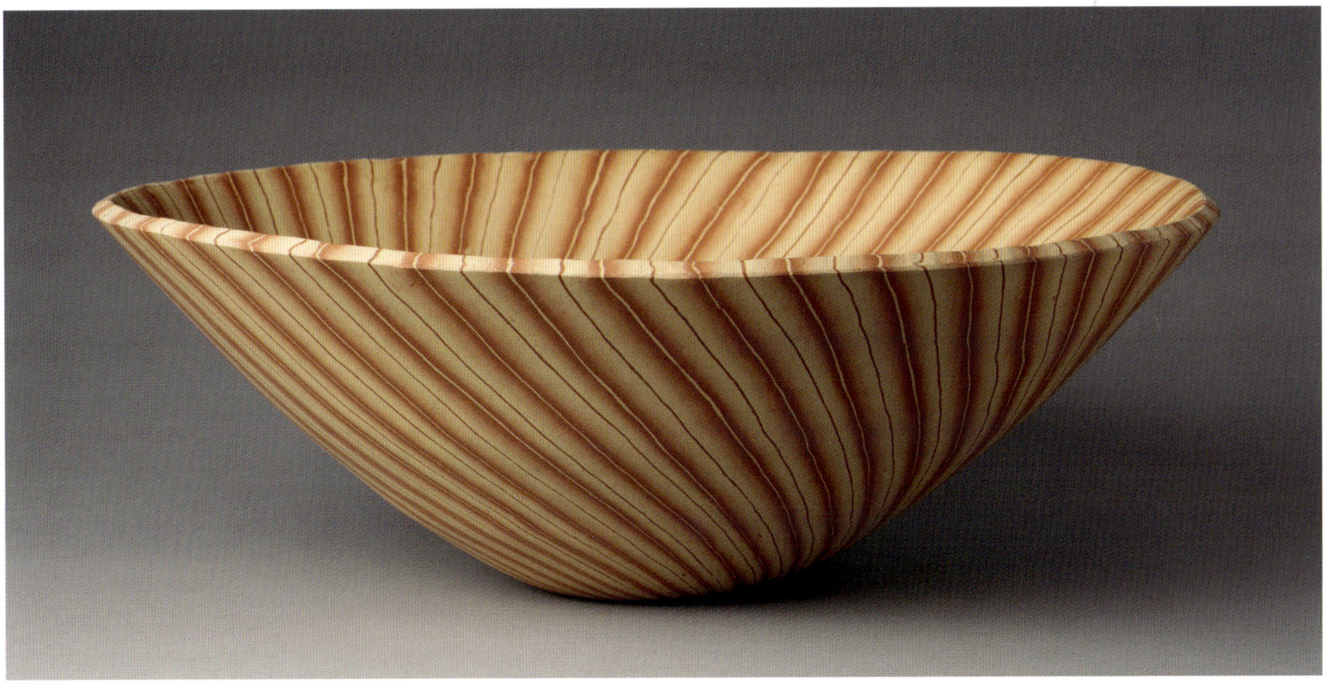

Large bowl with stripes, marbleized stoneware (neriage), 2016. Height: 6¼ inches, diameter: 17⅜ inches (15.9 x 44.1 cm)
Image: © The Metropolitan Museum of Art; image source: Art Resource, NY

▶ Cylindrical jar with flower patterns, stoneware, 2016. Height: 15¾ inches, diameter: 8⅝ inches (40 x 22 cm) *Image: © Itō Sekisui V*

▼ Square jar with flower patterns, stoneware, 2016. Height: 9⅞ inches, width: 5⅛ inches, depth: 5⅛ inches (25 x 13 x 13 cm) *Image: © Itō Sekisui V*

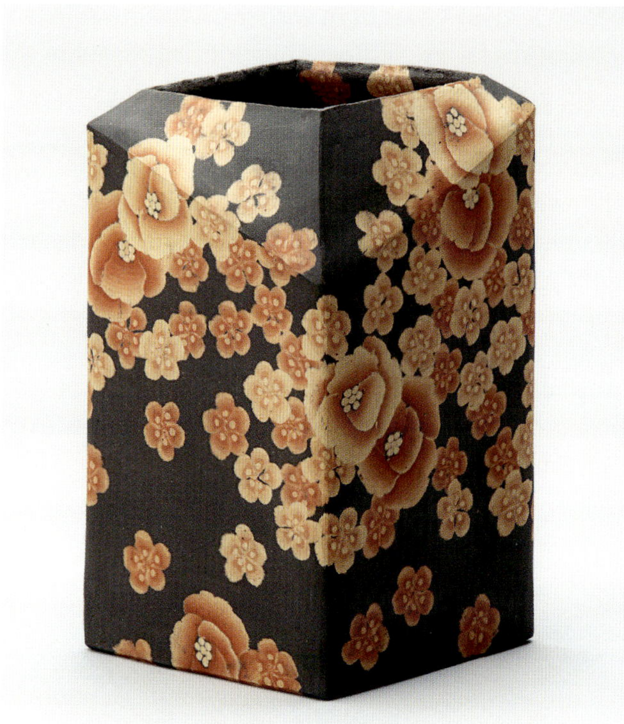

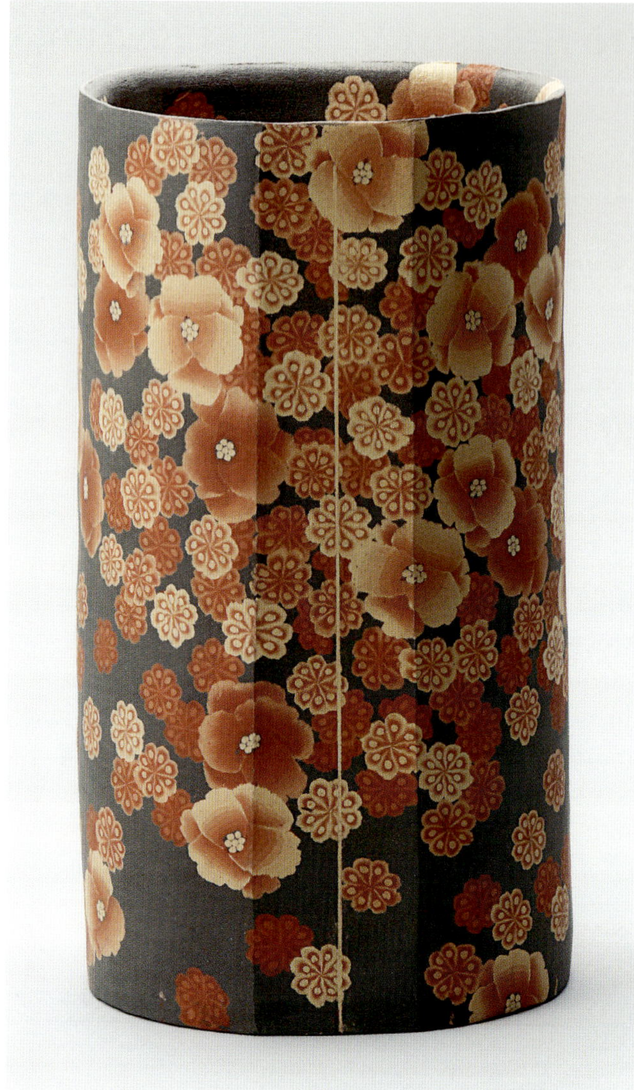

the result of the high iron content of the clay but the atmosphere inside the kiln is also partly responsible. Careful manipulation of the atmosphere leads to fired pieces with both vibrant red and areas of deep black which make for a striking and unique style called *yohen* ("kiln change"). Itō writes: "Depending on where potters place the works in the kiln, they can have a good idea of what kind of *yohen* will happen. However, it is not possible to do this with a precise level of control over the outcome. This is the interesting thing about pottery. With the color changes, I've added range to my work."

In addition to his traditional *yohen* work Itō Sekisui V is perhaps most famous for his highly skilled nerikomi vessels. This work started in the early 1980s. Using the red *mumyoi* clay along with a local yellow firing clay, as well as mixtures of the two, he has used the nerikomi technique to create beautiful cascades of flower petals and blossoms often against a deep black background which enhances the designs. He has also created mosaic, striped and gradated patterns with this technique. This work has set him apart from his ancestors although his use of the local clays ties this new work to the family, and more broadly, Sado traditions. His expertise in the traditional *yōhen* style of pottery as well as his innovative use of the *mumyoi* clay in the nerikomi pieces led to Itō Sekisui V being

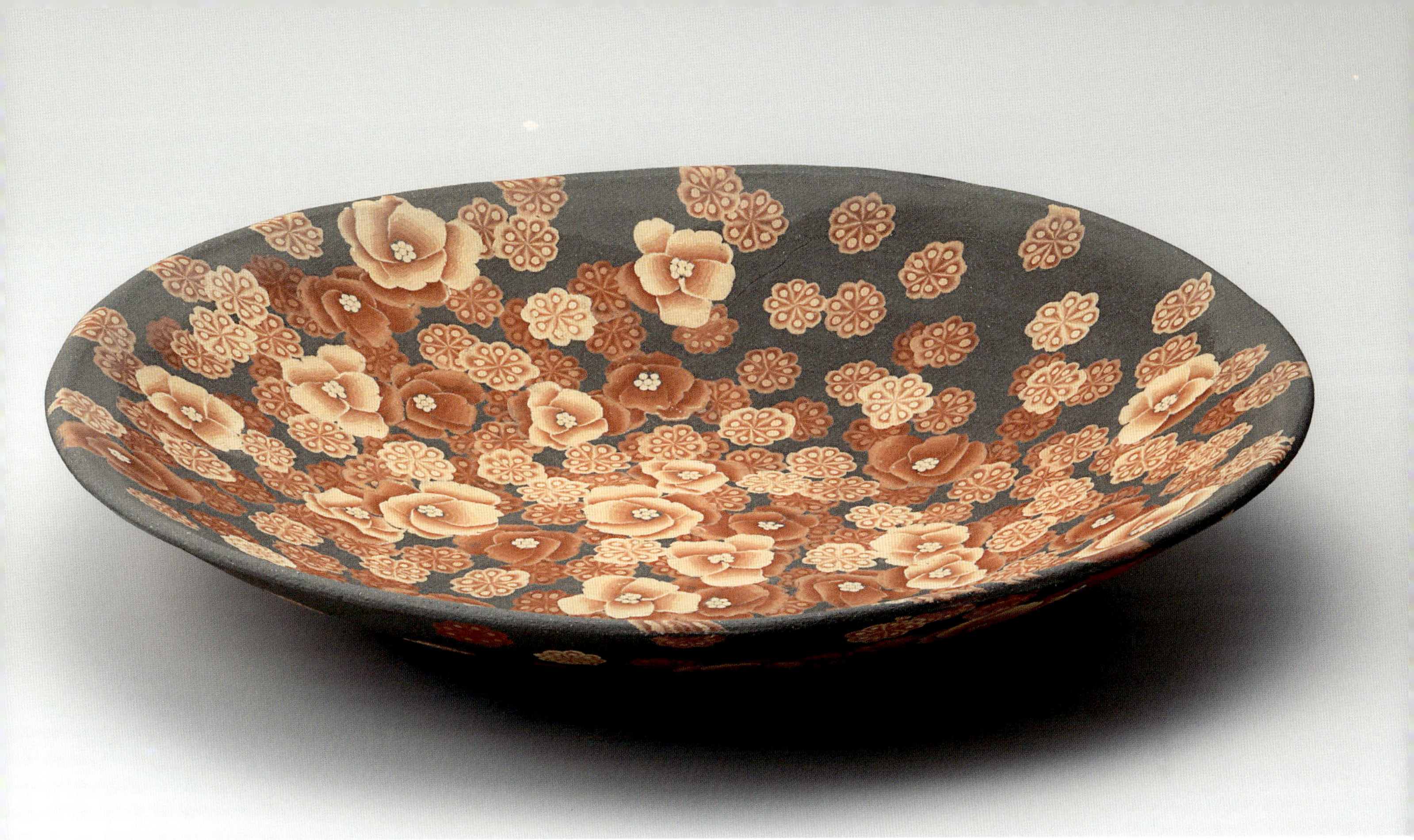

Plate with flower patterns, stoneware, 2016. Height: $3\frac{1}{8}$ inches, diameter: $16\frac{1}{2}$ inches (8 x 42 cm) *Image: © Itō Sekisui V*

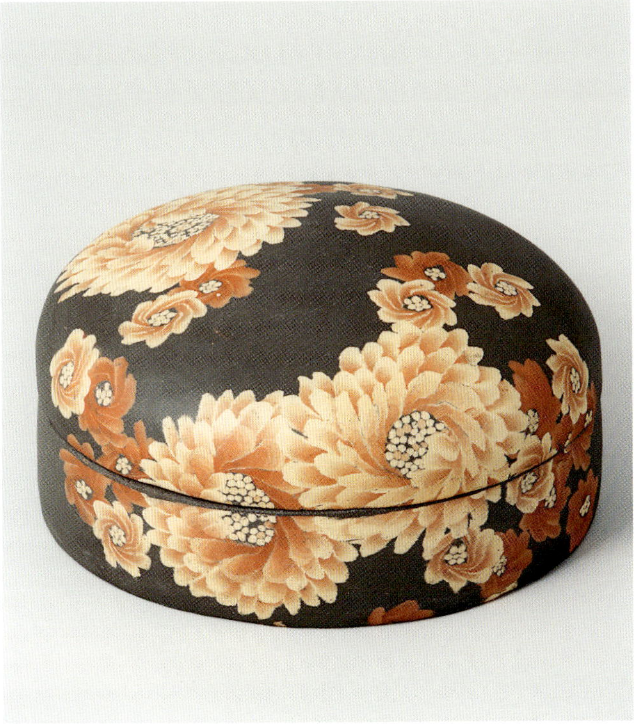

Case with flower patterns, stoneware, 2016.
Height: $5\frac{1}{8}$ inches, diameter: $9\frac{1}{2}$ inches (13 x 24 cm)
Image: © Itō Sekisui V

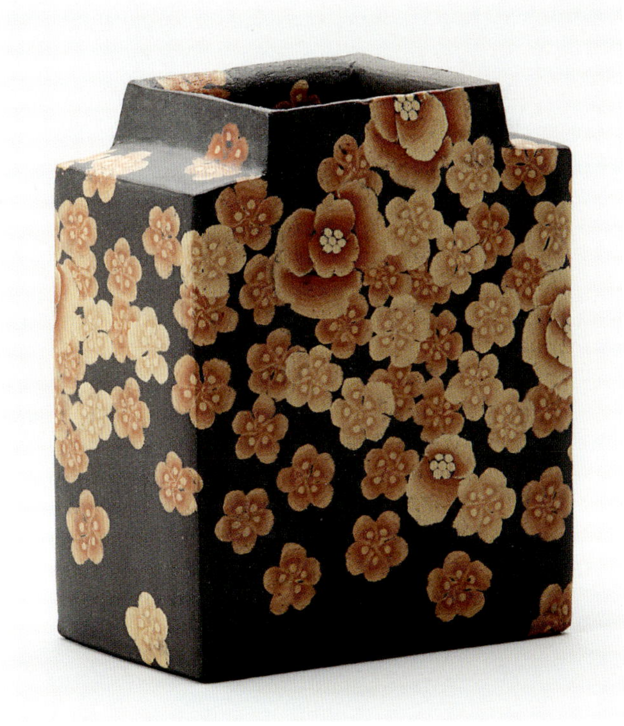

Square jar with flower patterns, stoneware, 2016.
Height: $7\frac{1}{8}$ inches, width: $6\frac{1}{4}$ inches, depth: $3\frac{1}{2}$ inches (18 x 16 x 8.9 cm) *Image: © Itō Sekisui V*

designated "Living National Treasure" in 2003. This honor would be considered by many as the capstone to a career but Itō has carried on in his drive to, as he states, "bring forth what has never existed, something new and attractive." His most recent work, *Sadogashima* of the "Sado Island" series, demonstrates his love of the potter's materials and the island's identity, incorporating a variety of crushed stones, including rhyolite and granite from the island which produce intriguing textures and a sense of ancient origins.

When questioned about Sado Island and its influence on his work Itō responds using the older name of the island:

> *When I think about what the island possesses to convey to the world, the answer has to be culture. Sadogashima has been an island of exile since ancient times. People who defied the central government were sent here, including Emperor Juntoku, the Buddhist priest Nichiren, and the playwright Zeami. But what was unusual was that they had to be self-sufficient. So long as they didn't leave the island, they were free to marry or apply their talents to business. As a result, the manners and customs of the nation's center were conveyed here, fostering a kind of highly cultured freedom, which took firm root.*

In 1972 Itō began a long history of participating in exhibitions throughout Japan with his acceptance into the Japanese Traditional Arts and Crafts Exhibition.

This was followed the next year by his Grand Prize in the 2nd Japan Ceramic Art Exhibition. In 1980 he was included in the Contemporary Japanese Ceramics Exhibition at the National Museum of History, Taiwan, and the following year was included in the Contemporary Japanese Ceramics Exhibition at the Smithsonian National Museum of Natural History (Washington DC) and Victoria and Albert Museum (London).

In 2003, the year Itō Sekisui V was designated as "Preserver of Important Intangible Cultural Property" (Living National Treasure), he also became an Advisor of the Japan Kogei Association. He had an audience with the Emperor and Empress presenting them with his *Mumyoi Neriage Flower Patterned Incense Burner* through the governor of Niigata and the Imperial Household Agency. Itō has been the recipient of many awards including the Grand Prize and Chichibunomiya Award at the 8th Japan Ceramic Art Exhibition in 1985 and the Praemium Imperiale Award at the 44th Japan Traditional Kogei Exhibition in 1997. In 2011 he received the Order of the Rising Sun from Emperor Akihito. 2017 brought his first international solo exhibition titled "Red Soil" at the Onishi Gallery in New York City. His work is included in the collections of many museums including The National Museum of Modern Art, Tokyo; the Ibaraki Ceramic Art Museum, Kasama City, Japan; the Sado International Art Museum, Sado Island, Japan; the Victoria and Albert Museum, London; the Smithsonian National Museum of Natural History, Washington DC; and The Metropolitan Museum of Art, New York.

Ogata Kamio (1949–2022)

Hokkaido, Japan's northern prefecture, was not primarily known for its ceramics traditions. The remarkable work of Ogata Kamio, however, has changed that perception. He was a self-taught artist with no ceramics background nor professional connections who was able to elevate the art of neriage to new heights throughout his career. In his thirties and already enamored with ancient Chinese neriage vessels, Ogata was inspired by a red-and-black vessel made by the famous potter Itō Sekisui V and began a journey to understand the mechanics of his techniques, ultimately moving beyond the common understanding of this work.

In a manner similar to Matsui Kōsei, Ogata's work is primarily a combination of neriage and nerikomi. Using the methods of nerikomi he layers hundreds of paper-thin sheets of colored clays in a precise order to create soft gradations of subtle color, often using grays, teal, and off whites. Thick slices of these pattern loaves are then formed into a cylinder which is placed on the wheel and thrown (neriage) to form his larger vessel shapes. At the leather-hard stage, trimming tools are used to carve the clay, disrupting the surface designs and adding an intriguing complexity. An early work in The Metropolitan Museum of Art's collection exhibits horizontally oriented carved knife-edged grooves which interact with the diagonal nerikomi pattern, creating an optical impression of movement reflected in its title, *Spiraling Vessel, Dizzy Shadings*.

Ogata continued his innovative search and went on to develop his classic style of hand-carving rippling surfaces into his thrown vessels, often bringing to mind the ripples on the surface of water. Ogata writes: "When thinking about wave patterns on the surface, what came to mind was the black river in Korin's byobu screen, 'Red and White Plum Blossoms.' I wanted to create a river in my works. In the process of making it, I also noticed an optical illusion."

Expanding on this idea of rippling surfaces he writes:

> *I'm fascinated by the "motion" of a design, something lacking in most designs using nerikomi. My personal theme, then, is "how to give motion" to nerikomi. Recently, my feelings have been changing: I'd like to give expression to motion in nerikomi to show that it, as a phenomenon, offers more than just delicacy and refinement. The outcome, I hope, will bring more emotion and feeling to the art of nerikomi.*

Ogata first set up his studio in 1983 in the Shinshinotsu village, Hokkaido. During a period from 1977 to 1994, he taught pottery at the Shinshinozu Institute for the Handicapped. By 1984 he was already exhibiting his work and a few years

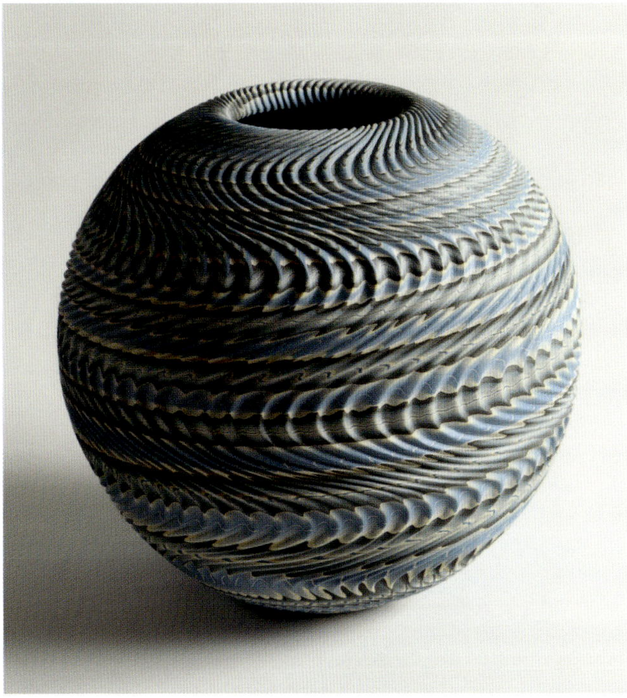

***Rinpa*, stoneware, 2019. Height: 9¼ inches, diameter: 9½ inches (23.5 x 24 cm)**

Round with blue, gray and white neriage (marbleized) vessel with carved, ridged surface and recessed mouth.
Image: © Courtesy of Joan B Mirviss Ltd

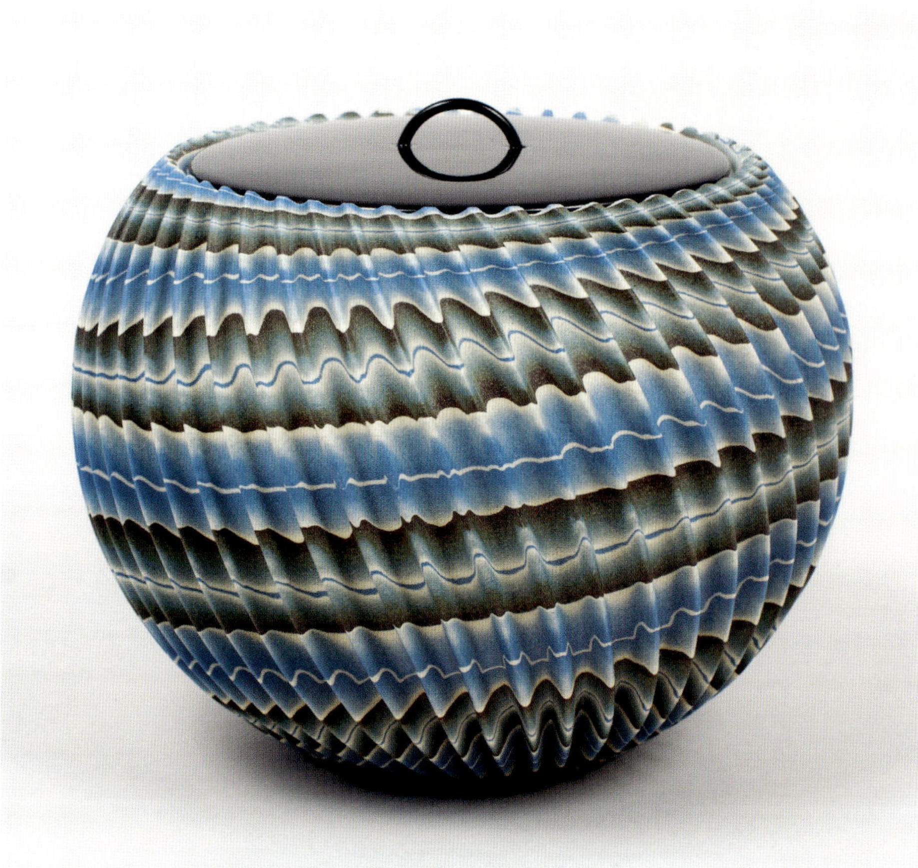

◀ **Covered fresh-water jar (*mizusashi*), stoneware, 2014. Height: 7¼ inches, diameter: 7⅝ inches (18.42 x 19.38 cm)**

The jar displays a design of marbleized and ridged surfaces. *Saint Louis Art Museum, William K. Bixby Trust for Asian Art 12:2015a-c; Image: © estate of Ogata Kamio*

▶ ***Abyss***, **stoneware, 2016. Height: 11⅝ inches, diameter: 12⅜ inches (29.5 x 31.4 cm)**

Standing, conical, neriage (marbleized) vessel in blue, gray, and white with carved, pleated surface and rolled-in mouth. *Image: © Kohiji Inoue*

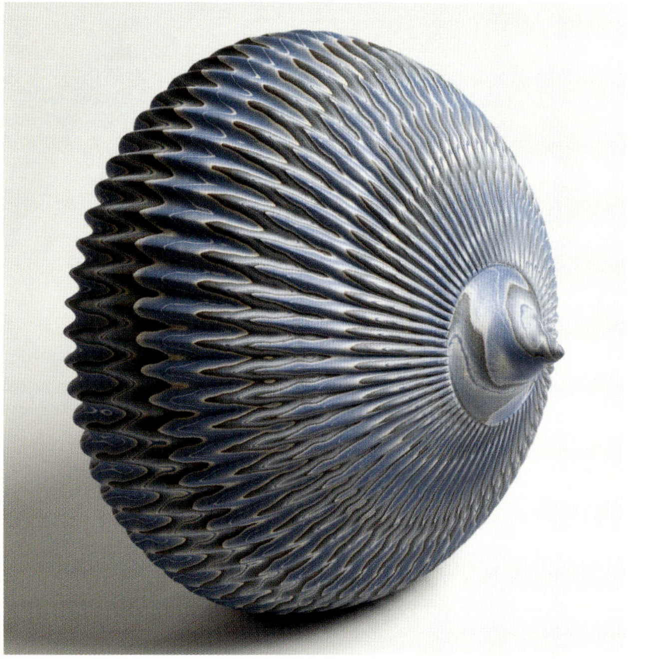

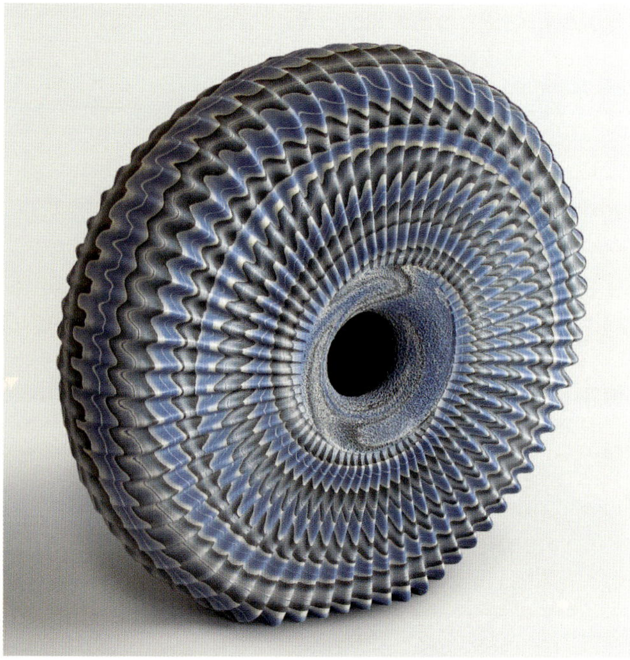

Two views of *Lunar Landing*, stoneware, 2017. Height: 9¼ inches, diameter: 9½ inches (23.5 x 24 cm)

Standing flattened round neriage (marbleized) top-like sculpture with blue, gray, and white pleated surface, with a pointed top on one side and a hole on the other. *Images: © Kohiji Inoue*

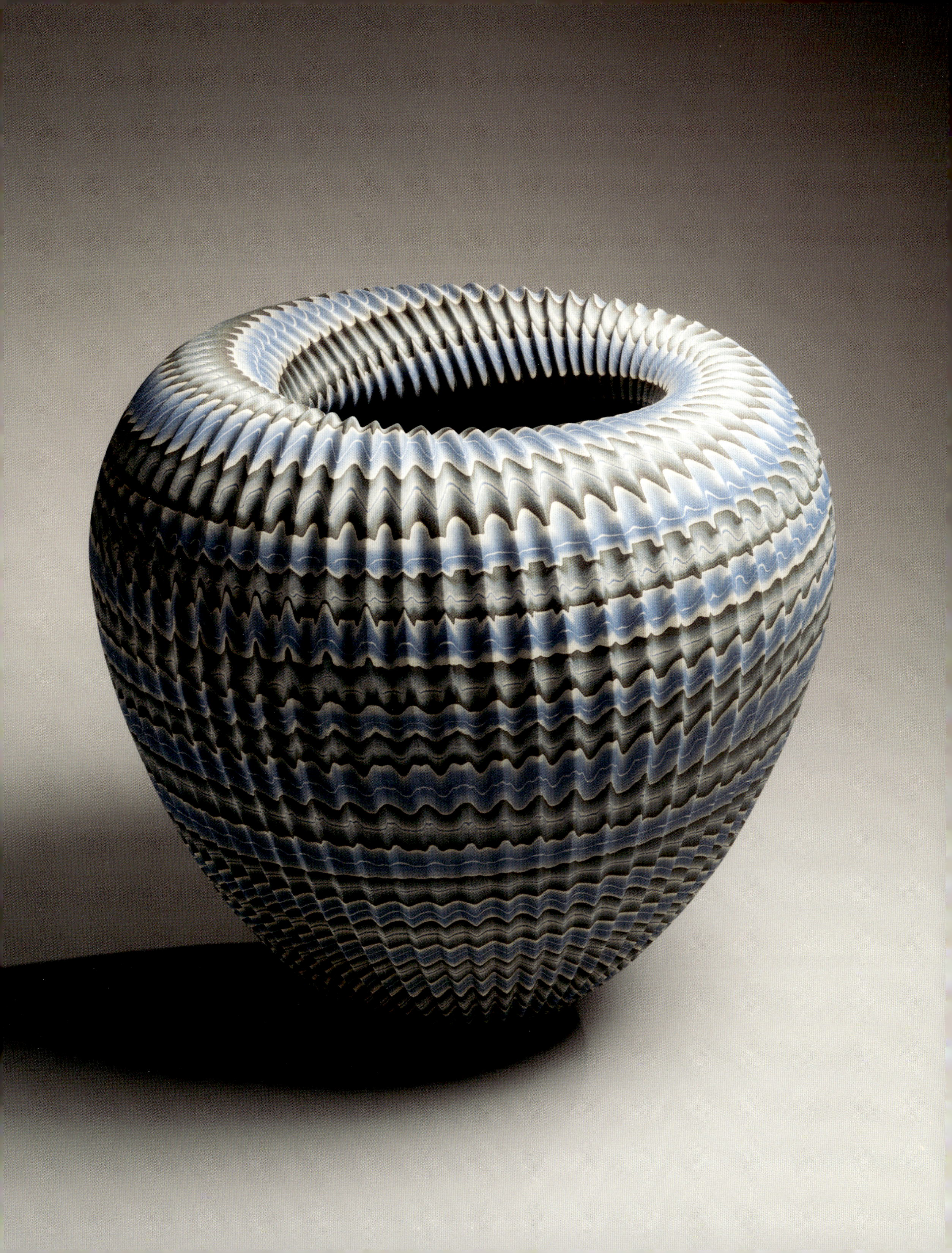

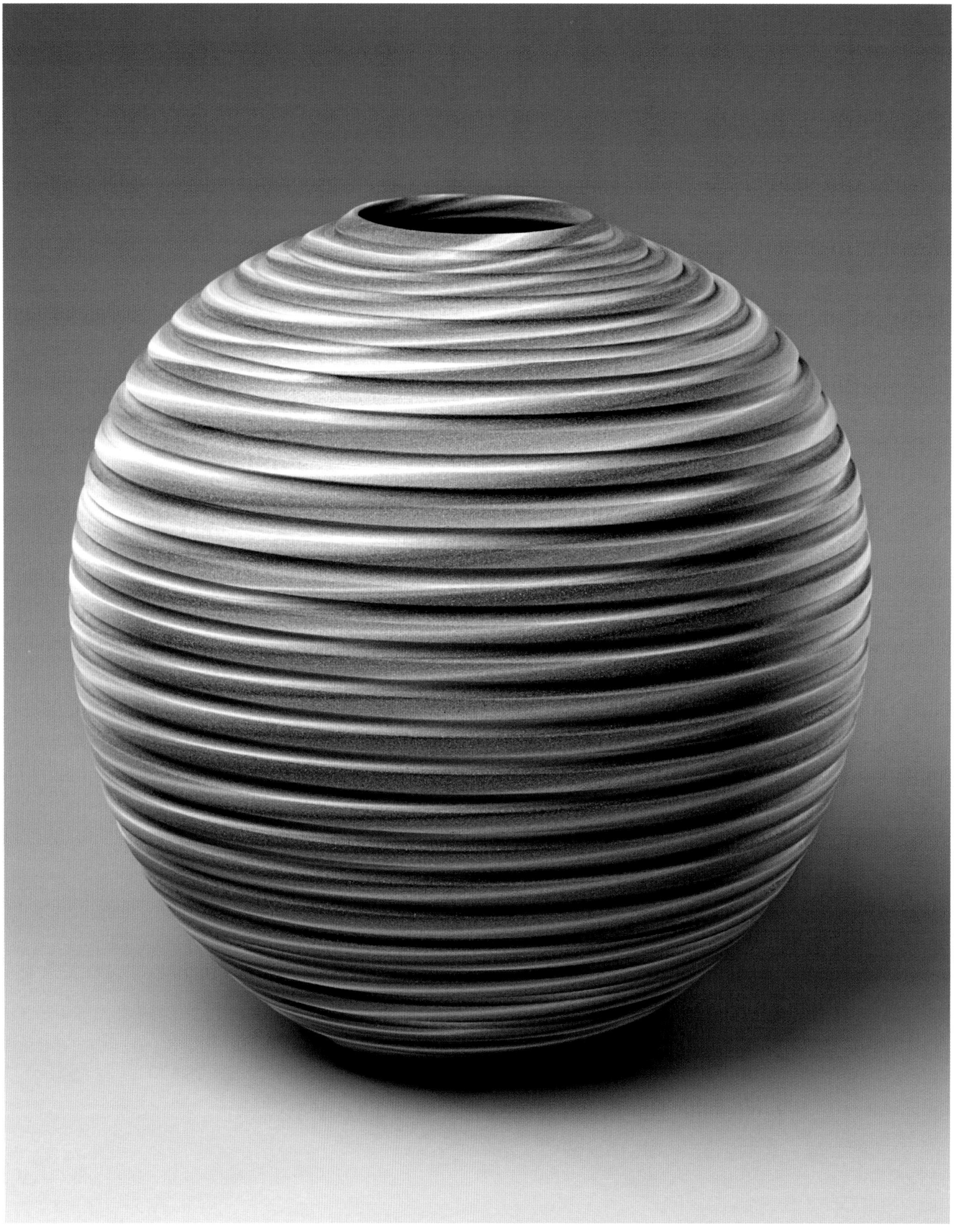

later began receiving prizes for his unique vessels. By the 1990s he was having regular solo exhibitions which continued throughout his career. His first solo exhibition outside of Japan was in 2019 at the Joan B Mirviss Gallery in New York City.

Ogata's remarkable achievements in the world of neriage and nerikomi are a reflection not only of his meticulous technique and tasteful aesthetic sensibility but also his boldness to constantly challenge himself, digging deeper into the relevance of his work in the rapidly changing twenty-first century. He writes: "I'm inspired to make new works by the combinations of the colored clay's gradations in color and shape, and their relevance to reality. Via the subtle changes in the clay, I'm able to express intangible emotions such as feelings of insecurity, instability, and struggle" – or put another way – "I try to express the spirit of our era, the gap between our physical and emotional selves, and how our daily lives are without focus. For me, this gap produces feelings of uncertainty and dizziness."

Ogata Kamio was a member of The Japan Crafts Association and the Hokkaido Art Association. His groundbreaking vessels have been presented in solo exhibitions at Kogei Murata, Ginza; Inui Gallery, Akasaka, Tokyo; Mitsukoshi, Nihonbashi, Tokyo; Nakamura Art Salon, Ibaraki; and Touch Ceramics Gallery, Hong Kong, among others.

He participated in numerous exhibitions throughout Japan and abroad and received many prizes and awards including the Oribe Silver Prize, Gendai chato-ten, Toki, Gifu (Contemporary Tea Ceramics Exhibition) (2005), and the Incentive Award, Japan Ceramic Society Hokkaido New Works Exhibition (2016). His work is found in many museums including several in the USA: The Metropolitan Museum, New York City; the Saint Louis Art Museum; the Asian Art Museum, San Francisco; and the Museum of Fine Arts, Boston, among them.

◄ ***Spiraling Vessel, Dizzy Shadings*, 2005. Height: 13 inches, diameter: 11⅞ inches (33 x 30.2 cm)**

Wheel-thrown and carved vessel, marbleized stoneware (neriage). *Image: © The Metropolitan Museum of Art. Image source: Art Resource, NY*

Kondō Takahiro (b. 1958)

Kondō Takahiro was born into a highly acclaimed ceramics family in Kyoto, known for their mastery of *sometsuke* or underglaze cobalt blue decoration. His grandfather, Kondō Yuzo (1902–1985), was given the title of Living National Treasure in 1977 for his *sometsuke* porcelain, and his father, Kondō Hiroshi, was also well known for this style of ceramics. His uncle, Kondō Yutaka who was also a ceramic artist and university teacher, lived nearby. Rather than immediately follow in his father's and grandfather's footsteps Kondō pursued a degree in literature at Hosei University. He later worked in business while pursuing his love of table tennis, ultimately becoming Japan's top-ranked table tennis player and traveling abroad to competitions. At the age of 26 he decided to return to his roots and went back to school to learn more about wheel throwing and glazing. After graduating from the Kyoto Ceramics Training School (1985) and the Kyoto Industrial Research Institute (1986) he became his father's assistant for three years, absorbing the family knowledge of *sometsuke*.

After an initial period of creating work in the *sometsuke* style, Kondō began exploring other styles and techniques, searching for a personal direction more in tune with his times and the state of the world. Slab building replaced wheel throwing and, never having studied this technique in school, he struggled through an initial period of trial and error. A major focus during this period was the relationship between form and decoration. Tomimoto Kenkichi, his grandfather's first teacher (named Living National Treasure in 1955) had impressed his ideas about this relationship to the entire Kondō family and Takahiro constantly strove to find a creative balance between the two. Over time he also made the decision to distance himself from the traditions of his ancestors and strike out on his own as an independent artist focusing on original work.

In 1990 when he was 32, Kondo's first exhibition helped to set him on a path of international exposure. The exhibition was at the São Paulo

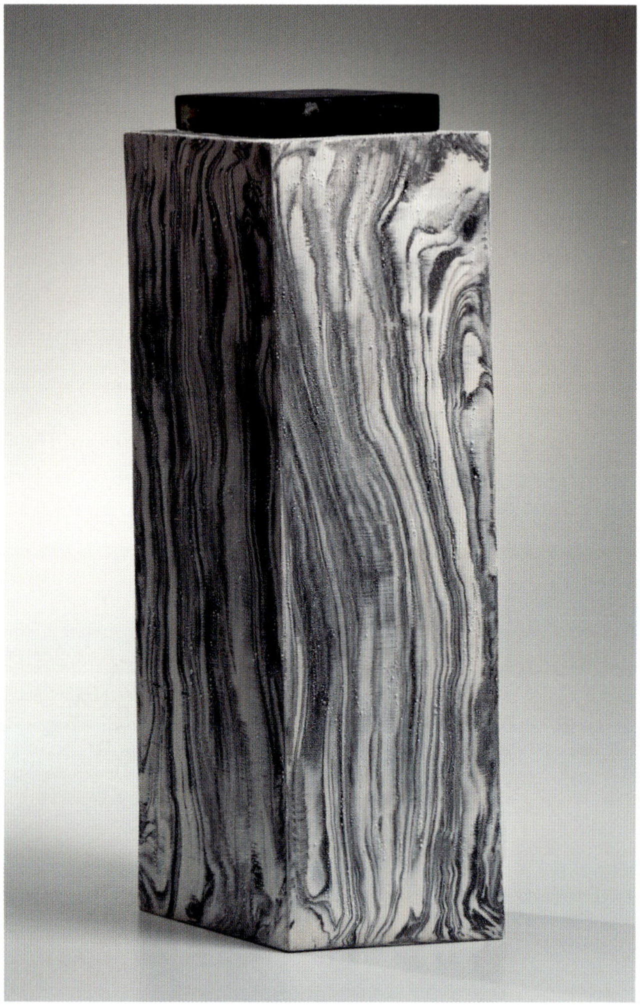

***Wave*, 2021. Height: 28⅜ inches, width: 13 inches, depth: 7⅛ inches (72 x 33 x 18 cm)**

Standing rectangular black-and-white marbleized form with *pâte de verre* cast clear-and-black glass cover and "silver mist" beaded glazing. Marbleized porcelain, "silver mist" overglaze, cast glass. *Image: Joan B Mirviss Gallery Ltd; © Richard Goodbody*

▶ ***Wave*, 2021. Height: 24⅝ inches, width: 9⅞ inches, depth: 7 inches (62.5 x 25 x 17.8 cm)**

Standing hexagonal black-and-white marbleized monolithic sculpture with "silver mist" beaded overglazing and small, slightly raised mouth. Marbleized porcelain, "silver mist" overglaze. *Image: Joan B Mirviss Gallery Ltd; © Richard Goodbody*

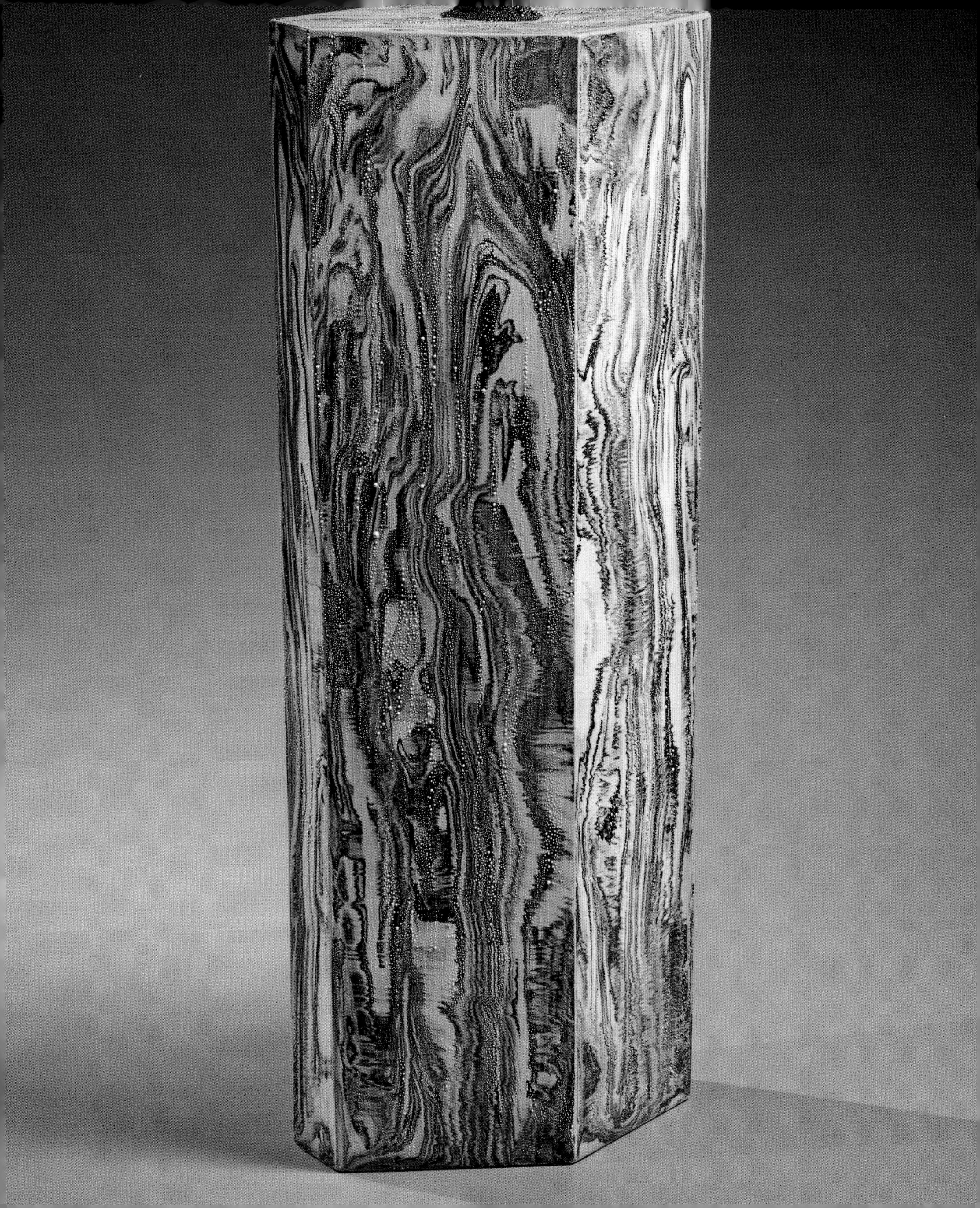

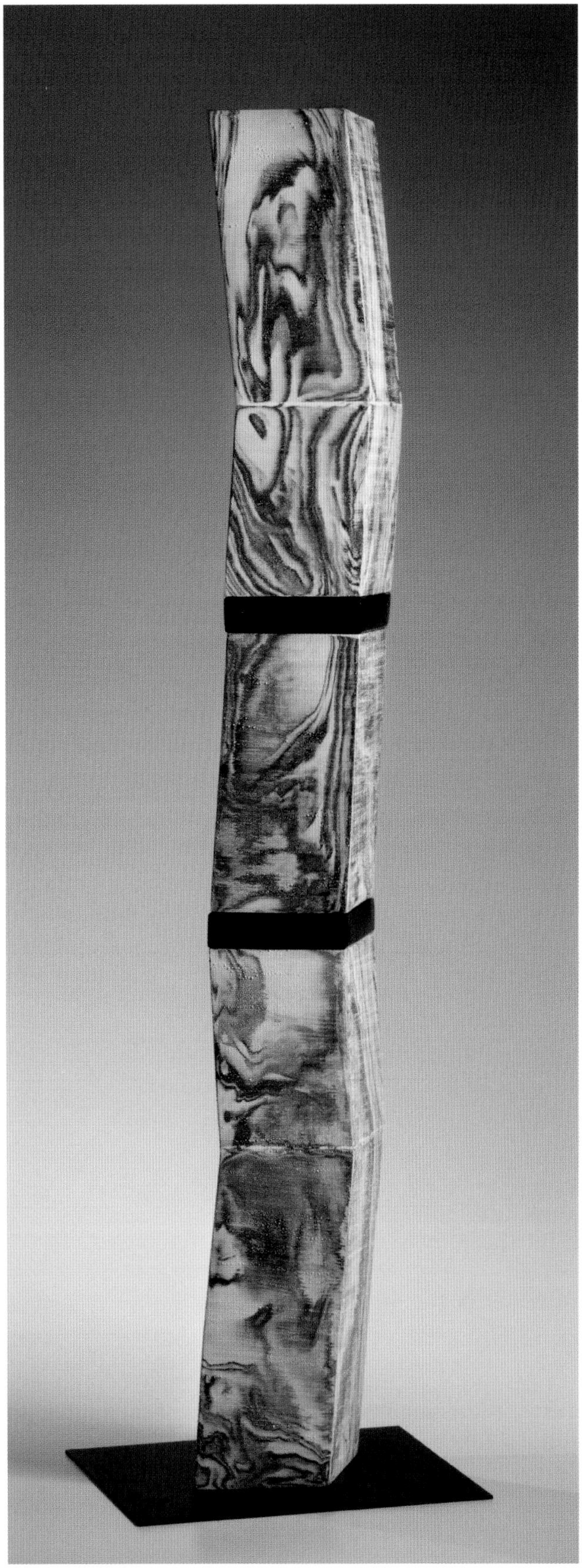

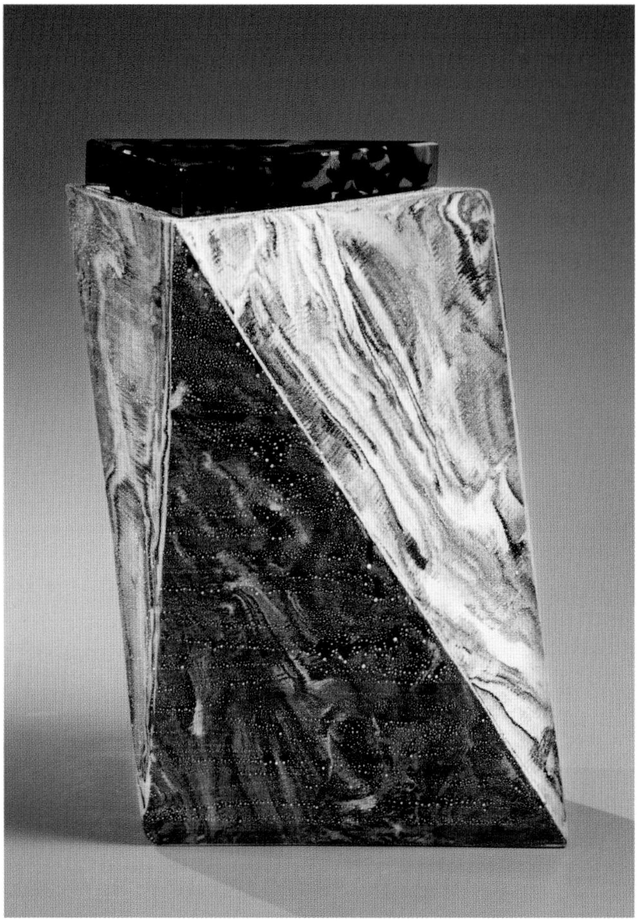

▲ **Wave, 2021. Height: 15⅛ inches, width: 9½ inches, depth: 6⅛ inches (38.4 x 24 x 15.5 cm)**

Standing black-and-white marbleized form with *pâte de verre* cast clear-and-black glass cover and "silver mist" beaded glazing. Marbleized porcelain, "silver mist" overglaze, cast glass. *Image: Joan B Mirviss Gallery Ltd; © Richard Goodbody*

◀ **Wave, 2021. Height: 44¾ inches, width: 7¼ inches, depth: 6¾ inches (113.6 x 18.4 x 17 cm)**

Standing zigzag black-and-white marbleized form with *pâte de verre* cast clear-and-black glass and "silver mist" beaded glazing. Marbleized porcelain, "silver mist" overglaze, cast glass, removable stainless-steel base. *Image: Joan B Mirviss Gallery Ltd; © Richard Goodbody*

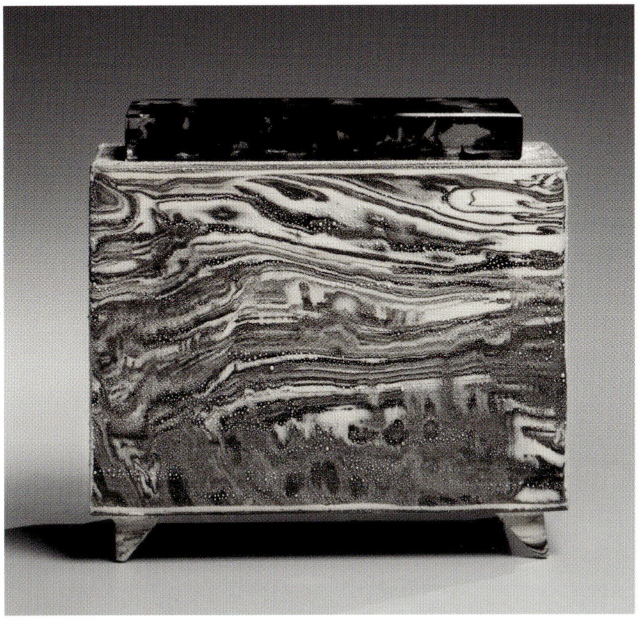

***Nami shizuku* (*Wave Droplet*), 2021. Height: 7¾ inches, width: 8½ inches, depth: 3¾ inches (20 x 21.6 x 9.5 cm)**

Standing rectangular marbleized box form on four legs with "silver mist" beaded glazing and *pâte de verre* clear-and-black glass cover. Marbleized porcelain, "silver mist" overglaze, cast glass. *Image: Joan B Mirviss Gallery Ltd; © Richard Goodbody*

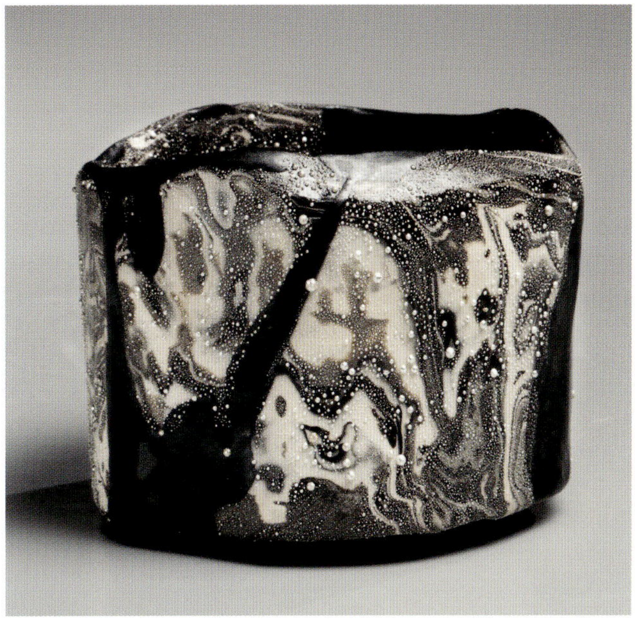

Ginteki wan nami (*Wave Tea bowl with Silver Mist Glaze*), 2020. Height: 3¾ inches, diameter: 4¾ inches (9.5 x 12 cm)

Rounded tea bowl in black, gray, and white marbleized clay with *gintekisai* (silver mist) overglaze and hand-brushed cobalt glaze decoration. Marbleized porcelain, "silver mist" overglaze. *Image: Joan B Mirviss Gallery Ltd; © Richard Goodbody*

Museum of Art in Brazil at a time when interest in contemporary Japanese ceramics was growing around the world. Exhibits in New York, Melbourne, and London soon followed and provided venues for Kondo's continuing artistic development.

Over time he began to focus on the concept of water and what it represents. Joan Mirviss writes of Kondo's concepts: "Water has been his principal theme for many years and he has explored not only its physical forms such as rain, ice, steam, and mist, but also its conceptual aspects as a source of life, purifier, natural resource, and, most currently, as a critical environmental issue in regards to the nuclear disaster in Fukushima."

Central to his investigation of water is Kondo's invention of his "silver mist" (*gintekisai*) glaze, which he calls "water born of fire," perfected in 1994 after years of experimentation and later patented. The glaze manifests as a beautiful coating of tiny droplets of silver, sometimes collecting as larger drops and even drips, while exposing the colored clay or other decoration beneath. Kondō has said of this glaze that the droplets appear to "take on an appearance that looks almost more like water than real water." The glaze is composed of a combination of platinum, gold, silver, and glass frit and has become one of his primary signature techniques, carrying with it the many associations and metaphors relating to water. The artist writes: "My work embodies the concept of '*utsuroi no bigaku*' or the art of impermanence. It is my intent to evoke that beauty found in the ephemeral through the waves and droplets on my surfaces that represent the cycle of nature, purification and rebirth."

Aways searching to broaden his artistic scope and challenge himself with new concepts, Kondō traveled to Scotland where he attained a Master's Degree of Design and Applied Arts at Edinburgh

College of Art in 2003. There he studied glassmaking techniques and became proficient in casting glass blocks, often with inclusions of bubbles or more opaque elements, which he now incorporates into his ceramic sculptures and box-like vessels.

While in Scotland he traveled to the Orkney Islands where he encountered the Ring of Brodgar, a prehistoric standing stone circle. Kondō writes of this as being "located in the midst of sky, rain, and wind at a point where land and sea horizons seem to meet: huge rocks connecting heaven and earth, life and death, past and future." This experience inspired his "Monolith" series of ceramic sculptural forms, larger than anything he had created before. It also marked a turning point where he began to prioritize both artistic concept and technical expertise. What followed was a wide range of sculptural experiments using a variety of materials as well as clay.

Perhaps Kondo's most famous sculptural work is his "Reduction" series, ceramic sculptures of seated figures meditating in a full lotus position. He made 21 versions between 2012 and 2017 from a life cast of his own body, the final works being smaller in size due to the shrinkage of clay as it is dried and fired. A wide variety of decorating and firing techniques were used so that each is quite unique. Some were made with nerikomi clay slices and finished with his signature "silver mist" glaze. This series was largely influenced by the devastating earthquake and ensuing tsunami of March 11, 2011 and the resulting Fukushima Daiichi nuclear disaster. Considered as a group, but also individually, they convey a strong spiritual feeling and elicit quite powerful emotions. Many are now in museum collections around the world, including four in the United States.

For his recent work Kondō Takahiro has used the traditional technique of nerikomi with white, dark brown, and light-brown porcelain in a loose gestural manner, again exploring his interest in the many forms of water. The flowing abstract designs have been described by Alain R. Truong as "reminiscent of seascapes in the tradition of Japanese ink painting (*suiboku-ga*)." The resulting nerikomi slices are used for both his slab-built sculptural forms and more intimate hand-formed tea bowls, the "silver mist" glaze adding a subtle and transformative layer of fascination to the work.

Kondō Takahiro's work has been exhibited in many solo exhibitions including "Making Waves" at the Joan B Mirviss Gallery, New York (2022); "Thoughts on Hands," Kahitsukan, Kyoto Museum of Contemporary Art (2017); "Seisui: Vessel Shifting," Setouchi City Museum of Art (2016); "Self Portrait," The Museum of Arts and Crafts, Itami (2010); "Metamorphose," Kyoto Art Center/Nariwa-cho Museum/Paramita Museum (2007); "Takahiro Kondo: New Blue & White," National Museum of Scotland (1995) and many others in Japan, the USA, Australia, Brazil, and England. His work has also been exhibited in numerous group exhibitions in Japan, Italy, Australia, the USA, France, Germany, England, the Netherlands, Taiwan, Singapore, China, and New Zealand. In 2020 he received the 33rd Kyoto Art and Culture Prize and in 2021 received the 39th Kyoto Prefecture Prize for Distinguished Service in the Arts.

Kondo's work is included in the collections of The Metropolitan Museum of Art, New York; the Brooklyn Museum, New York; the National Museum of Scotland; the Paramita Museum, Mie, Japan; the National Gallery, Victoria, Australia; and the São Paolo Museum, Brazil among many others in Japan, Northern Ireland, the USA, Scotland, France, and England.

▶ ***Reduction Wave*, 2017. Height: 33 7/8 inches, width: 25 5/8 inches, depth: 17 11/16. (86.1 × 65 x 44.9 cm)**
Marbleized porcelain with "silver mist" overglaze. *Asian Art Museum of San Francisco, Acquisition made possible by Dr. Phyllis A. Kempner and Dr. David D. Stein, 2020.3. Image: © Kondō Takahiro. Photograph Image: © Asian Art Museum of San Francisco.*

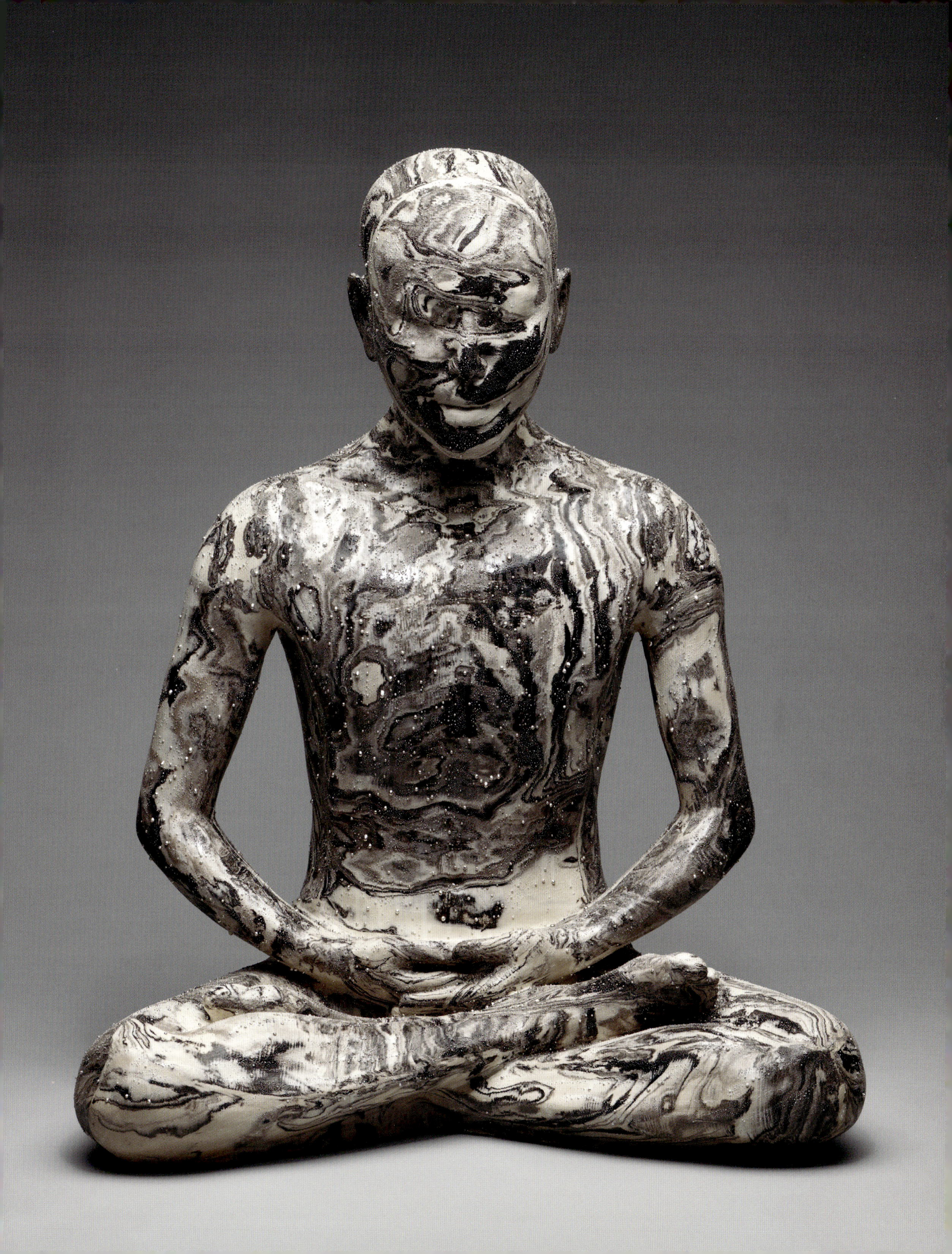

Nishi Koichi (b. 1948)

Born in Gifu in 1948, Nishi Koichi was inspired early on by the work of Shoji Kamoda, then one of the most renowned ceramic artists, and who had an attitude of continuously exploring new ideas. As he developed his pottery skills Nishi Koichi discovered and was inspired by the work of Matsui Kōsei, already well known for his work with colored clay using the neriage and nerikomi techniques.

What differentiates Nishi Koichi's work from that of others working with colored clay is his remarkable precision in creating meticulous patterns comprising subtle color differences incorporated into repeated geometric designs. His forming process involves making an inner mold with styrene foam and melting the styrene foam with thinner after molding. Another unique aspect in much of his work is the pebbly texture that he achieves through polishing the surface with a spoon or the like many times during the drying process and firing each piece without any glaze. This texture adds another dimension to Nishi's work and serves to make his precise designs even more astounding. He speaks of his work as "aiming to create works that cannot be expressed by any other method than kneading (working with colored clays)."

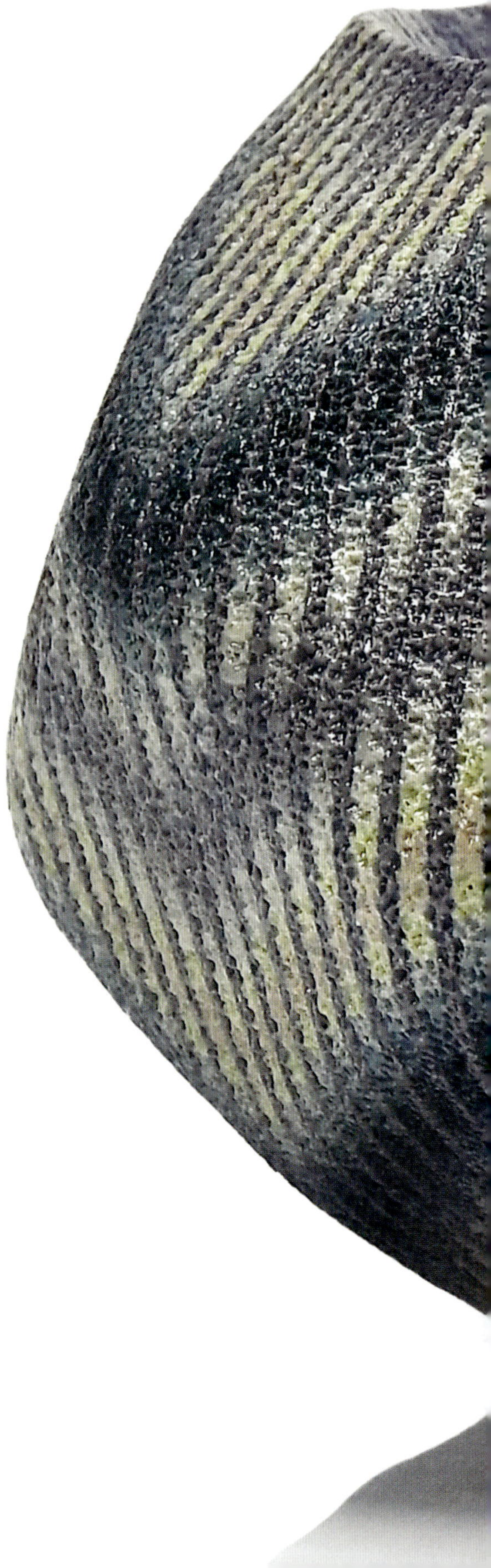

Nerikomi vase, 2022. Height: 5½ inches, diameter: 6¾ inches (14 x 17 cm) *Image: © The Stratford Gallery Ltd/2022*

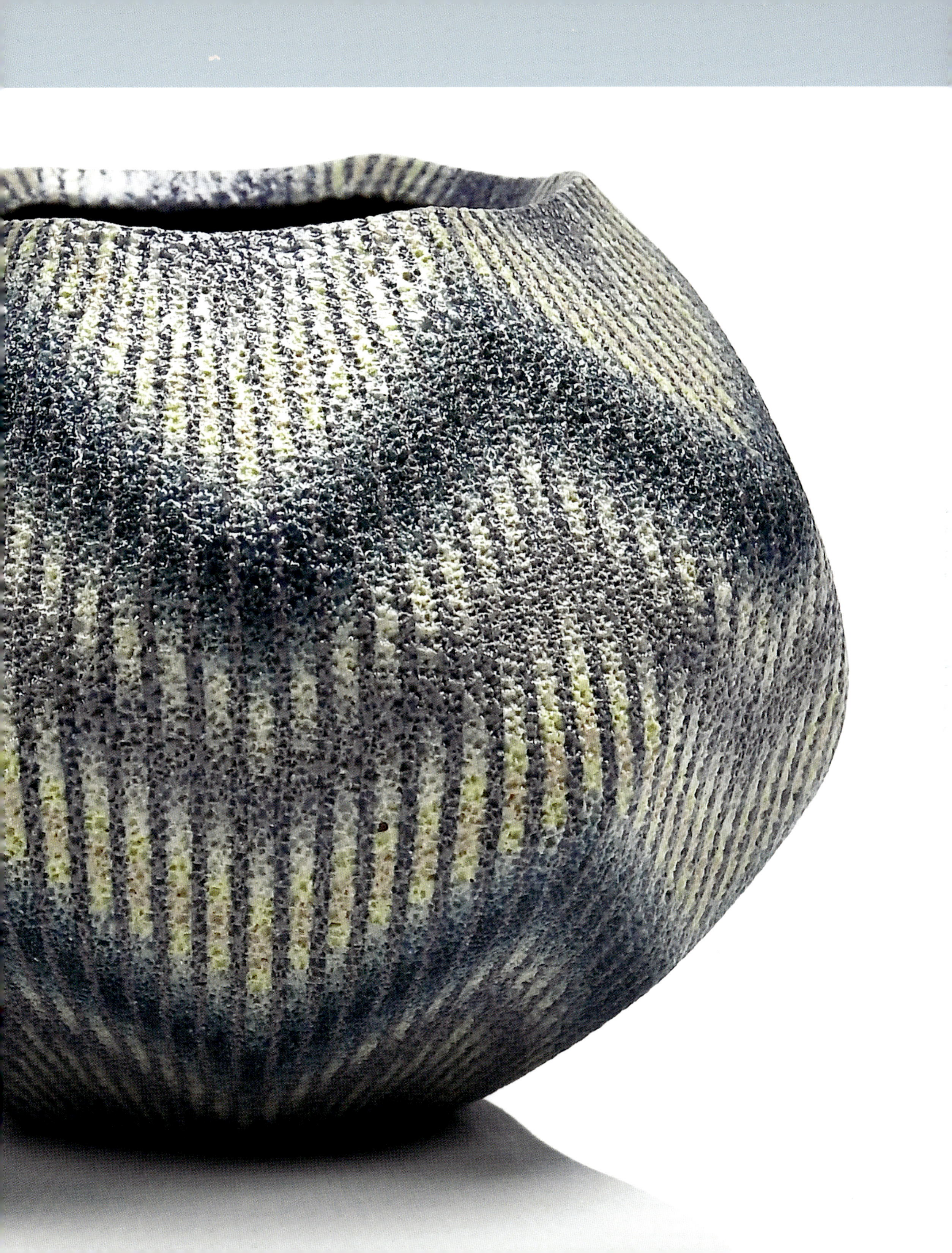

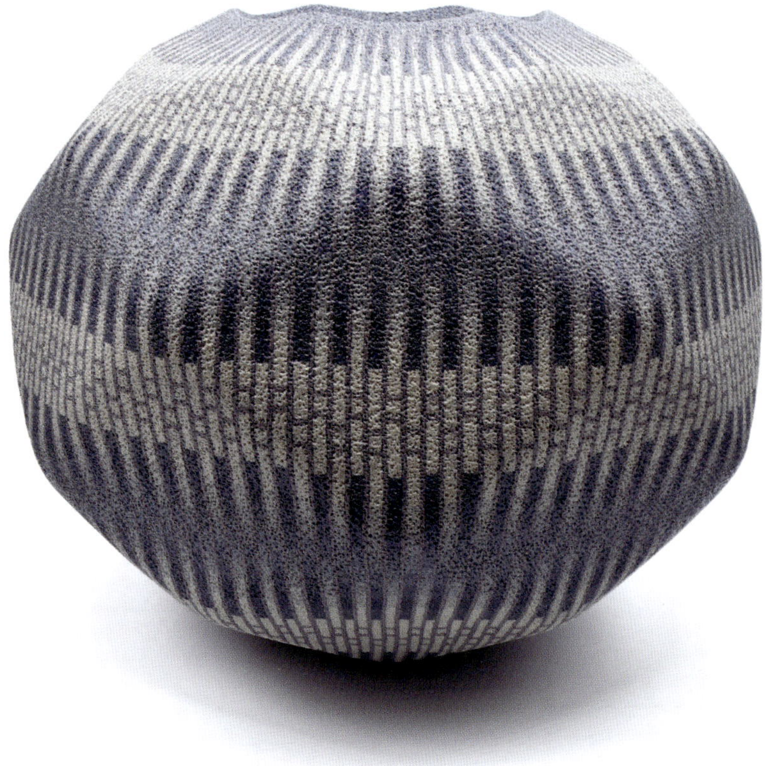

◄▼ **Two views of a large nerikomi vase, 2022. Height: 12⅝ inches, diameter: 15 inches (32 x 38 cm)**
Image: © The Stratford Gallery Ltd/2022

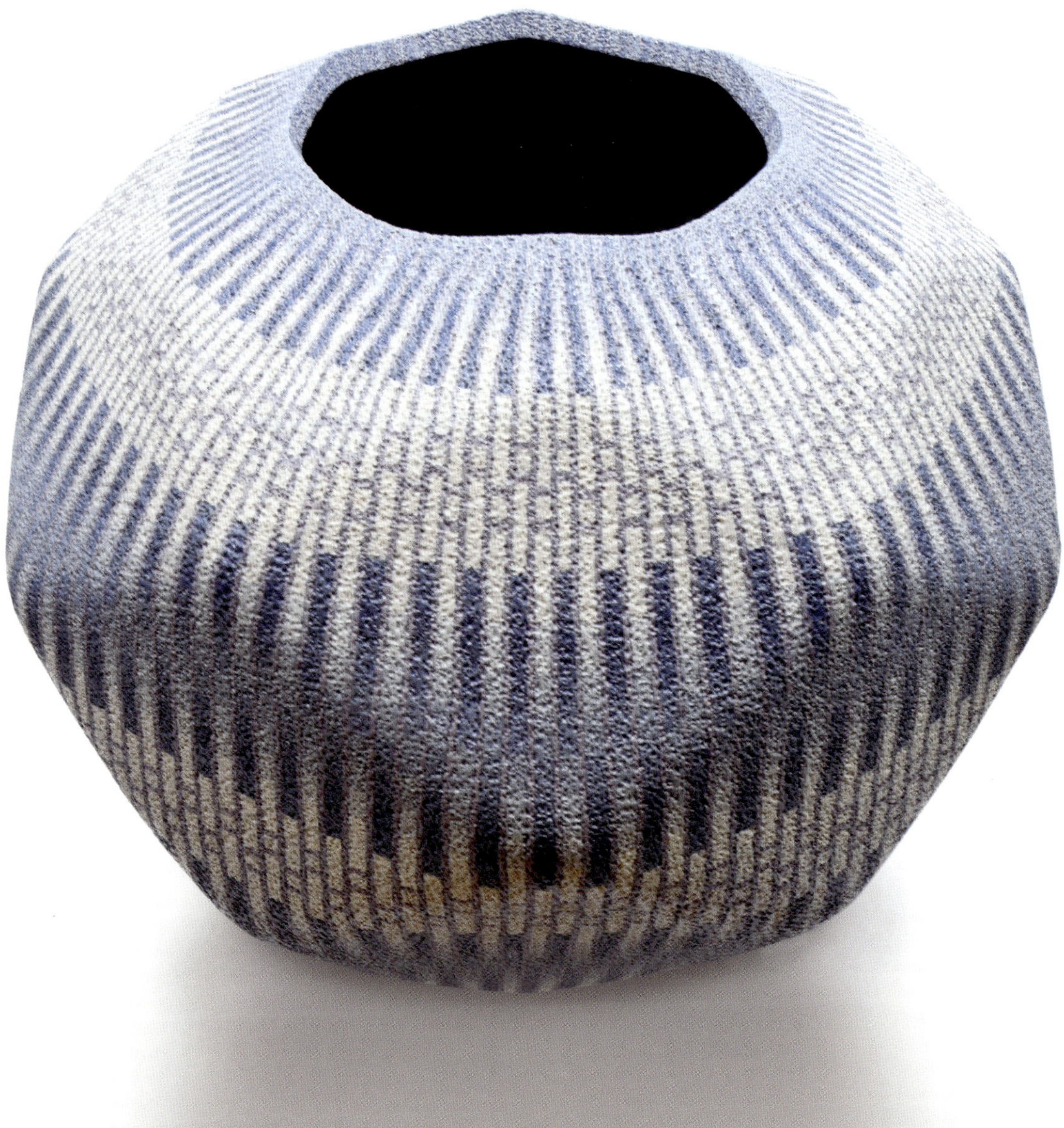

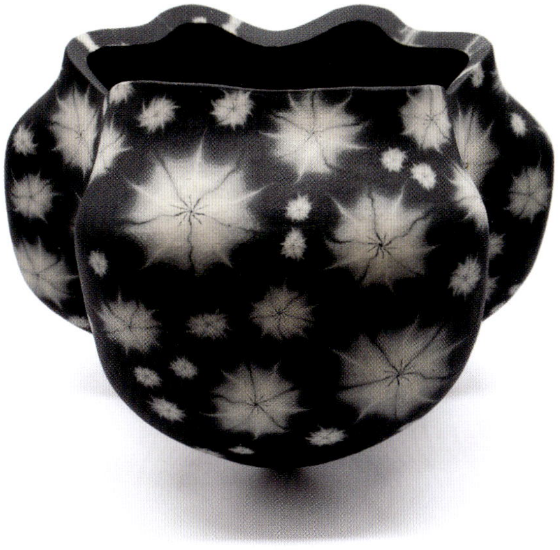

Nerikomi vase, 2022. Height: 7½ inches, diameter: 10¼ inches (19 x 26 cm) *Image: © The Stratford Gallery Ltd/2022*

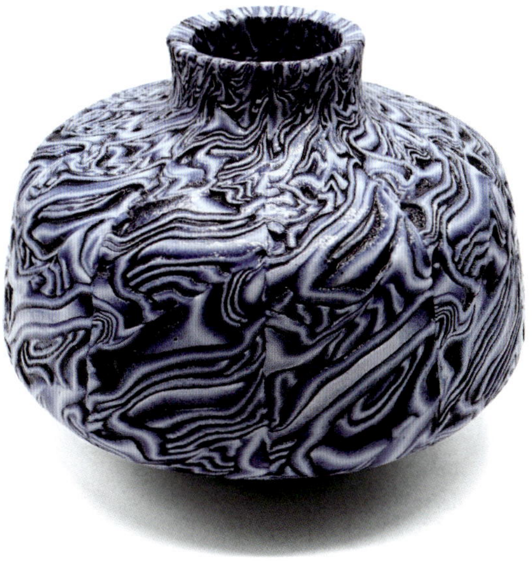

Nerikomi vase, 2022. Height: 8¼ inches, diameter: 9 inches (21 x 22.8 cm) *Image: © The Stratford Gallery Ltd/2022*

TIMELINE

1982 Selected for the Asahi Ceramic Art Exhibition (selected 7 times in total from this year onward)
1983 Selected for the Chunichi International Exhibition of Ceramic Arts (selected 5 times in total from this year onward)
1984 First Solo Exhibition at Gallery Moose (Seki City, Gifu)
1986 Selected for the International Ceramics Festival Mino '86 (selected 5 times)
1989 Artwork added to the collection of Mino Ceramic Art Museum, Tajimi
1993 Artwork added to the collection of the Museum of Fine Arts, Gifu
1994 Selected for the Japan Traditional Kōgei Exhibition (selected 15 times in total from this year onward). Holds many solo exhibitions from this year onward at Gallery Kohodo (Gifu), Nagoya Mitsukoshi, Takasaki Takashimaya, etc.
1999 NHK BS2 program "Yakimono Tanbo – Seeking the Possibilities of Neriage" broadcast
2012 Appears on NHK General TV "Kintoku – Yakimono Daihyakka [The Ceramics Encyclopedia]"
2013 Award by the Seki City Board of Education "Koichi Nishi Art Exhibition - The Neriage Challenge" held at Shinoda Toko Art Space Japan Ceramic Art Exhibition, 2 times

AWARDS

1986 Receives the Tokyo Governor Award at the Nihon Shin Kogei Exhibition
1987 Receives the Asahi Ceramic Art Incentive Award at the Asahi Ceramic Art Exhibition
1989 Received the Special Award at the Asahi Ceramic Art Exhibition
Received the Judge's Special Award (selected by Takuo Kato) at the International Ceramics Festival Mino '89
2006 Received the Grand Award at the Mino Togei Exhibition
2011 Award for Achievement of Traditional Culture Inheritence in Gifu Prefecture
2014 Received the 32nd Takuo Award

PUBLIC COLLECTIONS

The Museum of Fine Arts, Gifu
Mino Ceramic Art Museum

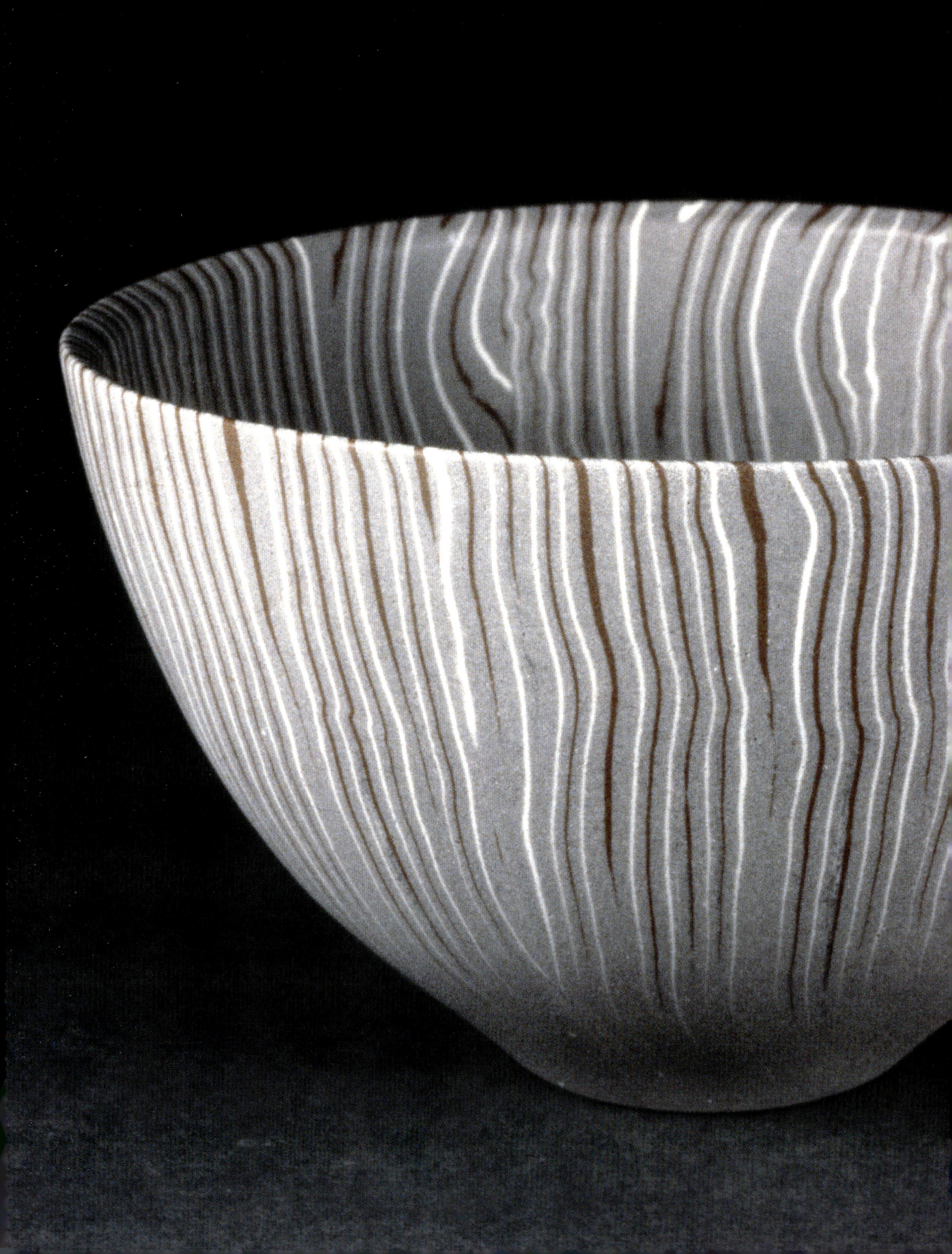

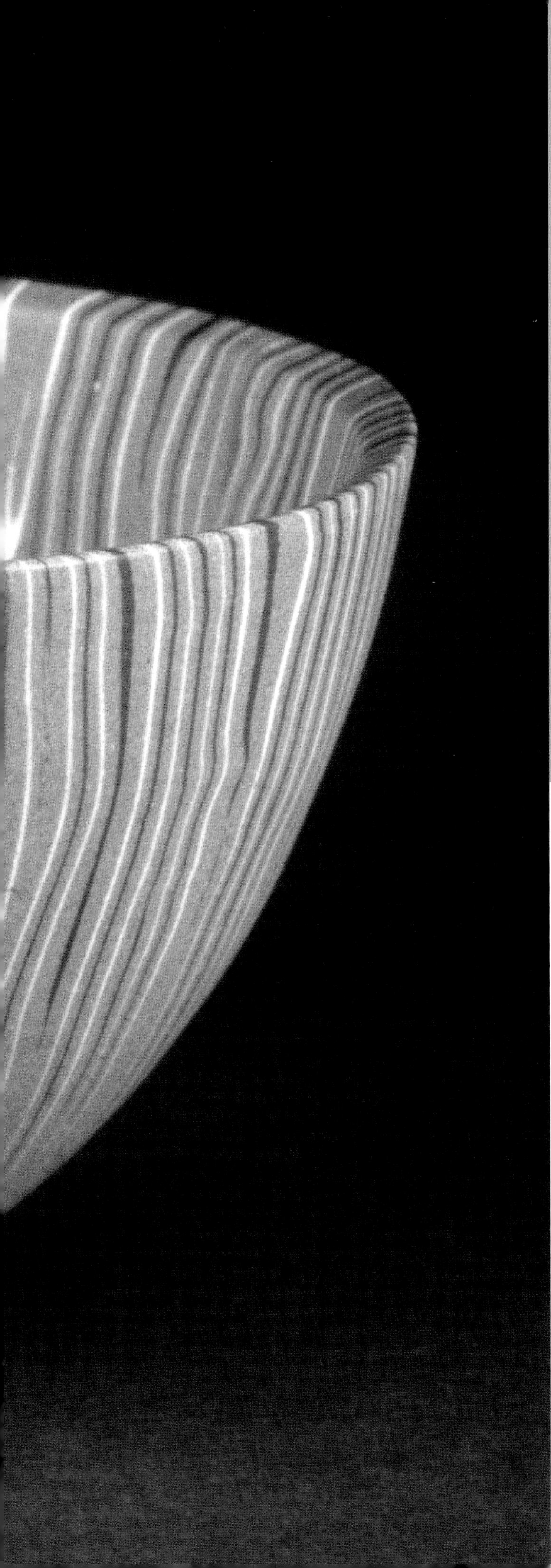

THE NERIKOMI TECHNIQUE

One Artist's Approach

THE CHOICE OF ONE'S PROFESSION can be quite simple, as in following in one's parent's line of work, a life-long journey in itself, or anything in between. For me it was a realization in my freshman year of college that an earlier interest in science wasn't quite right and I was instinctively being drawn in the direction of art, connecting again with my past. As a child, my siblings and I were treated to a wide variety of artistic experiences by our mother, who was an artist herself and full of creative ideas in many media (clay not being one of them). An early memory of pottery was watching a man at a large art festival in New Hampshire throw pots on a wheel. As most know, seeing the soft squishy mass of earth transformed in a matter of minutes into a beautiful symmetrical form is quite astounding. A similar memory was watching a wood turner create a bowl from a block of wood. Both transformations resulted in something pleasing to see but also demonstrated mechanical operations on materials with specific physical properties. This was both art and engineering, and it resonated with me. It just so happens that my father was an engineer by training, and as I grew up, both parents' interests and skills were passed on to me to some extent.

My high school experience took me deeply into the world of biology. The college transition to the arts took a course that may sound familiar to many. Photography was my initial focus but that gradually gave way to sculpture, with traditional abstract work and later more esoteric directions, influenced by the post-minimal work being done in the late 1960s and early 1970s.

The shock of reality of course hit me shortly before graduating, at which time I thought architecture might be an interesting career. I chose to both take a break from the academic world and learn more about architecture by getting an entry level job in a large firm in Cambridge, Massachusetts. As it happened, next door to the office was a small art gallery that exhibited Japanese ceramics and prints. I would pass it daily and soon was spending a great deal of time there studying the pots. This led to evening pottery classes and I was on my way.

Later on, I discovered a potter in Vermont, Malcolm Wright, who had studied in Japan under Takashi Nakazato, the fifth son of the famous potter Tarōuemon Nakazato XII, at Ochawangama in Karatsu, learning the traditional style of Karatsu pottery. Reading Malcolm's books on Japanese ceramics as well as seeing his two small colored clay objects were my first introduction to the nerikomi technique.

I studied with him for several months and later became his apprentice for a year, to be followed by a Master of Science degree in ceramics from Illinois State University. My wife, baby daughter, and I then moved to the Berkshires in Western Massachusetts, drawn by the natural beauty and rich cultural offerings, to start a career in clay.

I had been reintroduced to nerikomi by the Japanese potter Mitsuya Niiyama, who gave a workshop while I was in graduate school, and I had experimented with the technique for a while before moving on to explore other areas. A few years later, after one or two failed attempts at creating and selling a wholesale line of pottery, I was casting about for what to do next and was reminded of my earlier interest in nerikomi. About the same time a wise bit of advice was passed on to me. A colleague urged me to make the very best work that I could and that gallery representation and sales would follow.

Previously I had operated on the assumption that a professional potter ought to be making a line of functional pottery, letting the marketplace establish the format and then trying to be creative within that format. I had earlier ruled out nerikomi since the production time involved demanded high prices that, to my mind, were out of the range expected for functional pottery. My new attitude of doing what I was naturally attracted to, and what I did best, sent me back to nerikomi. My initial attraction to the technique had been the fascination of pattern and structure being one and the same. The "decoration" was made first, and was in the clay itself. Slices of that special clay were manipulated to create the forms.

In graduate school I had seen three basic patterns demonstrated: stripes, spirals, and checkerboards. I explored variations on these three patterns and gradually developed a style of my own. During this time, a friend showed me a Japanese book of work by "Living National Treasures." One, Matsui Kōsei,

◀ (previous) **Bowl, 1980. Height: 3⅜ inches, diameter: 5½ inches (8.6 x 14 cm)**

This small gray bowl was heavily influenced by one of Matsui Kōsei's pieces which I had seen in photographs. That piece had very fine parallel lines in dark clay against a white background, running from the lip down to the very bottom, converging miraculously at the center. I couldn't imagine the skill required to control the clay in such a beautiful way. Nevertheless, I tried. Not wanting to copy him outright, I chose a different form and decided to make my lines stop partway down the side of the bowl. By this decision I had unwittingly made a step forward into understanding the concept of a gradual fade from one color to another. By increasing the number of lines, through a "doubling" process, and by varying the length, I was ultimately able to create a fade from light to dark. This became more sophisticated later on but it started with this bowl. *Image: Private collection*

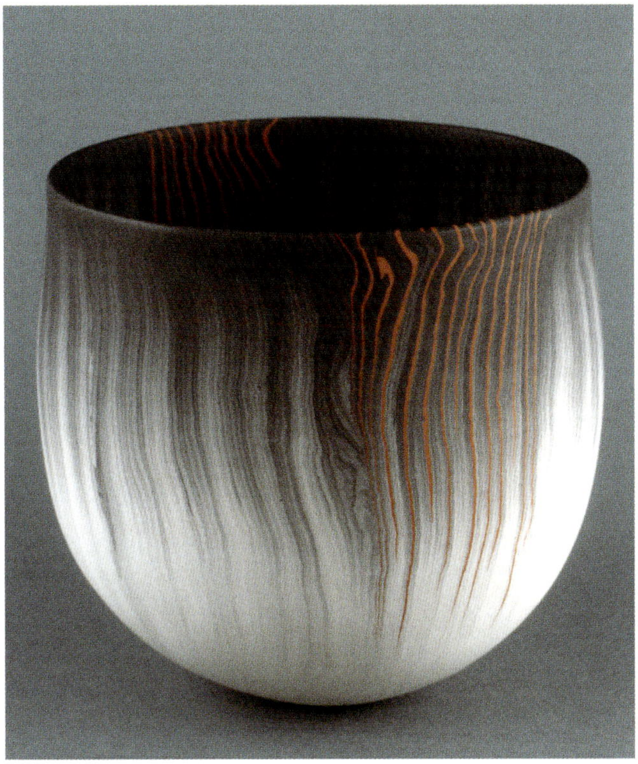

▲ **Bowl, 1985. Height: 5 inches (12.7 cm)**

This freeform bowl shows the further development of my color-fading technique along with contrasting pinstripes similar to previous work. Here the dark-to-light fade was created from black, white, and several shades of gray which, through the doubling process, began to overlap each other slightly, ultimately creating a soft fade. The number of layers in this type of design is in the thousands. I often speak of this doubling process as being similar to making puff pastry. *Image: Private collection*

▶ **Bud vase, 1980. Height: 8¾ inches, diameter: 2¼ inches (22 x 5.7 cm)**

This bud vase was made using a two-part plaster press mold and is an example of my developing technique which ultimately led to soft color fades. Here the pinstripes were added as a tapered slice of white clay embedded in the black, which was then multiplied many times through stretching, cutting, and re-stacking the loaf. Each time the stripes became thinner and the number was doubled. Since the tapered white slice was also cut at an angle, some stripes became shorter and others longer. The effect of this began to show a fading from an overall gray to black. *Image: Private collection*

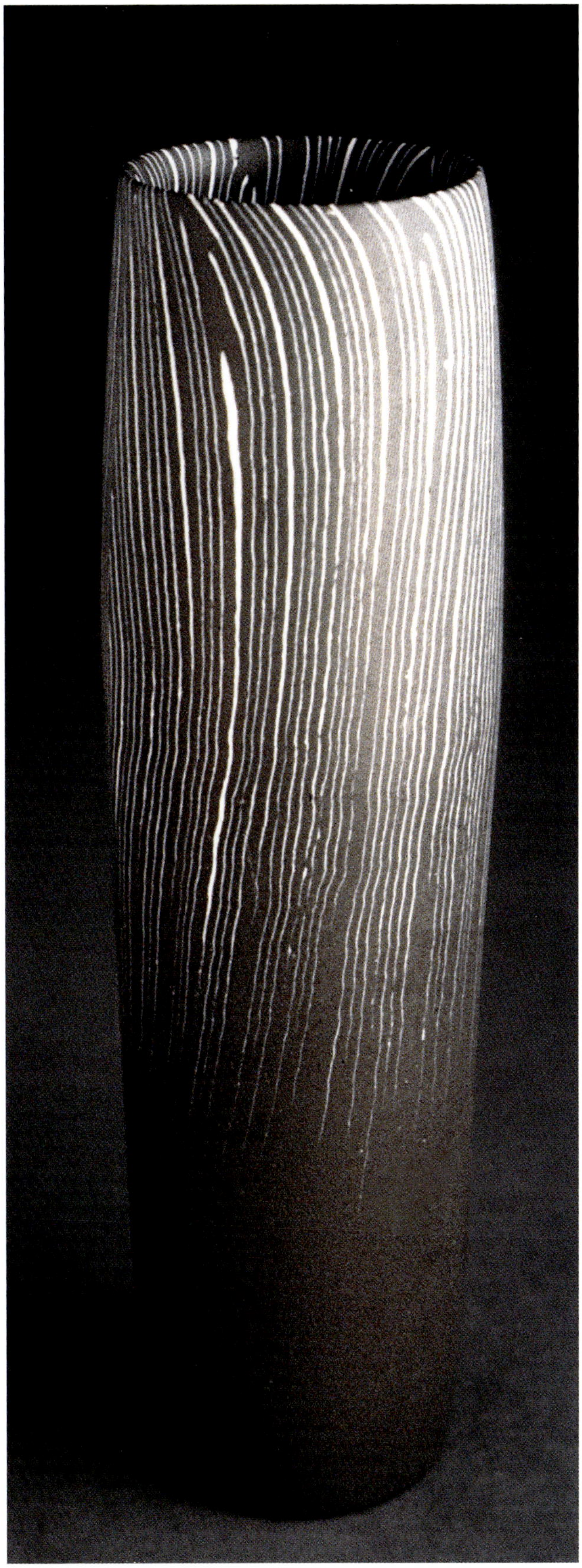

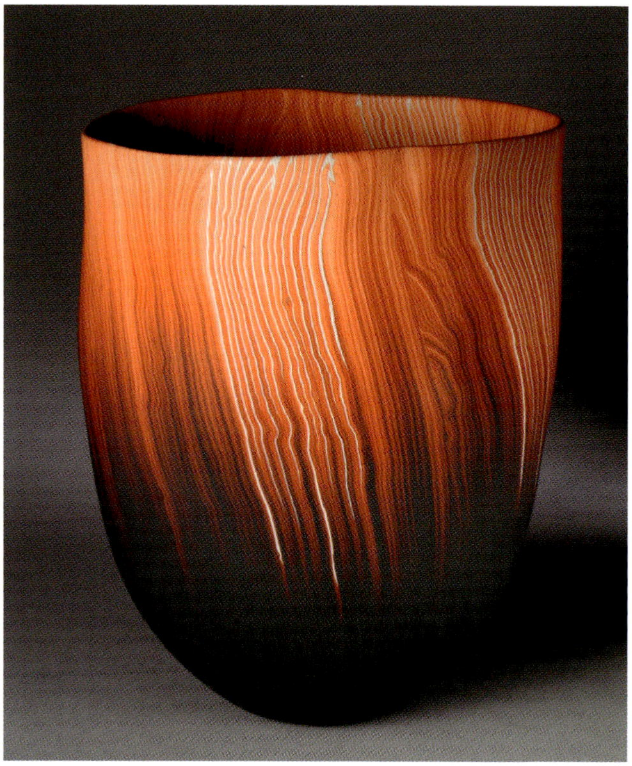

Vessel, 1985. Height: 7¾ inches, width: 6 inches, depth: 5 inches (20 x 15 x 12.7 cm)

This tall freeform vessel was constructed from four rectangular slices taken from a pattern block composed of several shades of brick-red porcelain along with black and light gray for the fine lines. The slices were joined together to form a long rectangle which was then draped around a cardboard tube with newspaper as a separator. The tube was placed upside down on a circular bat with a hole cut out in the center, allowing me to reach up inside the form. After removing the tube and newspaper, the black end of the cylinder was slowly pinched together to form what would become the base. Final shaping required one hand inside the form and the other, with a variety of tools, on the outside. Plastic was then used to enclose the red areas, keeping them soft, while the black third of the vessel stiffened up. When ready, the form was inverted, supported on a bat, and the upper two-thirds were expanded through gradual stretching to achieve the final form, the final step being to chamois the lip. At the leather-hard stage the entire piece was scraped inside and out, to clean up the pattern and create a more uniform surface. *Image: Los Angeles County Museum of Art, gift of Howard and Gwen Laurie Smits (M.87.1.63) photo © Museum Associates/LACMA*

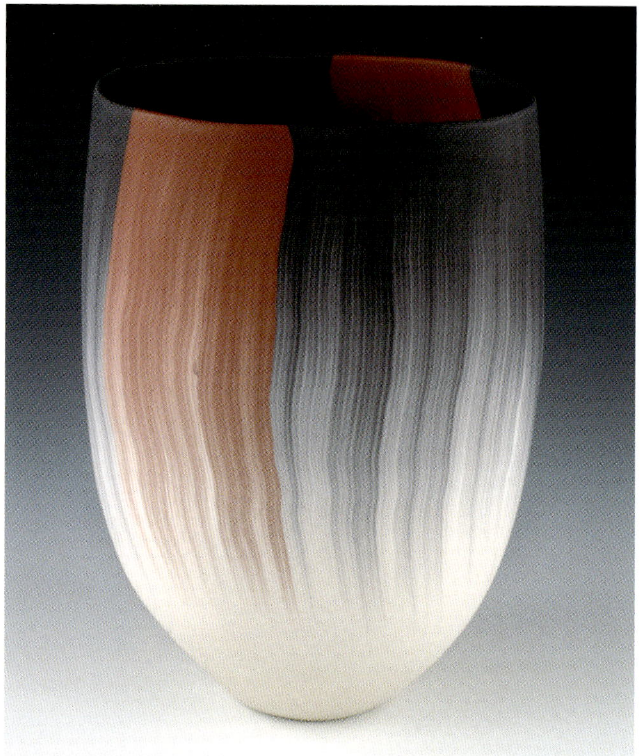

Vase, 1988. Height: 7⅞ inches, width: 5⅛ inches, depth: 4⅛ inches (20 x 13 x 10.5 cm)

This tall vase form shows further development of the fade technique, especially as it relates to the vessel's form. Two fade blocks were created with similar amounts of doubling, resulting in similar levels of blurring between the graded steps of color intensity down to white, the black/gray block being three times the weight and size of the terracotta/white block. They were then joined together and two slices were taken from the large block and joined side by side. Construction of the form was similar to the description of the previous image. *Image: Delaware Art Museum, Gift of Donna Turner Petersen and Robert E. A. Petersen, 2005*

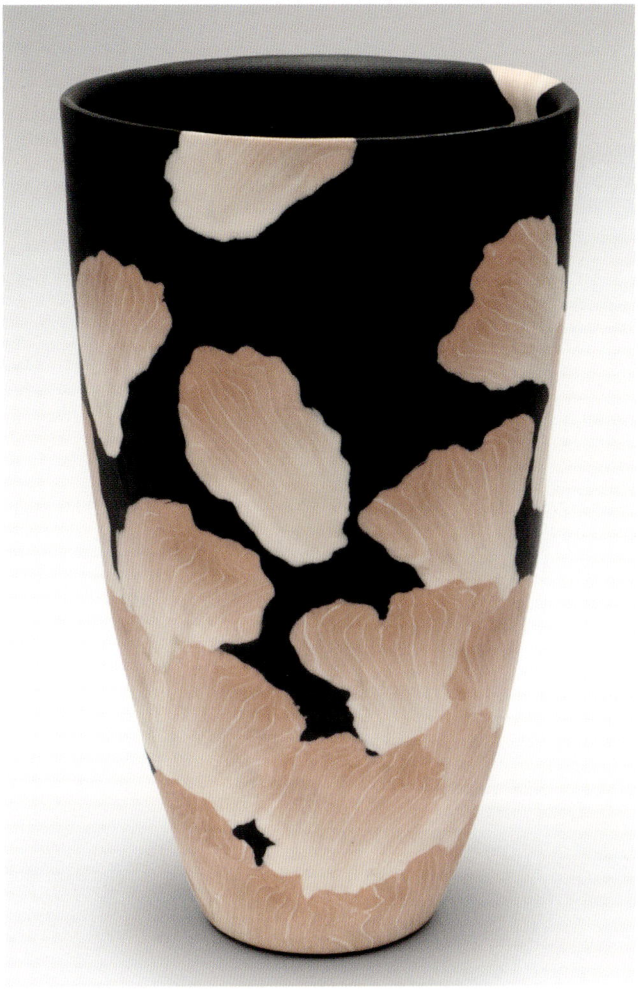

Vase, 1981. Height: 9½ inches, diameter 5⅜ inches 24 x 13.6 cm)

This open vase form was made using a one-part plaster press mold. Slices were taken from a pink fade-pattern loaf and torn at the edges to create petal-like shapes. These were pressed into the mold and joined where they overlapped with scoring and dampening with water and a toothbrush. When the petals were all placed, slices of black were added, reaching into the mold and filling the gaps. Press-molded forms were left to dry in the mold until they could hold their own weight without deforming. Slow drying was then accomplished in a "damp box" over many days. *Image: In the collection of the Philadelphia Museum of Art*

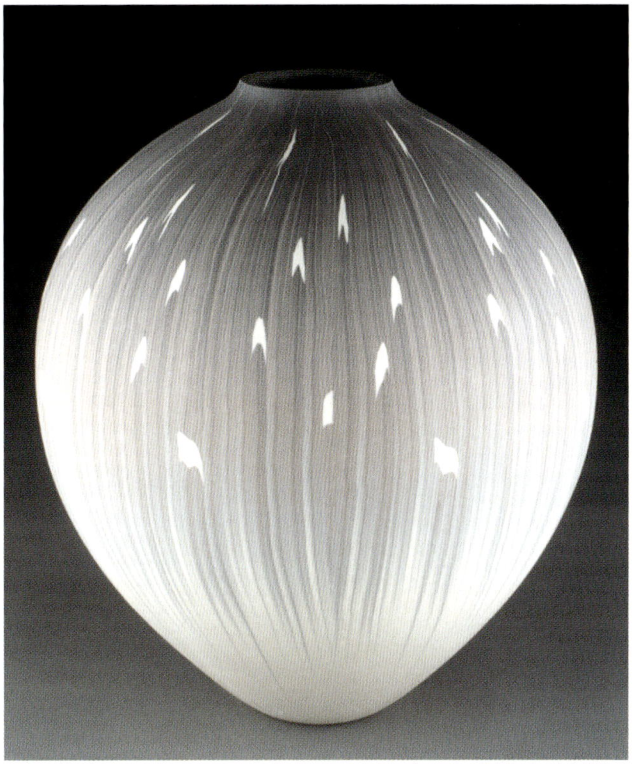

Jar, 1983. Height: 12 inches (30.5 cm)

My 12-inch jars are the second to the smallest of my press-molded jars, the molds being cast from wheel-thrown forms. For this type of design, a large pattern block was made, incorporating a four-color fade with flecks of white. The flecks started out as a thin slice of white porcelain inlaid into a diagonal cut through the fade, which was then doubled a few times resulting in the scatter effect. The block or loaf was incrementally stretched to the proper length and formed into an elongated leaf shape, eight slices of which were joined in the mold – four for each side. This particular jar form has an opening that is too small for a hand to be inserted so a long, curved tool was carved that could reach inside and be manipulated to compress the final joint between the two halves. These larger jars were dried very slowly, still inside the mold, with the help of a hair dryer (cool setting) to get the inside drying at the same rate as the outside, a bit of a trick. The piece was removed from the mold after several days, followed by more drying in a damp box. *Image: Cover image for* Ceramics Monthly *magazine, September 1983; private collection*

worked with colored clay. His pots astounded me and I recognized that he had taken nerikomi to another level. Interestingly, he called his work "neriage" rather than "nerikomi," I think partly because the final shaping of his vessels was done on the wheel, after the initial nerikomi pattern-making. I was inspired by the images I saw and spent lots of time trying to figure out how he did what he did. I tell friends that to work with nerikomi one has to think backwards in three dimensions. How do I create a block of clay which, when sliced open, will reveal the pattern I have in mind? Many steps are involved and the sizes and shapes of each colored clay addition to the block must be thought out in advance.

As in most endeavors, mistakes lead to new discoveries and I experienced many setbacks as I explored the seemingly endless possibilities of this technique. I ended up with a large palette of designs rendered in several color palettes as well. As I worked it became evident to me that with this technique, the plasticity of the soft porcelain could actually be seen. Stacked slices of clay start out as straight lines, when viewed in cross section, but through the manipulation of the pattern block or loaf, those lines become undulating and can be made to taper down to a hair's breadth. I think of my patterns as being a collaboration between my imposed structure and the clay's wise alteration of that structure.

I continue to be fascinated by the beauty of nature, especially when examined close up. The intricate designs are endless. When I work with soft clay, itself part of nature, I feel that I am participating in this flow of design creation. I start with conscious intent but the forces of nature, that is, the physical properties of the clay, seem to take over and move my simple ideas to another level. When I cut the clay blocks in cross-section I am always taken aback by the transformation that has occurred.

My early work, starting in the 1980s, was made in the traditional way using plaster molds into which I pressed slices of patterned colored porcelain. The molds were cast from wheel-thrown bisque-fired forms. Open bowl and vase shapes were formed in single molds while closed jar forms and narrow bud vases required two-part molds. With the forms standardized I had the freedom to delve into new patterns and color combinations, expanding my repertoire and experimenting with new ideas. Over time, I became tired of the few shapes I could make and more mold-making wasn't very exciting for me. Instead I experimented with freeform shapes made without molds, or at the most, a simple drape mold to form the base of the vessel. This has been my style of construction for many years.

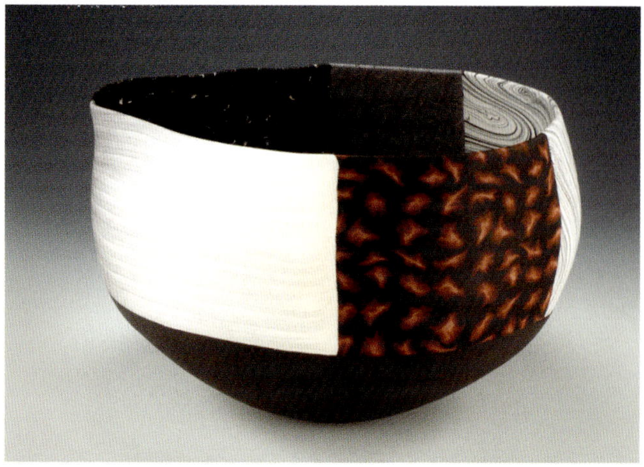

Large bowl, 1993. Height: 7⅛ inches, width: 9¼ inches, depth: 10⅜ inches (18 x 23.5 x 26.4 cm)

1993 was declared the "Year of American Craft" by President George H. W. Bush near the end of his presidency. When the Clintons moved into the White House, First Lady Hillary Rodham Clinton took it upon herself to organize fitting celebrations. Along with decorating the White House Christmas tree with hand-crafted ornaments from all 50 states, she conceived of an exhibit to be shown in the White House, on the period furniture in many of the public rooms, consisting of craft works in many media. Michael W. Monroe, then curator-in-charge of the Renwick Gallery, which holds the Smithsonian Institution's craft collection, was asked to curate the exhibit, and this large bowl which I made in 1992 was among the 72 works chosen for the exhibit. We were thanked personally by both the President and the First Lady at an artists' reception during the Christmas season, something that I will always remember. The exhibit, called "The White House Collection of American Crafts," then went on to travel the country, and later toured internationally. The collection is now held at the William J. Clinton Presidential Library and Museum in Little Rock, Arkansas. *Image: © Thomas A. Hoadley*

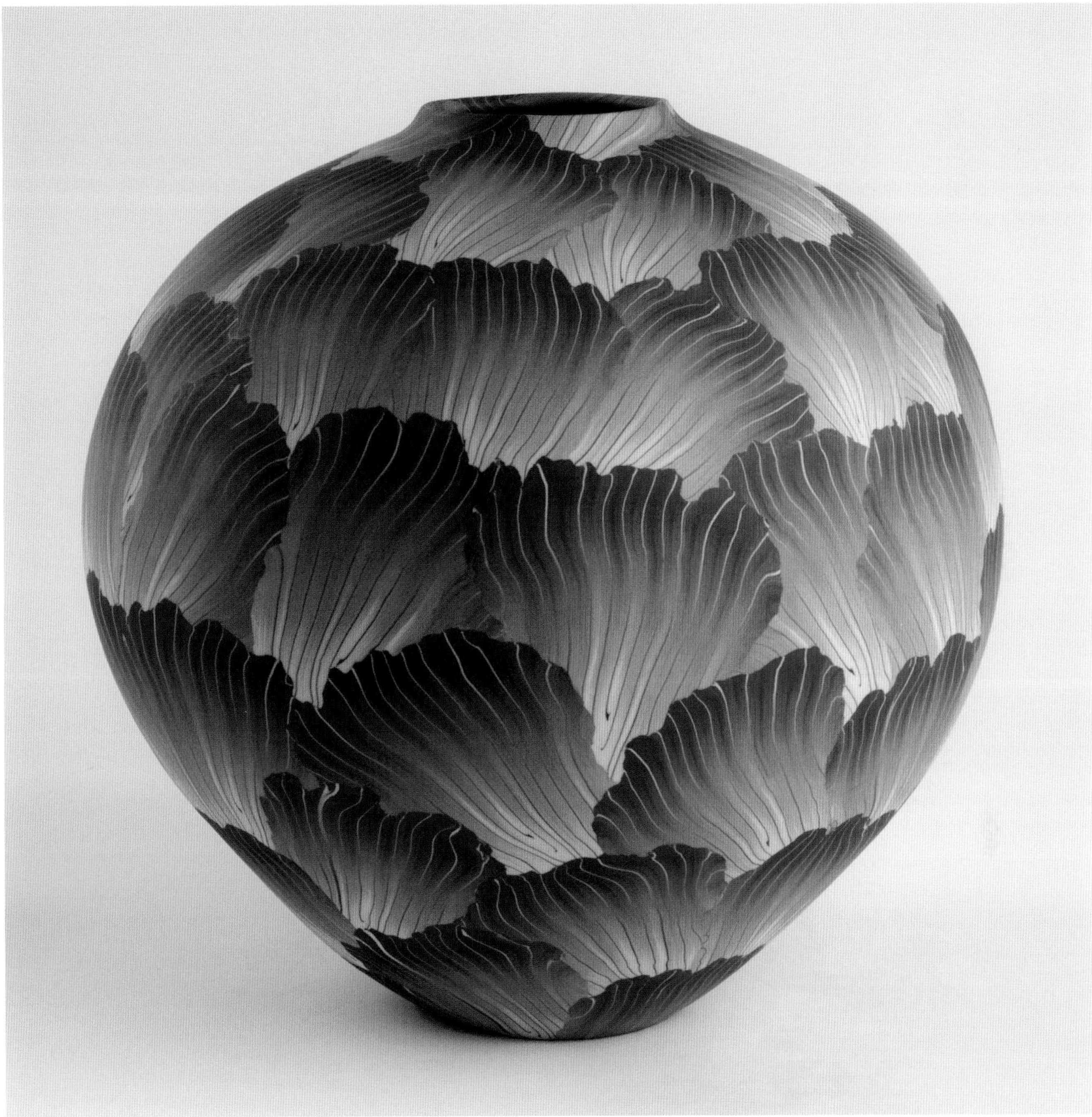

“Petal jar,” 1984. Height: 13 inches, diameter: 13 inches (33 x 33 cm)

I refer to this type of design as a “petal jar,” seen here on the second to the largest of my jar forms. Using my fade technique, in this case incorporating pinstripes of light blue, the piece was assembled in a two-part plaster press mold. Slices from the pattern block were first torn along the edges and then pressed into the mold starting from the bottom. The edges were scored and dampened with a toothbrush and overlapped by the next petal, working this way up to the top. The petals were placed just short of the mold’s edge on one side and placed on the opposite side in such a way as to go beyond the edge of the plaster. This was done for both halves so that, when put together, the overlapped petals from one mold would fill the gaps left on the edge of the other mold, thus disguising the side seams. *Image: In the collection of the Delaware Art Museum, Gift of Donna Turner Petersen and Robert E. A. Petersen, 2005. Photo © Thomas A. Hoadley*

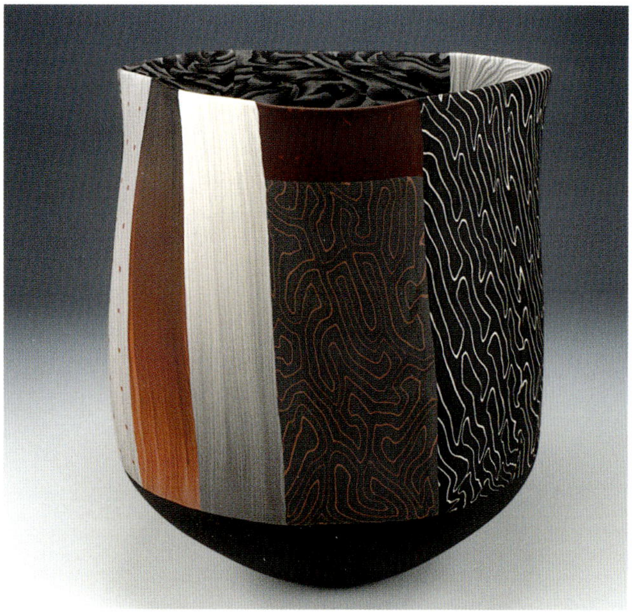

Nerikomi vase, 1995. Height: 10 inches, width: 8 inches, depth: 9 inches (25.4 x 20.3 x 22.8 cm)

I was honored to receive a Bronze Award in Ceramic Arts at the 4th International Ceramics Competition 1995 Mino, Japan for this large nerikomi vase. Well into my career at that point, my preferred color palette remained black, gray, and white with accents of brick reds, perhaps influenced by the natural colors of native American pottery, the colors of raw clay itself, and the colors resulting from primitive firing techniques. I still find these to be quiet and soothing, but have also created vessels with a brighter palette. *Image: Museum of Modern Ceramic Art, Tajimi City, Japan*

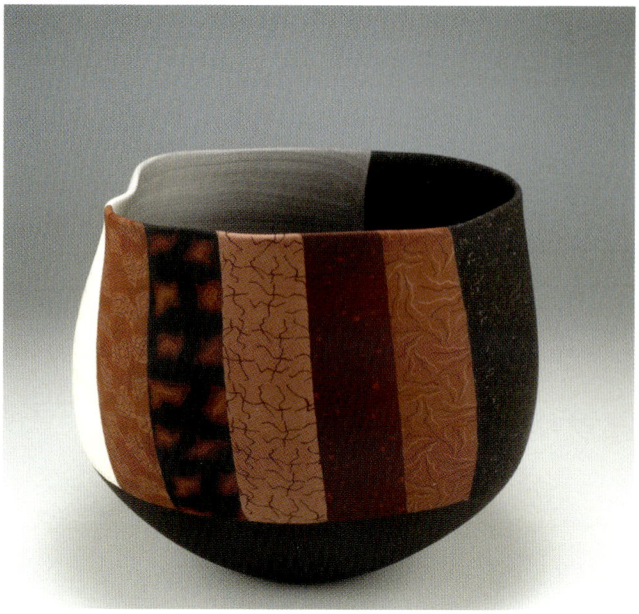

Bowl, 2003. Height: 6 inches, width: 6⅞ inches, depth: 6¾ inches (15 x 17.5 x 17 cm)

My interest in contrast is clearly seen in this small bowl from 2003. I have enjoyed pairing both light vs. dark as well as color vs. black and white. Often I have composed my pieces to present a visually stimulating section as foreground with quieter areas behind. In this case the soft gray-to-white fade and the black granite texture present a quiet background for the five bands of intricate patterns in shades of red. *Image: Private collection*

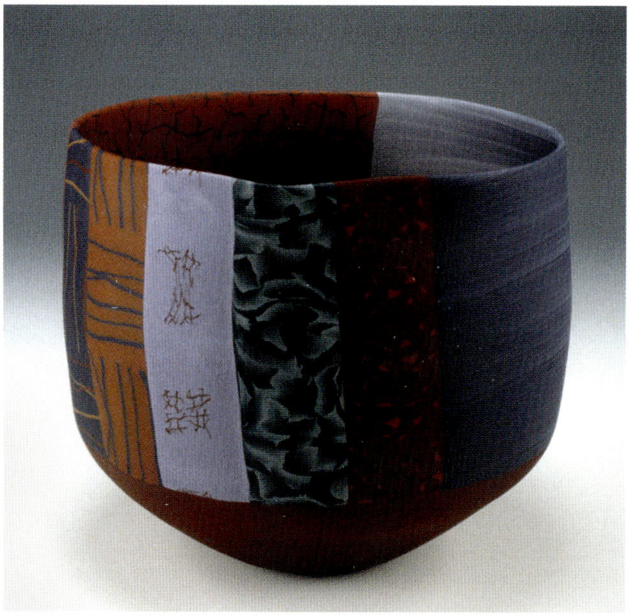

◀ **Bowl, 2005. Height: 6 inches (15 cm)**

More colorful palettes were explored in the early 2000s, here incorporating a dark-red base. This style of bowl starts with a disk of deep-red clay being draped and pressed over a bisque-fired wheel-thrown form. When stiff enough to hold its own shape, it is removed from the mold and placed right side up on a wooden bat with clay supports beneath the edge. Meanwhile, slices of various patterns are joined together flat on a cloth and later added to the edge of the base. This can be a precarious operation, especially when the form is tall as seen in the vase above left. All joints are made with scoring and dampening the beveled edges and then overlapping the bevels and pressing together. Bamboo tools are used to smooth over the joints both inside and outside the piece. *Image: Private collection*

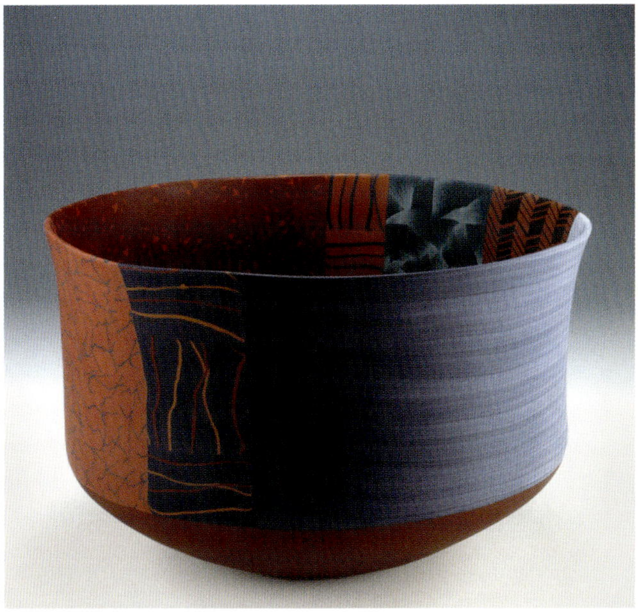

Bowl, 2006. Height: 6 inches (15 cm)

This is a rare nearly symmetrical form composed from my multicolor palette where the soft blue fade and deep reds balance the more vibrant colors and designs. As with all my work, the piece is dried slowly and, when leather-hard, scraped with a metal rib to remove the smeared clay and even out the form. Further drying includes covering the piece with a single thickness of a dry-cleaning bag which is removed, inverted, and reapplied to the piece daily. This slowly removes moisture which condenses on the inside surface of the plastic. When bone dry the work is sanded with two grit sizes of "grit cloth" in a booth with continuous exhaust to remove the clay dust. This is followed by a biscuit firing, wet sanding, high firing, and final wet sanding. *Image: Private collection*

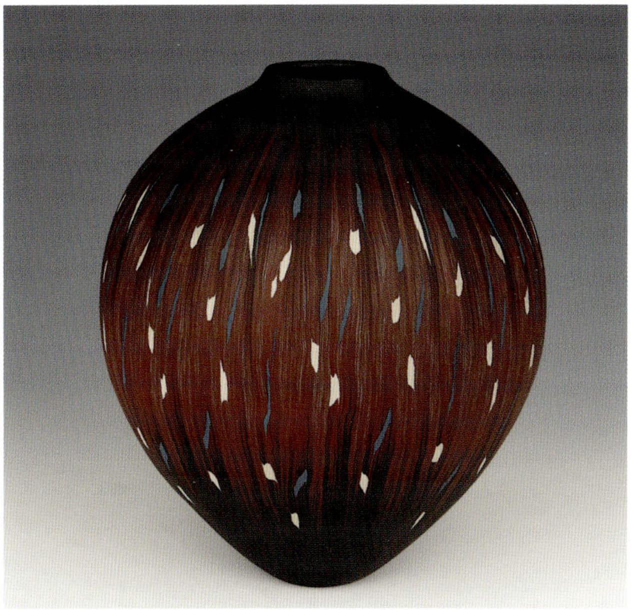

Jar, 1984. Height: 12 inches, diameter 9½ inches (30.5 x 24 cm)

This deeply colored jar was made in the same mold as the one on page 55. White flecks were introduced into the block at an early stage and doubled several times so that in the final piece they show up faintly as numerous thin lines of white clustered in groups. Additional white as well as light-blue were later added and doubled a few times to create the scatter effect in the final piece. *Image: Private collection*

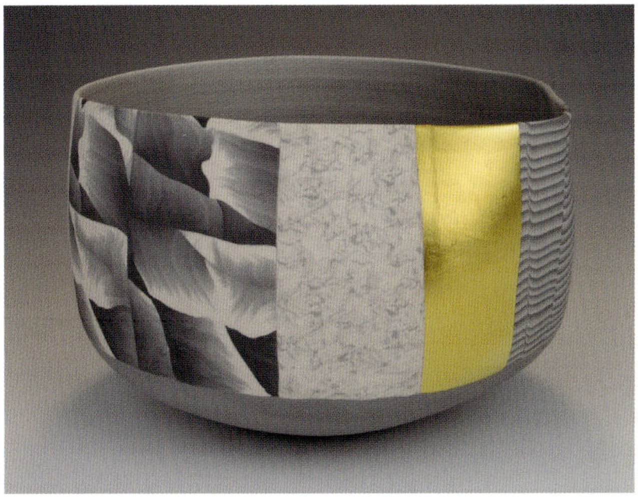

◄ **Medium bowl, 2022. Height: 5½ inches, width: 8½ inches, depth: 7½ inches (14 x 21.6 x 19 cm)**

In recent work I have reduced my color palette to lighter shades of gray, perhaps looking for a more quiet or calming mood. Forming methods are similar to those described in the images above left and opposite bottom. I have also been using gold leaf as an added contrast, both in material and color for several years. This is added after the final firing by brushing on a sizing material and later carefully draping extremely thin sheets of gold leaf onto the sized area. Excess is brushed off with a soft watercolor brush. This is followed by a second coat for extra depth of color. *Image: Private collection*

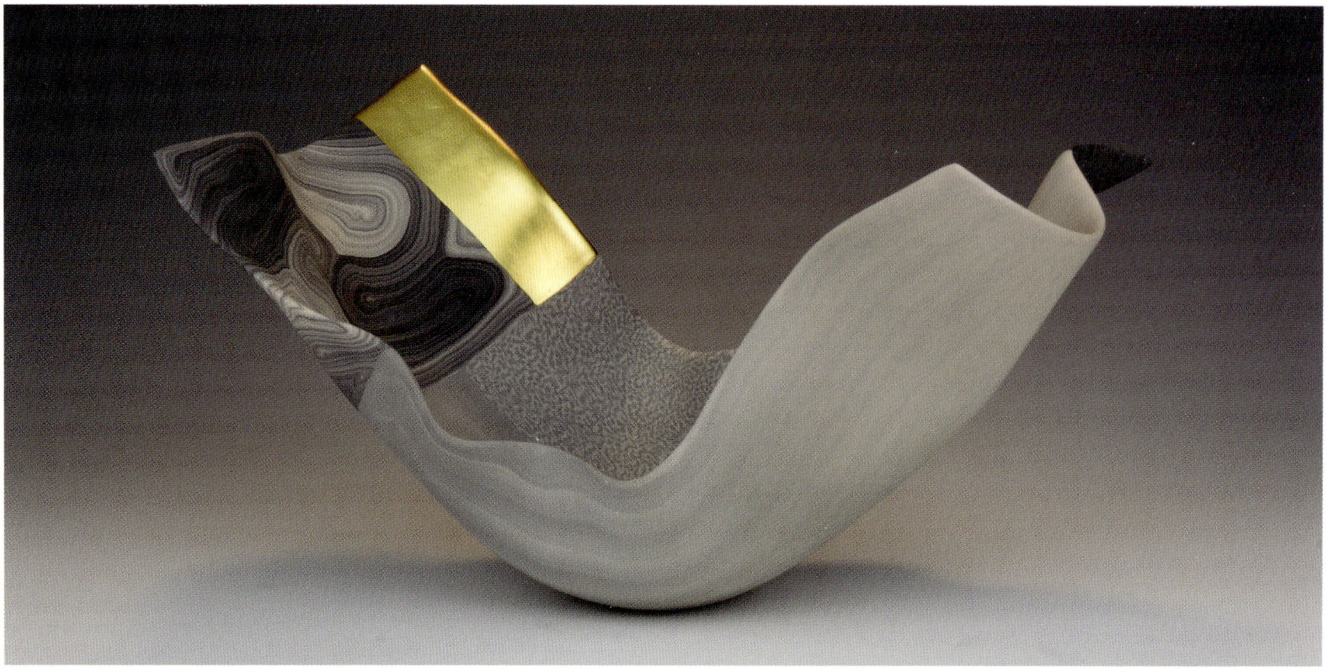

▲ **Large arabesque, 2018. Height: 8 inches, width: 14½ inches, depth: 8¼ inches (20.32 x 36.83 x 20.96 cm)**

I call this shape an arabesque, after the ballet position by the same name. Slices of varying dimensions from several pattern blocks are joined together on the work table to form a large rectangle. This is then draped over a biscuit-fired mold, now being upside down, and covered with a half-inch-thick foam sheet which has been cut in such a way as to fit closely around the curved form. This is then again covered by a second mold, this time made of thin plaster and the whole ensemble is flipped over so the form is now upright. The original biscuit mold is removed to reveal the inside of the arabesque which is supported beneath by the foam-and-plaster mold. At this point the shape is adjusted with the use of small foam wedges which support the undulating edge during the drying cycle. I see the final flowing form as a frozen version of the type of shapes that soft slices of porcelain naturally make when handled. The upward lifting shape balanced on a small base gives a sense of weightlessness. It is still a vessel, in the broadest sense, but also aspires to being sculpture.
Image: Private collection

The general steps are as follows: pattern-making, construction of the vessel, slow drying, dry sanding in a booth with both coarse and fine "grit cloth," bisque firing (943°C [1730°F] – cone 08 electric kiln), wet sanding (340 grit waterproof sandpaper), high firing (1188°C [2170°F] – cone 5 electric kiln), final wet sanding, gold leaf application (most pieces) in two coats, and mineral oil (rubbed on and washed off).

The sequence shown on pages 61–3 is an example of a system of pattern-making which I developed through trial and error. These photos demonstrate what I call a "loose checks" design but the general steps of doubling are the focus here. I often use this doubling system with other designs, the main idea being to create a pattern in a large loaf and then multiply it through doubling, gradually decreasing the size of the original design and multiplying the number of copies. I usually work with 10lb loaves which yield about ten slices, 3⁄16 inch (48 mm) thick, that are 6 inches x 9½ inches (15.2 x 24.1 cm). These are each checked for gaps or weak sections, compressed, and stored between sheets of plastic food wrap, the whole group then wrapped in many layers of dry-cleaning bag plastic and sealed in an airtight plastic box along with a small sponge soaked in water. I have been able to keep usable slices for up to two years or more using this method.

This sequence demonstrates the "doubling" technique I use to create my "loose checks" pattern.

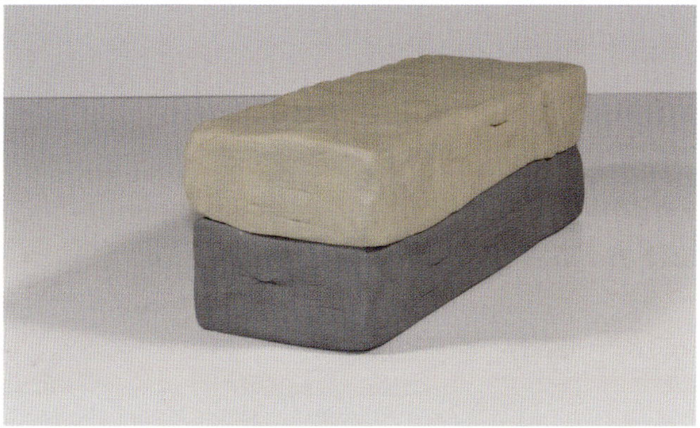

1. The 5lb blocks of white and gray porcelain are stacked to form a square cross-section.

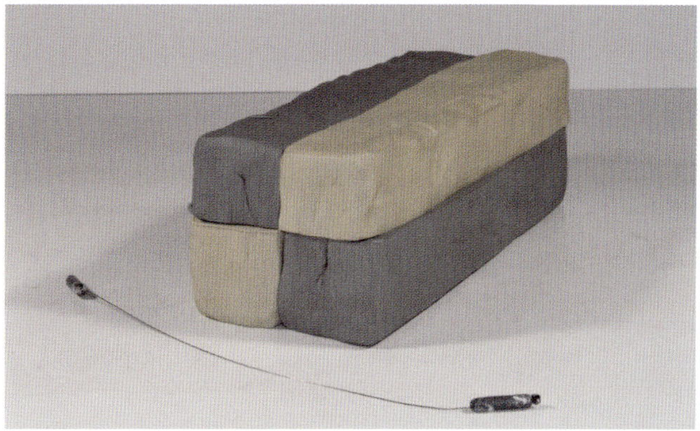

2. The block is turned 90° and is wire-cut in half. The top half has been lifted, reversed, and replaced.

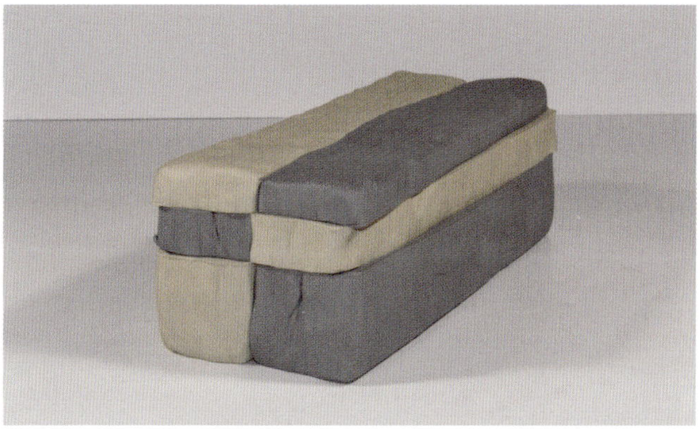

3. The top half is wire-cut and, again, the top slice is lifted, reversed, and replaced.

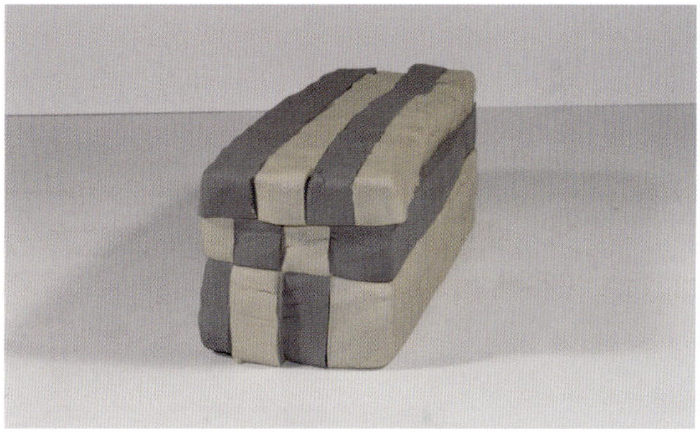

4. The bottom half is cut again and reversed. The block is now turned 90° and the top half is wire-cut, the slice lifted, reversed, and replaced.

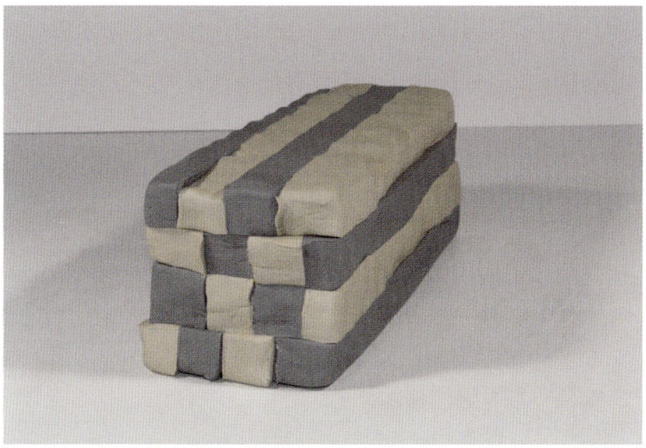

5. The same procedure is made on the bottom half. Notice how the checkerboard design is already "slipping." This image represents the basic pattern which will then be doubled several times. Many other patterns seen in my work were multiplied in this same manner.

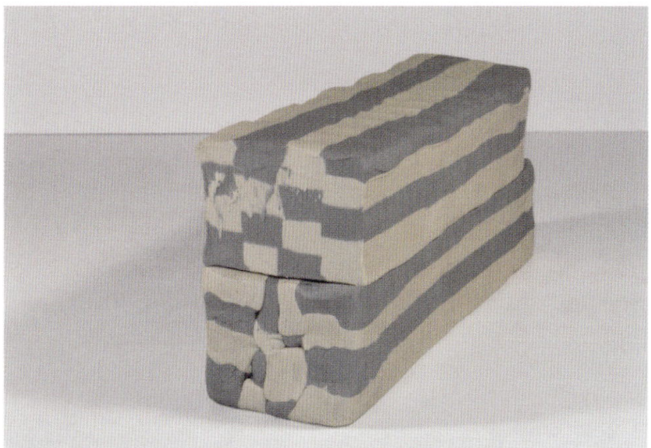

6. The block has been stretched through repeated throwing-down at an angle on the work table, turning it 90° each time, roughly doubling the length. It is now cut in half and, after smoothing the surfaces with a wide spatula, the two halves are stacked together.

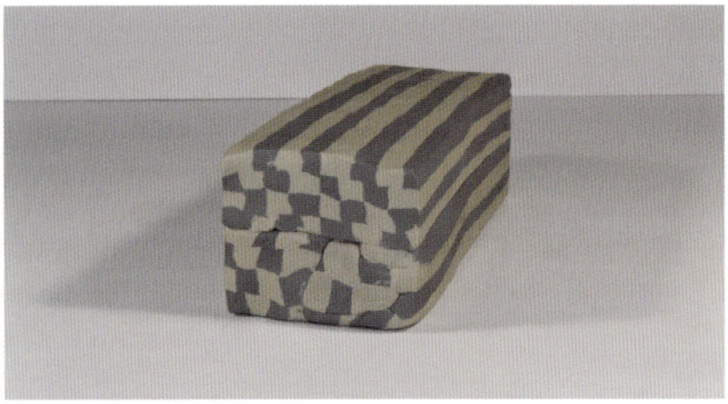

7. The now "doubled" block is doubled again, resulting in a square cross-section again. With each doubling, the pattern becomes half the size and there are twice as many as before.

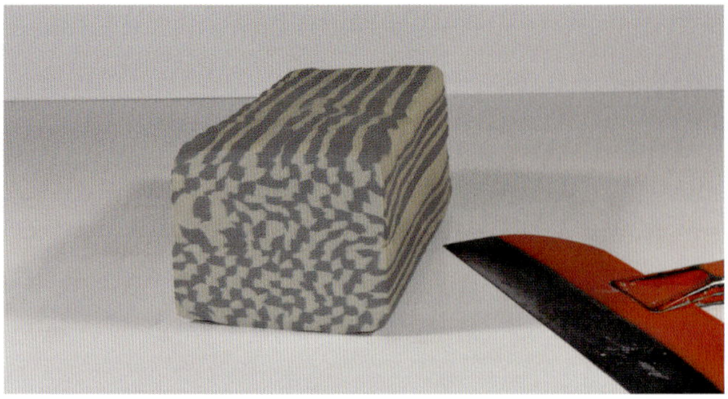

8. The block has been doubled twice more.

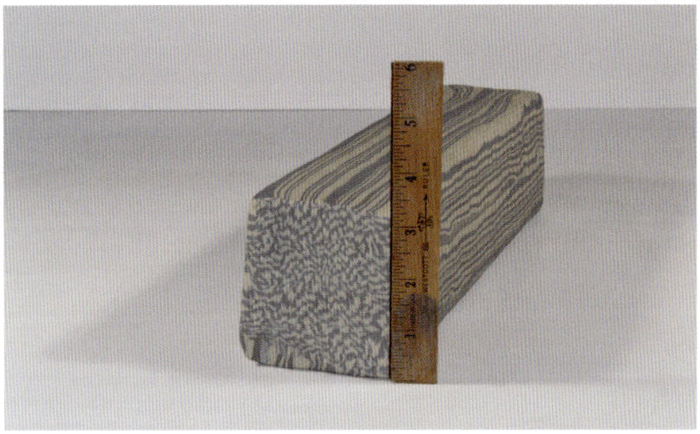

9. Again, the block has been doubled twice more. At this point preparations are made anticipating the ultimate size of the finished block, which will be 6 inches x 9½ inches (15.2 x 24.1 cm), a standard size for most of my pattern blocks. The block has been stretched a bit, through repeated slapping and turning, so the cross-section is 3 inches (7.6 cm).

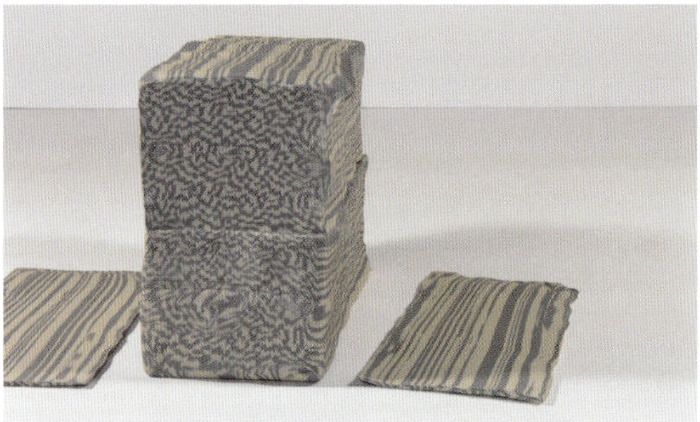

10. The 3-inch loaf is cut in half and, after smoothing the surfaces with the spatula, a thin slice is wire-cut at the bottom, leaving a slightly roughed-up surface which will bond with its double more effectively than if the two outside smoothed surfaces were to be joined. The action of the spatula serves to align the microscopic clay platelets. If two such surfaces are slapped together, chances are they will separate easily later on. The roughing up of the clay by wire-cutting helps the two surfaces to mesh together more effectively – just one of several strategies to reduce cracking later on.

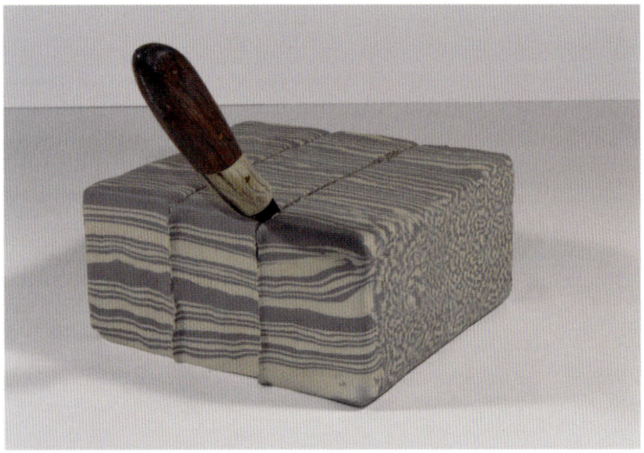

◀ 11. The block is now placed on its side and knife-cut into thirds.

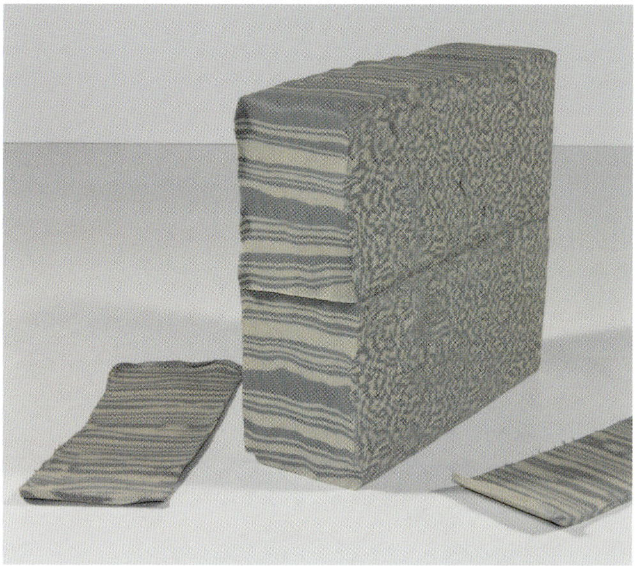

12. Again, thin slices are wire-cut from the surfaces to be joined. Here the first two are stacked together.

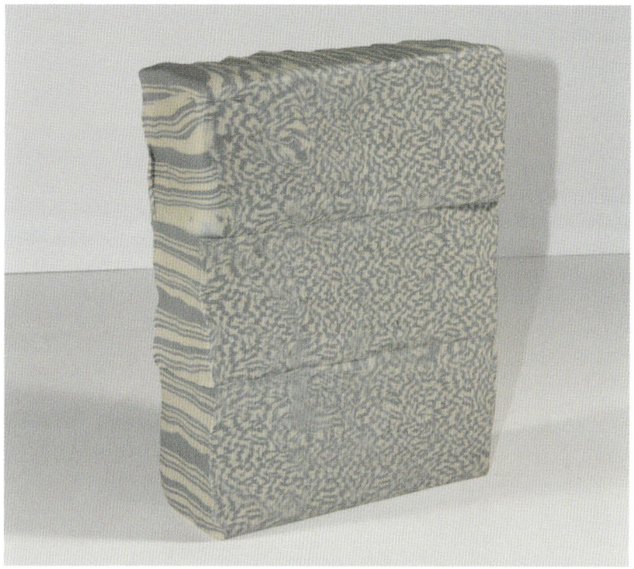

13. The final third has been added and the entire block is pounded carefully against the table, turning in all directions, to consolidate the entire 10lb block.

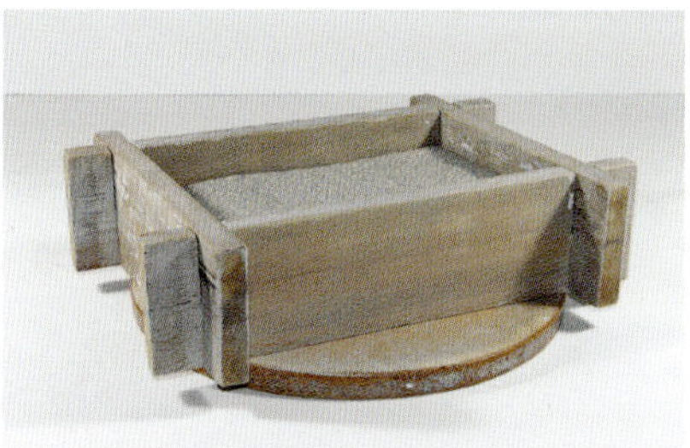

14. The block is placed on a circular bat and enclosed by a homemade wooden frame used to further consolidate it, all for the prevention of future cracks.

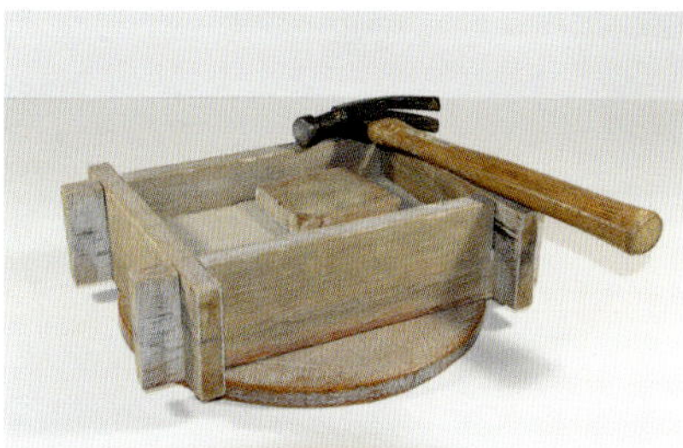

15. A piece of artist linen is placed on top of the clay and a small square of wood is rhythmically pounded with a hammer, moving it across the entire surface, to compress the top part of the block. A thin knife is used to cut around the perimeter, releasing the clay from the frame and a second bat is placed on top. The entire stack is flipped, the first bat removed, and the condensing procedure is performed on the other side of the block. Again, the clay is released from the frame by cutting the perimeter and the frame is removed.

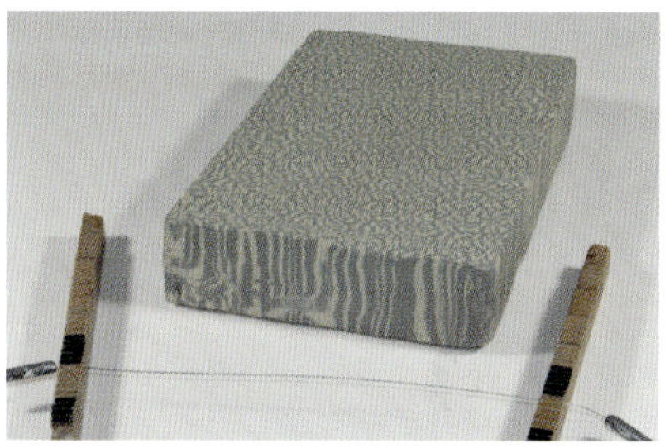

16. The pattern loaf is ready for slicing. This is done with Japanese-style cutting sticks and a wire. Each slice is examined for slight defects or weak spots by careful bending, and repairs are made. Then, each slice is compressed between two layers of cloth with a smaller spatula, attempting to keep the thickness of the slice close to the originally cut dimension. I stack the finished slices between pieces of plastic food wrap and, along with several layers of dry-cleaning plastic bags, they are stored in tightly sealed plastic food container boxes – along with a small water-soaked sponge. This way they can actually stay soft for a few years.

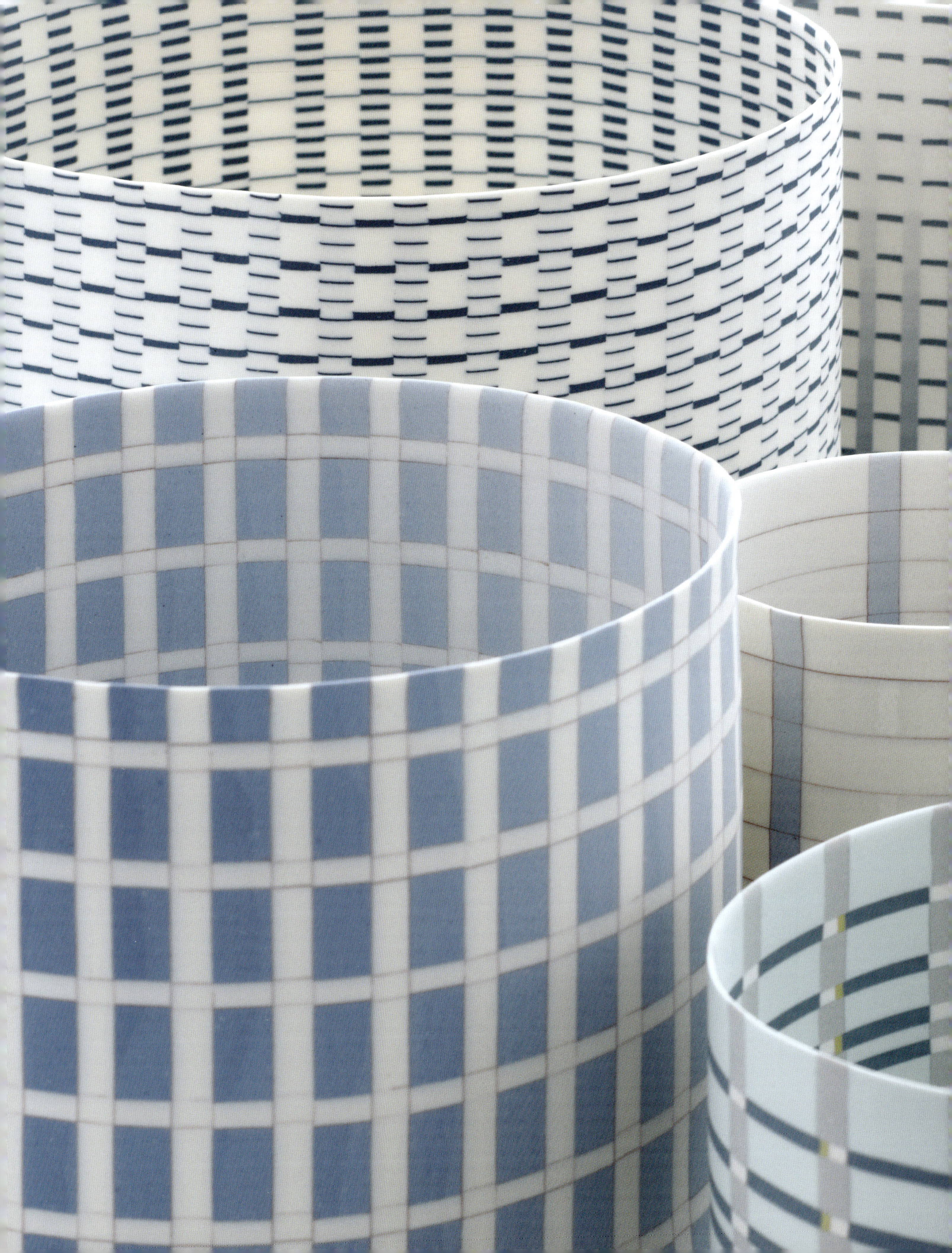

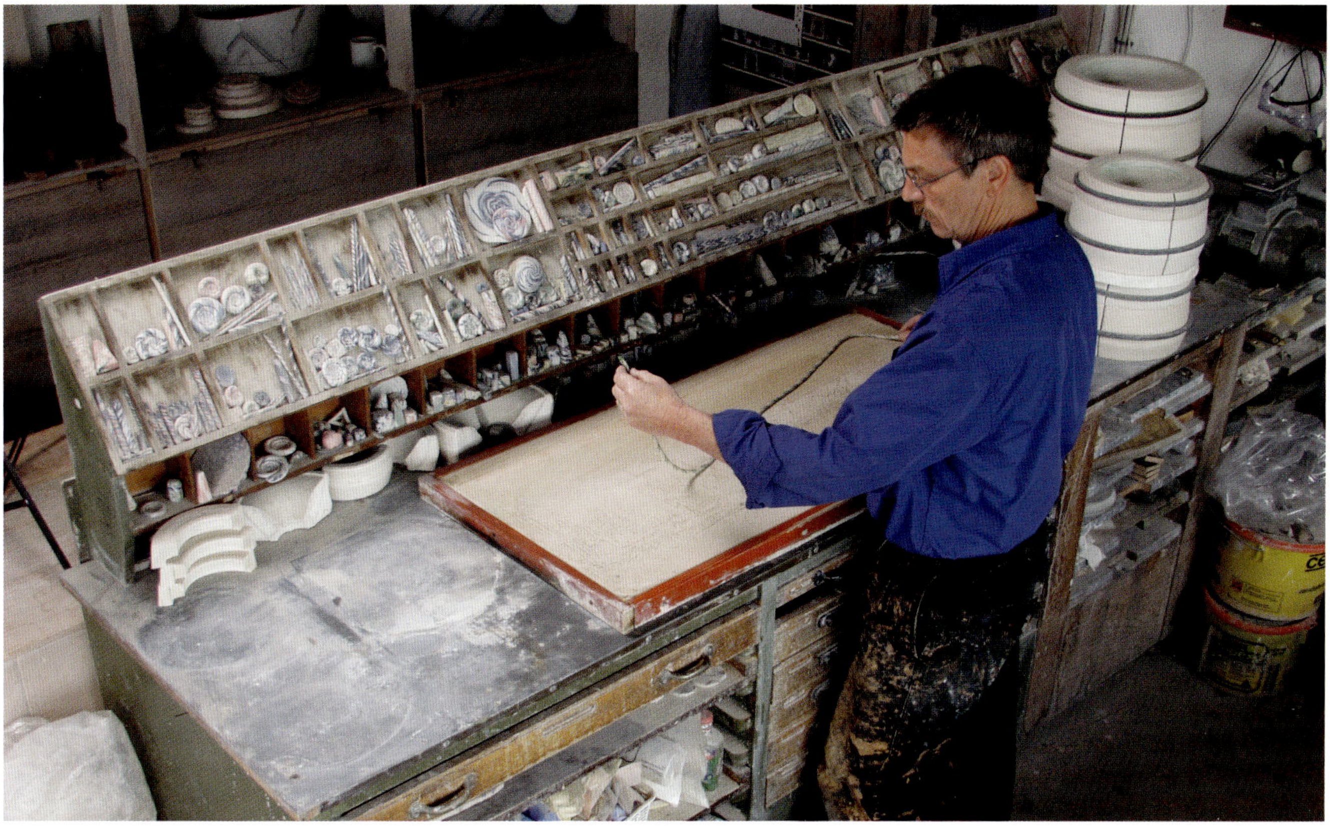

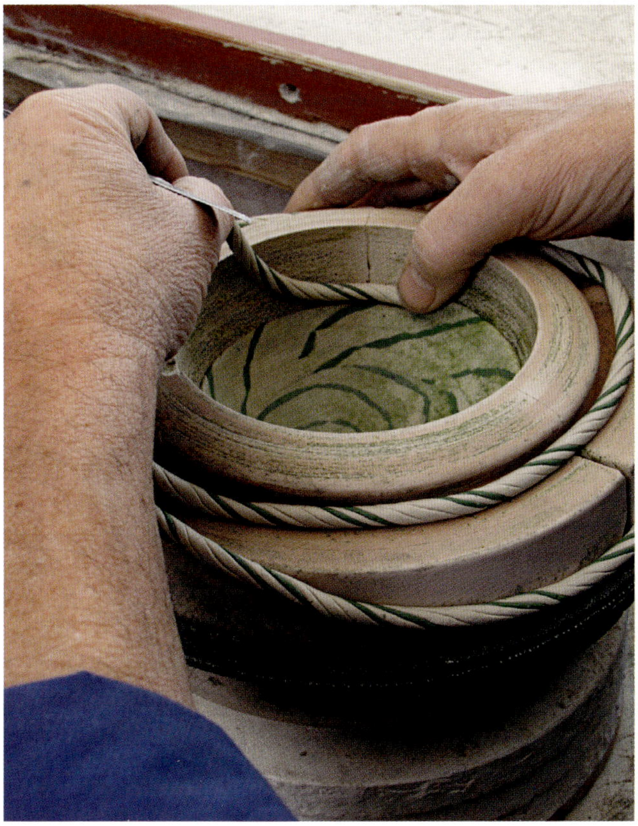

▲ Making coils and twists on a plaster slab.

◀ Building up a bowl inside a plaster mold.

AWARDS AND EXHIBITIONS

Hans Munck Andersen has exhibited widely, won numerous awards, and his porcelain bowls and sculpture are included in 34 museums worldwide, including the Museum of Modern Art KUNSTEN, Aalborg; the Kunstmuseet Trapholt, Kolding; the CLAY Keramikmuseum, Middelfart; the Designmuseum Denmark, Copenhagen; and Bornholms Art Museum – all in Denmark. He has also exhibited in other museums in Denmark, Sweden, Norway, the Netherlands, Germany, Hungary, Latvia, the USA, Japan, Taiwan, China, Turkey, Lapland, Iran, and South Korea.

Janny Baek, USA

The colorful sculptures of Janny Baek are unique in their combination of organic forms with vibrant color which is inherent in their structure. Baek's work would be included in the broad sense of the term nerikomi but her approach lies apart from the traditional understanding of the technique. She often takes coils of carefully prepared colored clay and twists them together to create a candy cane-like coil with which to build her forms.

Baek, born in South Korea and raised in Queens, New York, received a BFA in ceramics from the Rhode Island School of Design. This was followed by employment as a sculptor in the animation and toy world, after which she completed a Master's Degree in Architecture at Harvard University. She was a founder of the firm McMahon-Baek Architecture in 2014 along with her husband. By 2019 Baek had returned to ceramic sculpture and was selected as an Emerging Artist by *Ceramics Monthly* in 2021.

ARTIST'S STATEMENT

In choosing my forms, patterns, and colors, I pursue an expression of fantasy, hope, and pleasure. I would like my pieces to appear both familiar, in referring to known forms such as vessels and natural objects, and at the same time occupy a more ambiguous territory of strange or unfamiliar, occupying some state of dynamic imbalance. I intentionally choose an abundance of rich colors and a lot of change in the pieces because I think the expression of something "other" (the foreign, the complex, the alien), and the possibilities of growth and transformation, are some of the most hopeful ideas we have and this is primarily what drives my work.

A main theme in my work is change and growth, and it's important that my pieces reflect this idea in both form and surface. Many of my forms begin as vessel-like forms that are captured in a state of transition, whether they are one becoming many, or many combining into one, or changing from one type of thing to another. For this reason, I began using techniques for the surface that would also have a gradual change as it moved across the form, rather than a repeating pattern as in a more traditional nerikomi technique.

My work explores the strangeness and pleasure of growth as a way of asking questions of ourselves both as individuals and as a species. As a woman artist of color, immigrant, and mother, I am invested in mining the space between the familiar and the alien. My pieces exist in a captured moment of transformation, or in an ambiguous state of in between, which I think is fertile ground for reflection, change, progress, and curiosity. I use colorful and vibrant forms as a joyful and hopeful entry point to contemplate our perceptions and assumptions about ourselves, our world, the future. Through my work, I advocate for the strange, uncategorized, undefined, changeable, hybrid, multiple, alien, and pleasurable.

My forms and surfaces are informed by references to familiar traditions in art and design: organic and figurative forms, geometry in nature, patterns in textiles and architecture, textures that expose the material qualities of clay. Though I use very specific techniques and considerations when working, my pieces intentionally reside in uncertain, ambiguous states of in-between: unity/multiple, soft/hard, solid/intricate, familiar/alien, hidden/revealed. My intention in leaving the work in this state of flux is to create the opportunity for subjective interpretation, open to questions of shifting perceptions and assumptions. From my experience working in animation and toys, I bring an interest in the playful, imaginary, and sci-fi, and my experience in architecture has trained me in materiality and design. I view the making of my art as an act of speculation that reflects on the present, while intervening to influence the future.

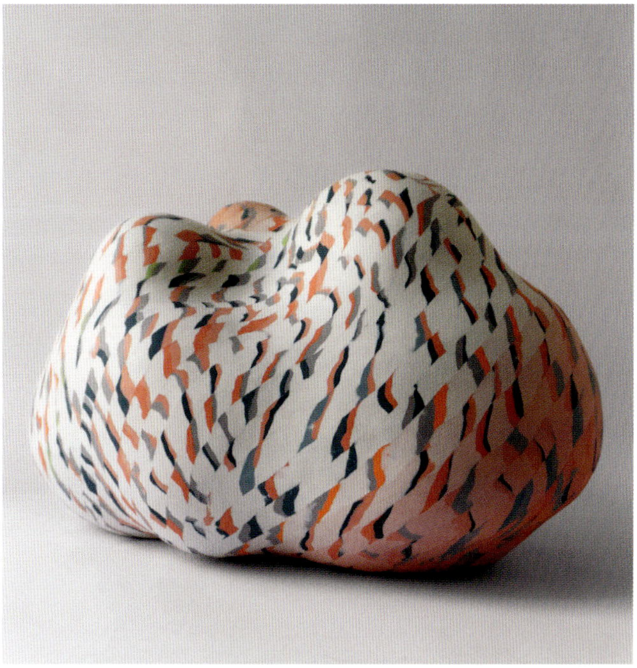

Barbe à Papa, 2021. Height: 12 inches, width: 20 inches, depth: 12 inches (30.5 x 50.8 x 30.5 cm)

These images all © Janny Baek

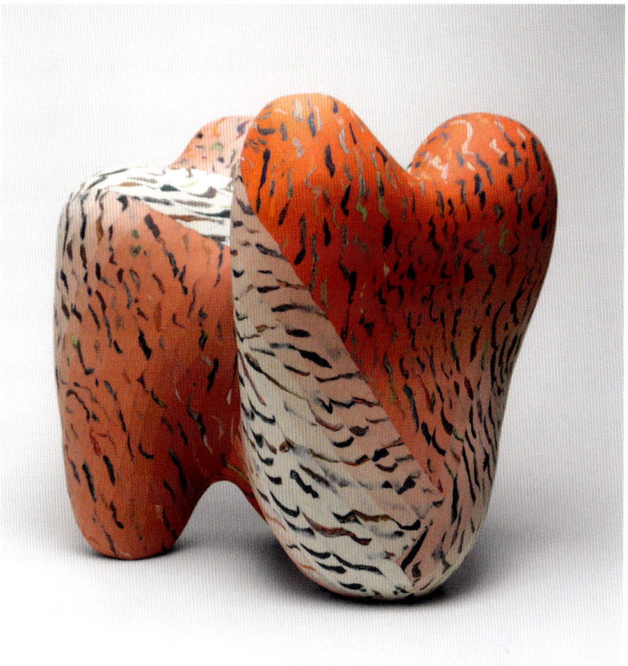

Burning Sensation, 2021. Height: 15 inches, width: 17 inches, depth: 13 inches (38 x 43 x 33 cm)

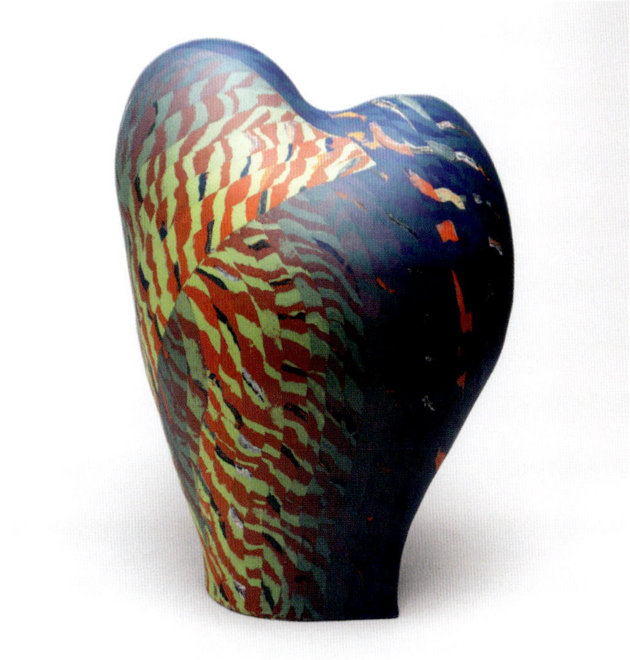

Tidal, 2021. Height: 16 inches, width: 12 inches, depth: 12 inches (40.6 x 30.5 x 30.5 cm)

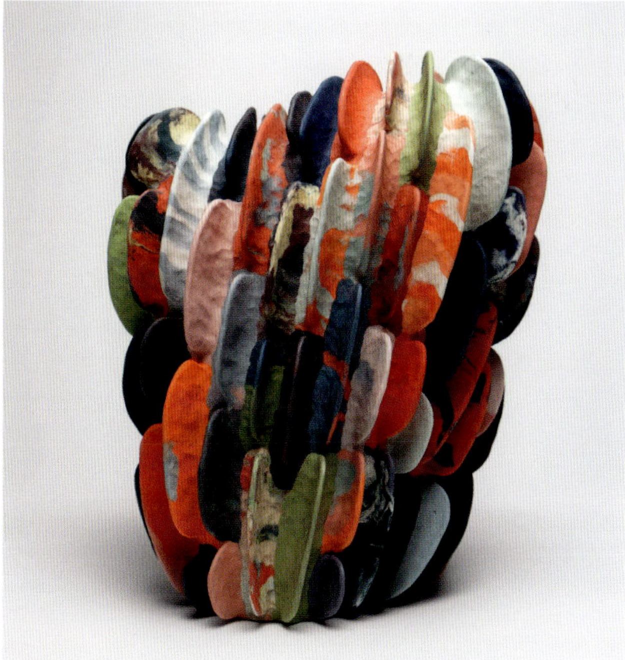

Feathered vessel, 2022. Height: 11 inches, width: 8 inches, depth: 7 inches (28 x 20.3 x 17.8 cm)

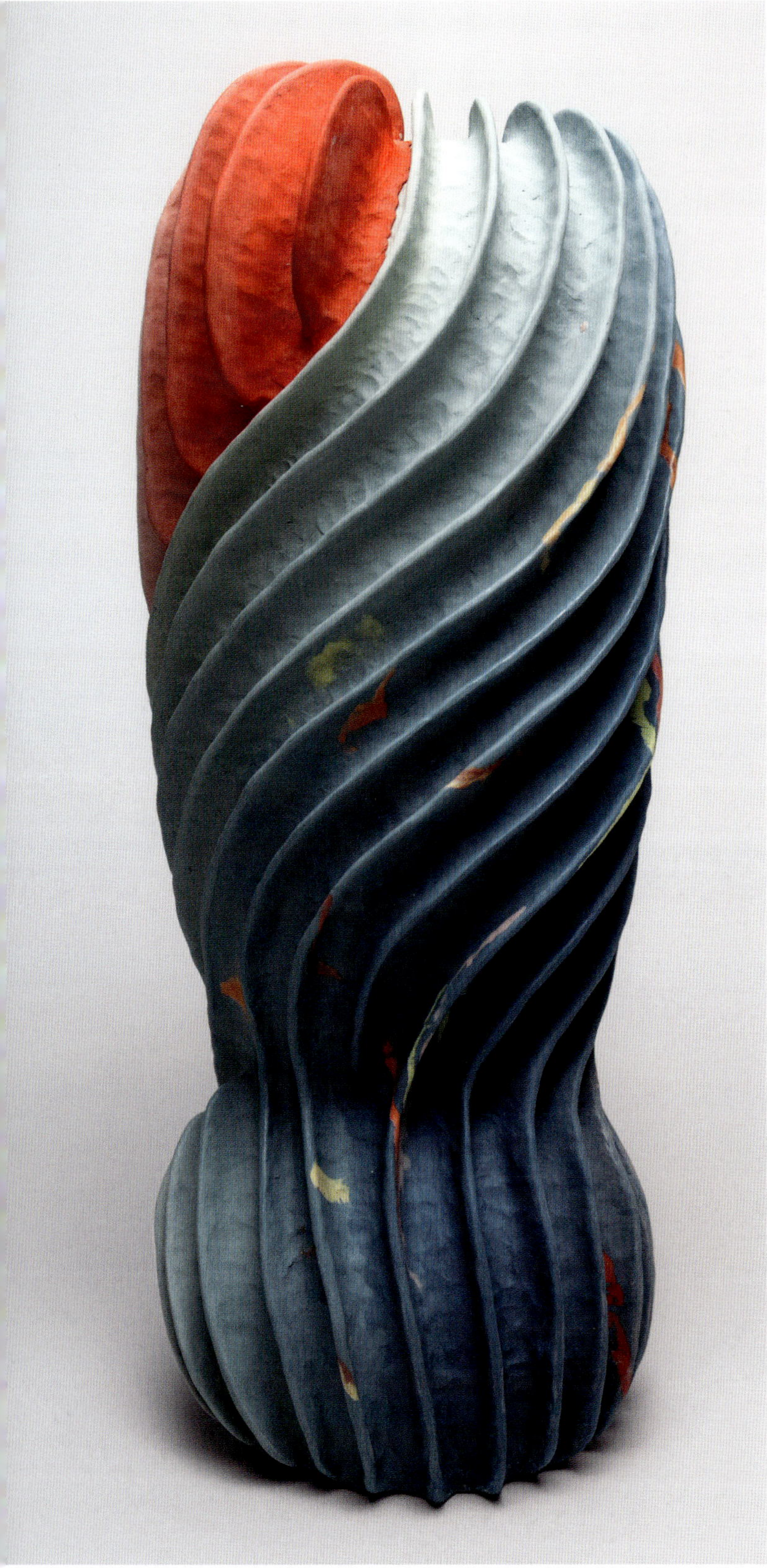

Ramune, 2022. Height: 16½ inches, width: 7 inches, depth: 7 inches (42 x 17.8 x 17.8 cm)

Baek considers drawing to be of prime importance to her practice, devoting a short period of time each morning to non-judgmental sketching in her sketchbook. This regular practice can be a useful tool to tap into the subconscious and in times of indecision she can browse through her pages to rediscover seeds of ideas, yet to be developed. Aside from sketching she also recharges by playing music, usually on the piano or cello. She now sees this as another essential practice that ultimately helps her creativity.

Baek uses mid fire (1200°C [2192°F] – cone 6) porcelain and commercial stains.

AWARDS AND EXHIBITIONS

Janny Baek's first solo show was "The Pleasure of Growth" at Culture Object, New York, 2023 and she has exhibited her works at the Houston Center for Contemporary Craft and the Lighthouse Art Center, Tequesta, Florida, as well as participating with Kasper Contemporary Gallery at "Market Art + Design," Bridgehampton, New York; Galerie Joseph in "1000 Vases," Paris Design Week, Paris, France; and "Factory Made" at the Old American Can Factory, Brooklyn, New York. She has demonstrated her approach with clay at Gasworks NYC, and has been published in "Studio Talk, Process and Perspectives in Clay," *Ceramics Monthly*, May 2022; *The Pinch Journal*, fall 2022, Volume 42, Issue 2, University of Memphis; and "Reliquary: Alive With Hope," *Galerie Magazine*, March 2022.

Janny Baek's development of a form.

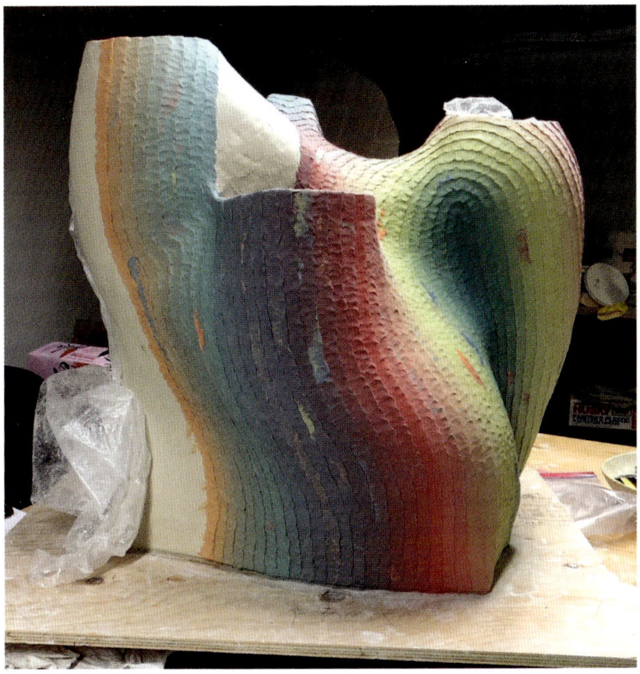

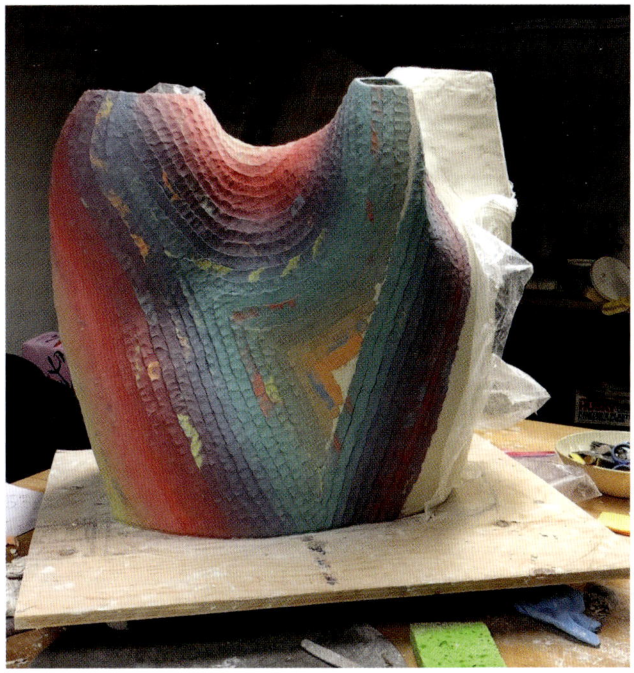

After the base form is made using regular coil-building techniques, the surface is added in strips of color-stained porcelain, accentuating the changing curvature of the surface.

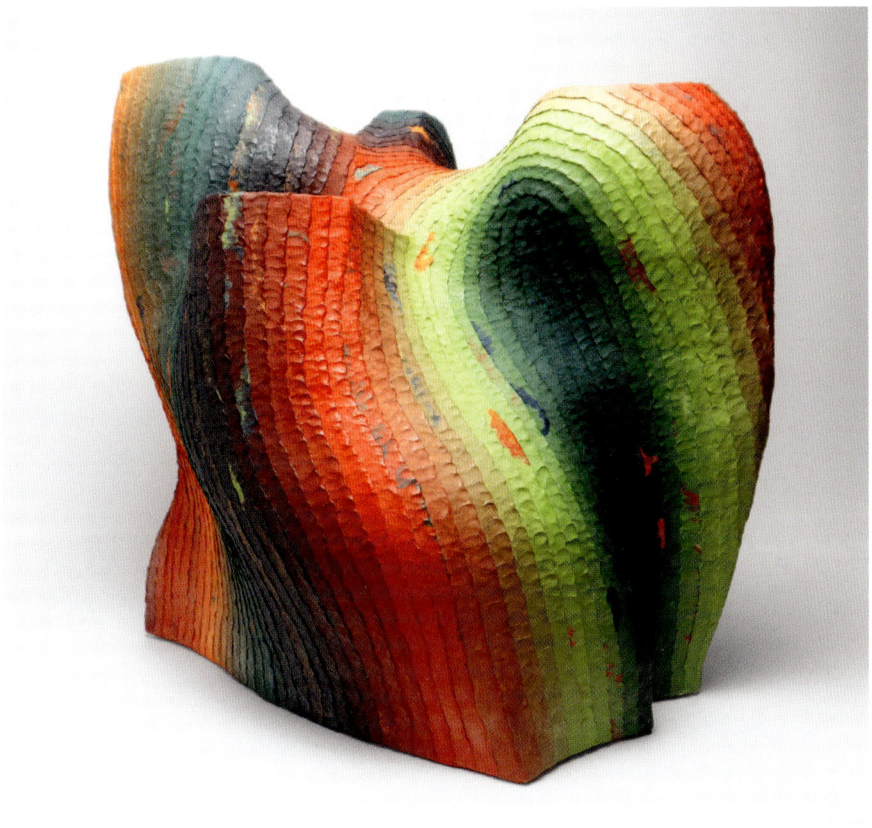

◀ Finished and fired *Mineral*, 2022. Height: 18 inches, width: 19 inches, depth: 14 inches (45.7 x 48.3 x 35.5 cm)

Curtis Benzle, USA

There are many examples of aesthetic beauty in ceramics, but few hand-made vessels possess the ability to touch the innermost sensations of the heart as those made by Curtis Benzle.

These are the words of Robert Piepenburg, a prominent ceramics artist, teacher, and writer. He is rightly focusing on the uncommon quality of Benzle's vessels, their translucency and ability to capture, hold, and transmit light in a beautiful spectrum of color, and the way the human spirit is touched by the light's magic.

ARTIST'S STATEMENT

The primary purpose of my art is to embrace the illusive, emotional content of traditional beauty. Form, color, pattern, and light are the bones of this pursuit but, in the end, the driving issue comes down to my desire to convey the poetry of a quiet moment – to create a visual sanctuary.

I struggle with difficult materials, invent frustrating techniques, and tolerate relentless failure. My goal is both elusive and seductively possible. I have an unfailing desire to give form to feeling – to connect myself and my audience with a better place and to express some of what is so wonderful in our world.

In the early 1970s Curtis Benzle explored the field of color dynamics of glass, but soon moved on to ceramics to take advantage of clay's inherent malleability. A wise tip from Rudy Staffel, a well-known ceramics artist and teacher, set him on a path toward a new translucent porcelain body. Staffel advised Benzle to "leave the feldspars and flint but remove the majority of clay from the recipe and replace it with an organic plasticizer." This last ingredient would burn out in the firing, leaving a porcelain body that would be approaching the properties of glass, but the trade-off is a clay body that is a real challenge to work with.

Benzle's porcelain vessels are extremely thin and unusually difficult to assemble, refine, and fire. Partial solutions to these problems include creating refractory firing saggars that are used to both assemble the pieces and support them during the firing as they essentially begin to melt.

Construction of a typical Benzle vessel often begins by brushing porcelain slip onto a piece of fabric. Thin slices from small pattern loaves created through nerikomi techniques are then placed on the slip background. These may be small fish forms, leaf shapes, or abstract intricate nerikomi designs. Additional areas of multicolored slips may be added to enrich the color and unify the various parts of the design. The mishima technique of filling controlled depressions in the clay with a contrasting color is often used, creating an overall screen of fine dots.

The final result of these layers, when seen in the fired vessel, is a transparent blending of unique colors and designs in such a way that one is always finding something hidden or unexpected. The interior is usually quite different than the exterior, but close examination reveals complex blending through the thickness of the wall as light unifies these multiple layers.

After many years of perfecting his design and forming techniques, Benzle discovered the Japanese ceramic repair tradition of kintsugi, the filling of cracks (results of firing stresses or accidental breakages) with gold. Although the cracks appear to be filled with pure gold the technique traditionally makes use of fine lacquer to fill the gaps and, later, gold powder which adheres to the lacquer and is

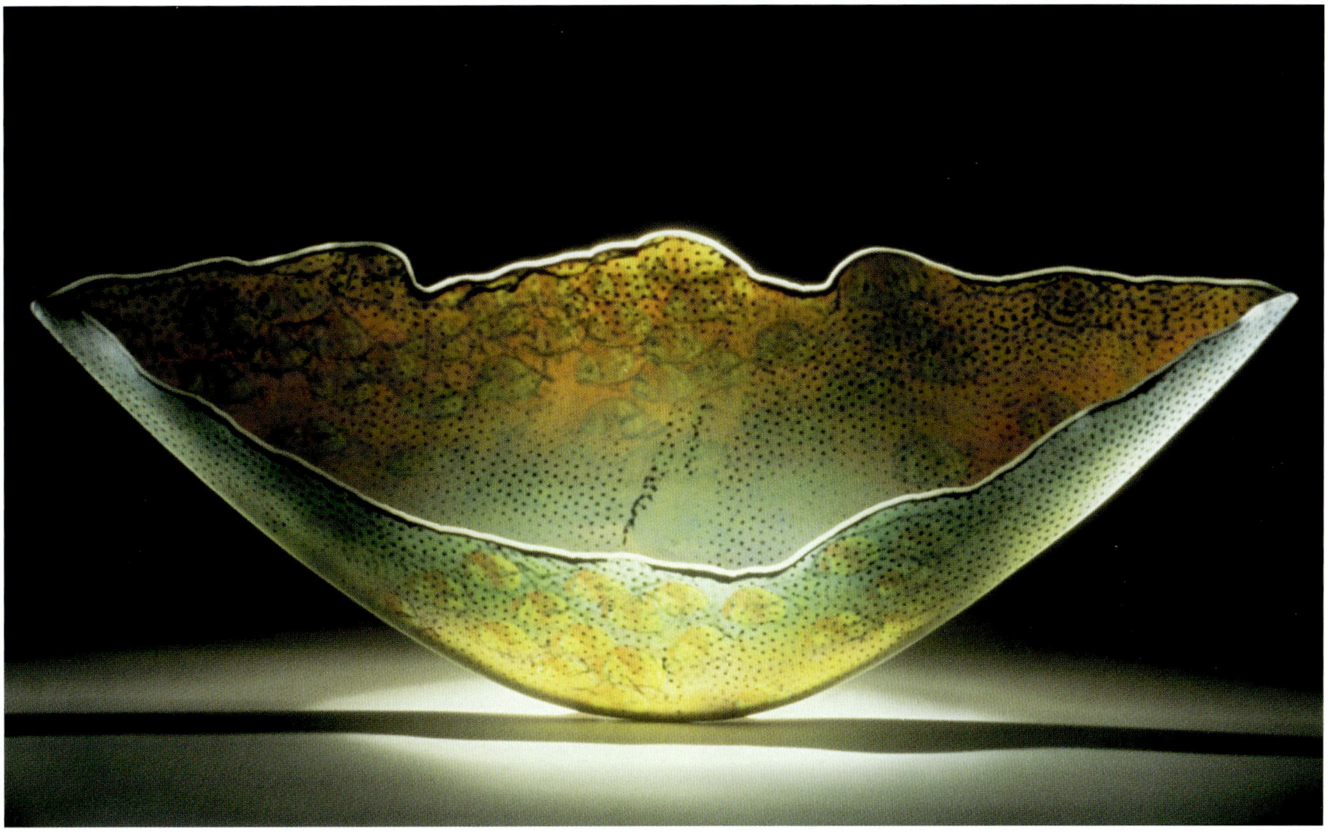

Migration, 1986. Height: 8⅔ inches, width: 17⅓ inches, diameter: 4 inches (22 x 44 x 10 cm)
These images all © Curtis Benzle

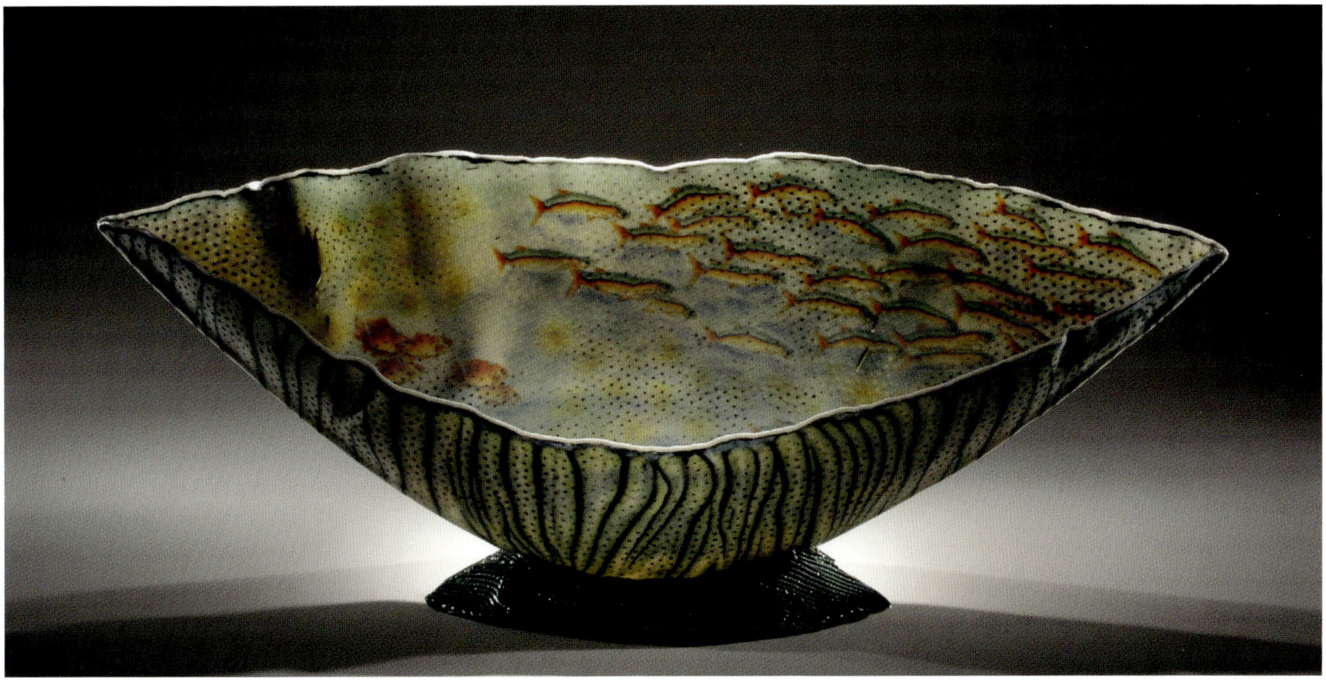

Softly, As In a Morning Sunrise, 1997. Height: 7⅘ inches, width: 16 inches, depth: 3 inches (20 x 41 x 7.6 cm)

then burnished. These techniques have often been modified using more contemporary materials like epoxy and gold leaf. These time-consuming techniques not only result in a repair of the obvious defect but also a decorative enhancement to the beauty of the piece which also alludes to time, vulnerability, repair, and a new life made richer through difficulties.

In her book *Kintsugi: The Poetic Mend*, author Bonnie Kemske writes of "kintsugi's greatest strength: an intimate metaphoric narrative of loss and recovery, breakage and restoration, tragedy and the ability to overcome it."

Benzle's vessels are truly made richer by the practice of kintsugi.

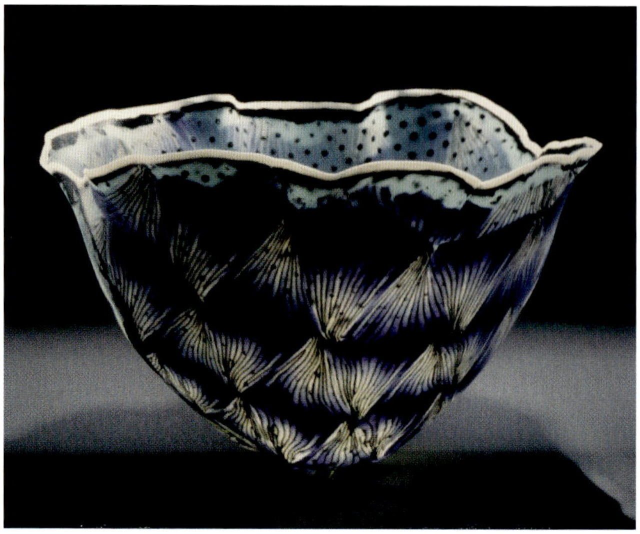

Crosshatch, 1996. Height: 3½ inches, diameter: 4 inches (9 x 10 cm)

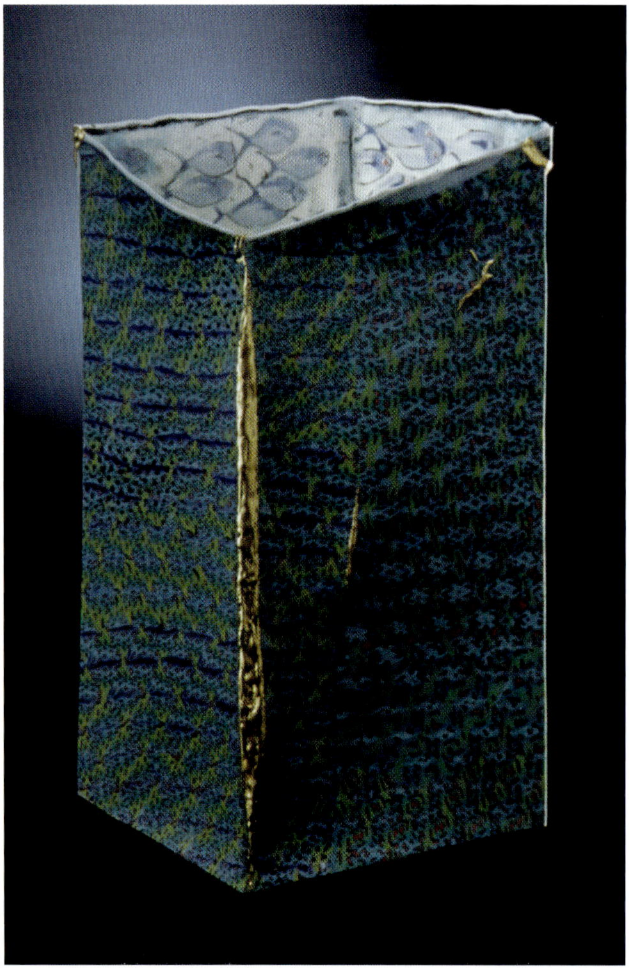

Seto Gratitude, 2001. Height: 11⅘ inches, width: 6⅓ inches, depth: 4⅓ inches (30 x 16 x 11 cm)

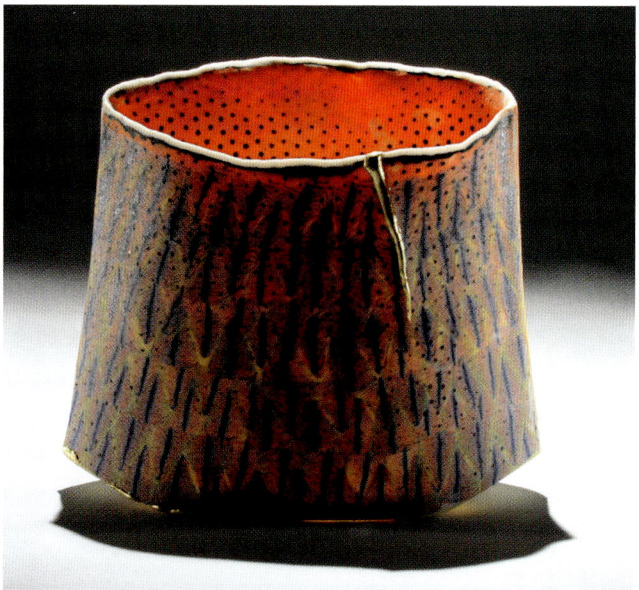

Dificilisimo, 2017. Height: 4⅓ inches, width: 4¾ inches, diameter: 2¾ inches (11 x 12 x 7 cm)

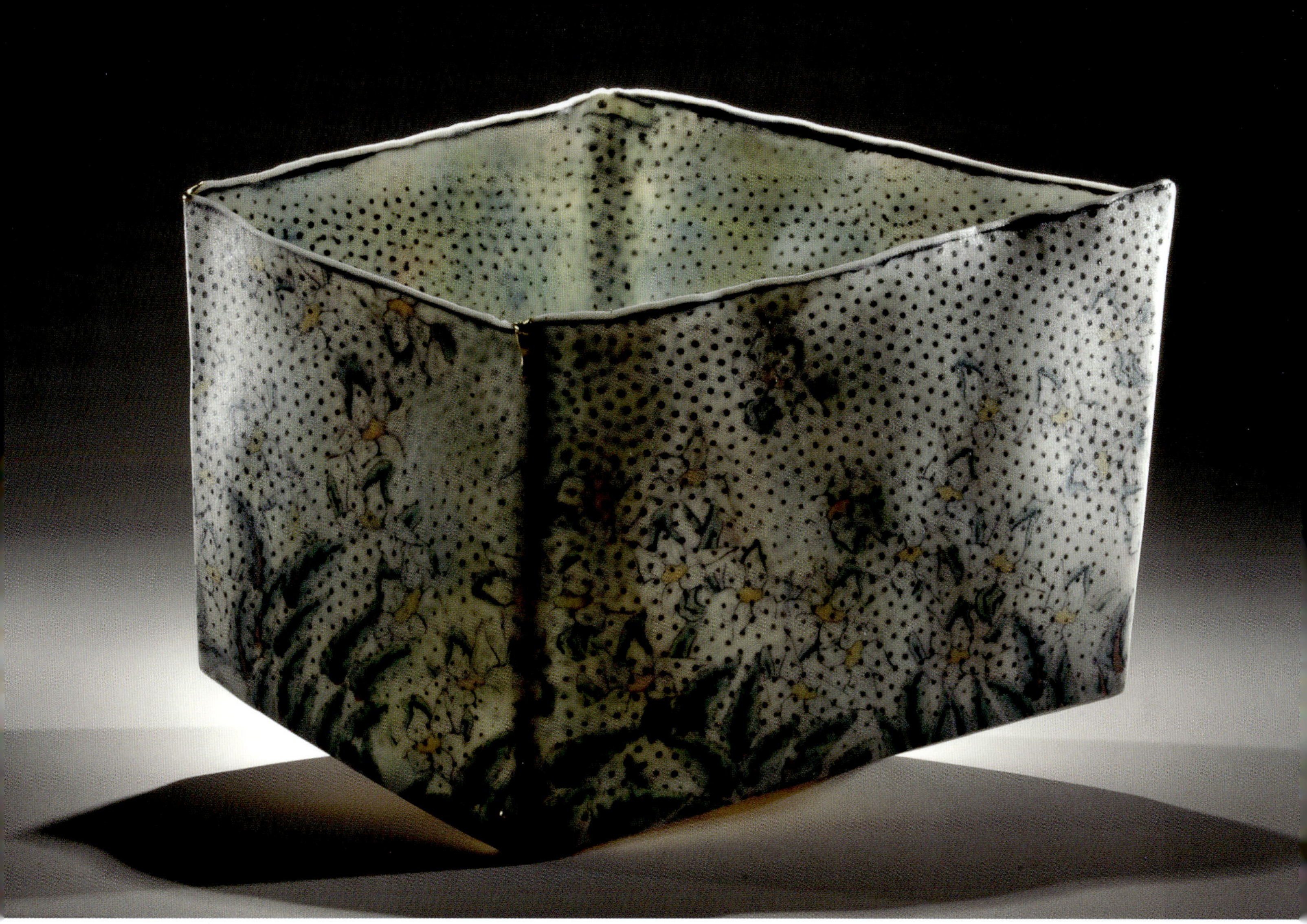

After You're Gone, 2005. Height: 7⅘ inches, width: 11⅖ inches, depth: 4 inches (20 x 29 x 10 cm)

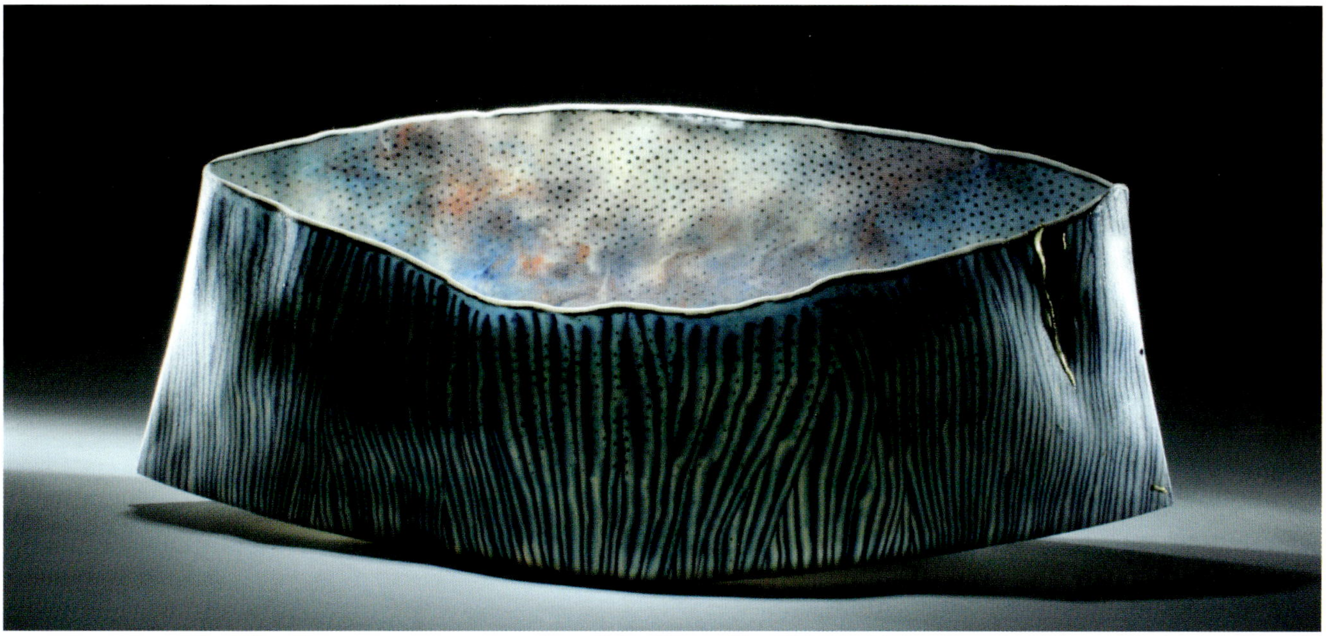

Vieques, 2017. Height: 6⅔ inches, width: 15¾ inches, diameter: 4¾ inches (17 x 40 x 12 cm)

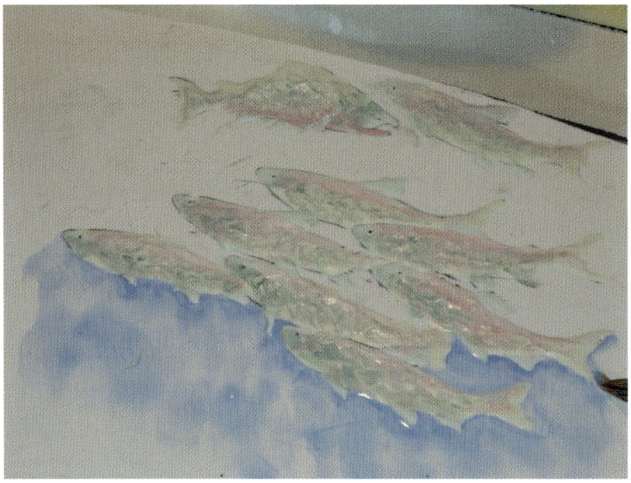

Painting with slip

This is a shot of painting with slip. Early on, I created two pieces using fish made using the nerikomi process but here, opted for the painting technique because it gave me greater latitude in developing the individual fish. The slip I use is the same Benzle #89 porcelain clay used in the entire vessel – stained and watered down to slip consistency.

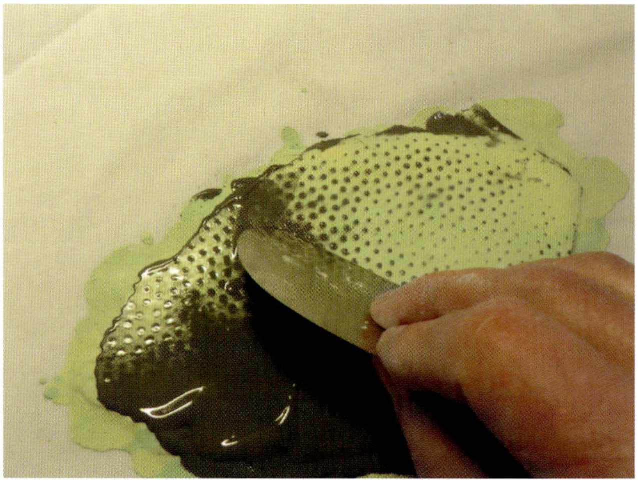

Mishima process

This is a shot of filling impressions in the clay during the mishima process. The black slip is the same Benzle #89 that I use throughout the vessel-making process. The normal slip stain percentage is 3–5 percent, but the black is increased to 10 percent in order to provide greater color intensity.

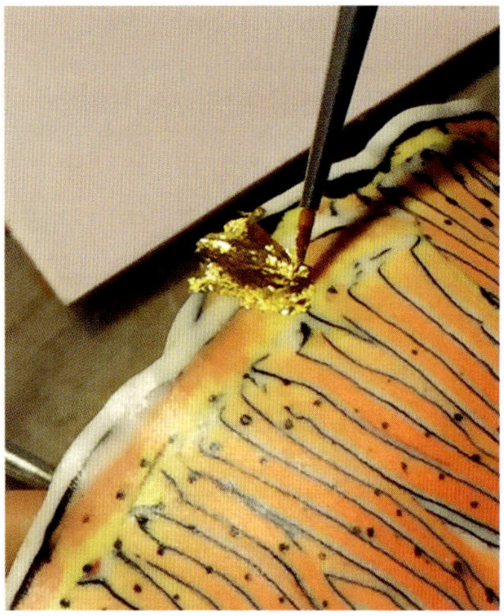

Kintsugi process

This is an illustration halfway through the kintsugi process. Because Benzle #89 porcelain is prone to cracking, I decided after about 15 years of repairing cracks to use the Japanese technique of kintsugi to repair and accentuate the cracks. Cracks would occur in either of two ways; one – natural stresses in the clay body during firing or two – careless or accidental breakage after the pieces were completed. In both cases, I find that the repaired and accentuated cracks enliven the surface and indicate interesting hints concerning the nature and provenance of the work.

AWARDS AND EXHIBITIONS

Curtis Benzle is Professor Emeritus of Columbus College of Art and Design and has taught extensively in the US and abroad. He has participated in numerous exhibitions worldwide and his works are in the collections of the Metropolitan Museum of Art (New York); the National Collection of American Art/ Smithsonian Institution; the White House Collection of American Crafts (William J. Clinton Library and Museum); the Los Angeles County Museum of Art; and the American Museum of Ceramic Art; as well as museums in Ohio, New York, Wisconsin, Arizona, Indiana, and North Carolina. Internationally, Benzle's work is held in museum collections in China, Taiwan, Canada, Japan, Italy, and Belgium. He is also the recipient of several grants and fellowships including a National Endowment for the Arts Craftsman Fellowship in 1980.

Angela Burkhardt-Guallini, Switzerland

Born in Switzerland in 1953, Angela Burkhardt-Guallini has established over the course of more than 40 years an international reputation for her functional yet highly artistic bowls made with colored porcelain. She opened her first studio in 1978 and, starting in 1984, has worked exclusively with the nerikomi technique. This change of focus was the result of a chance meeting in Basel with the Japanese calligrapher, author, and ceramist Mitsuya Niiyama who gave her a small porcelain slab made from colored clays in the nerikomi technique. Instantly entranced, Burkhardt-Guallini commenced a period of research and experimentation through which she taught herself the basics of this ancient technique.

Her early nerikomi work made use of vibrant colors and geometric patterns but, over time, the physical characteristics of porcelain itself became an underlying influence which can be seen in the flowing lines of her signature bowls. The reduced palette of white, black, and red (or sometimes blues or green with white) draws our attention but also allows the fine lines and modern design to softly speak to us.

ARTIST'S STATEMENT

Ever since I have worked as a ceramist, I have been fascinated by the nerikomi technique. This highly demanding technique offers an infinite variety of possibilities. It is more than a fascination; it has become my passion. When I am looking for new patterns, relentlessly striving for perfection, I am guided by emotion.

So many components like form, pattern, and color need to be balanced in order to produce a harmonious effect. Nature is my teacher and an inexhaustible source of inspiration. Everywhere I come upon form and structure. In addition, my creative abilities are stimulated by architecture.

The nerikomi technique enables me to express my feelings through my own individual style. It allows me to break new grounds in the field of ceramics, to explore the very limits of colored porcelain.

The nerikomi technique leads me to new ways of interpreting an ancient technique. I see the combination of traditional handicraft with contemporary design as a great personal challenge. My bowls want to depict the diversity of cultures. I use white transparent porcelain colored with different pigments, which gives an interesting interplay.

To me it is much more than work: it is my mission in life.

For Burkhardt-Guallini the specific type of porcelain clay is of utmost importance. In 2005, having won a Bronze Award in the ceramics competition in Mino, Japan, she discovered a porcelain body from Seto, Japan, that is known for its pure white color and beautiful, almost shining surface. She ordered a tonne to be delivered to Switzerland, and after a challenging period of getting to know the idiosyncrasies of this special porcelain, she developed an admirable mastery in bringing out its beauty.

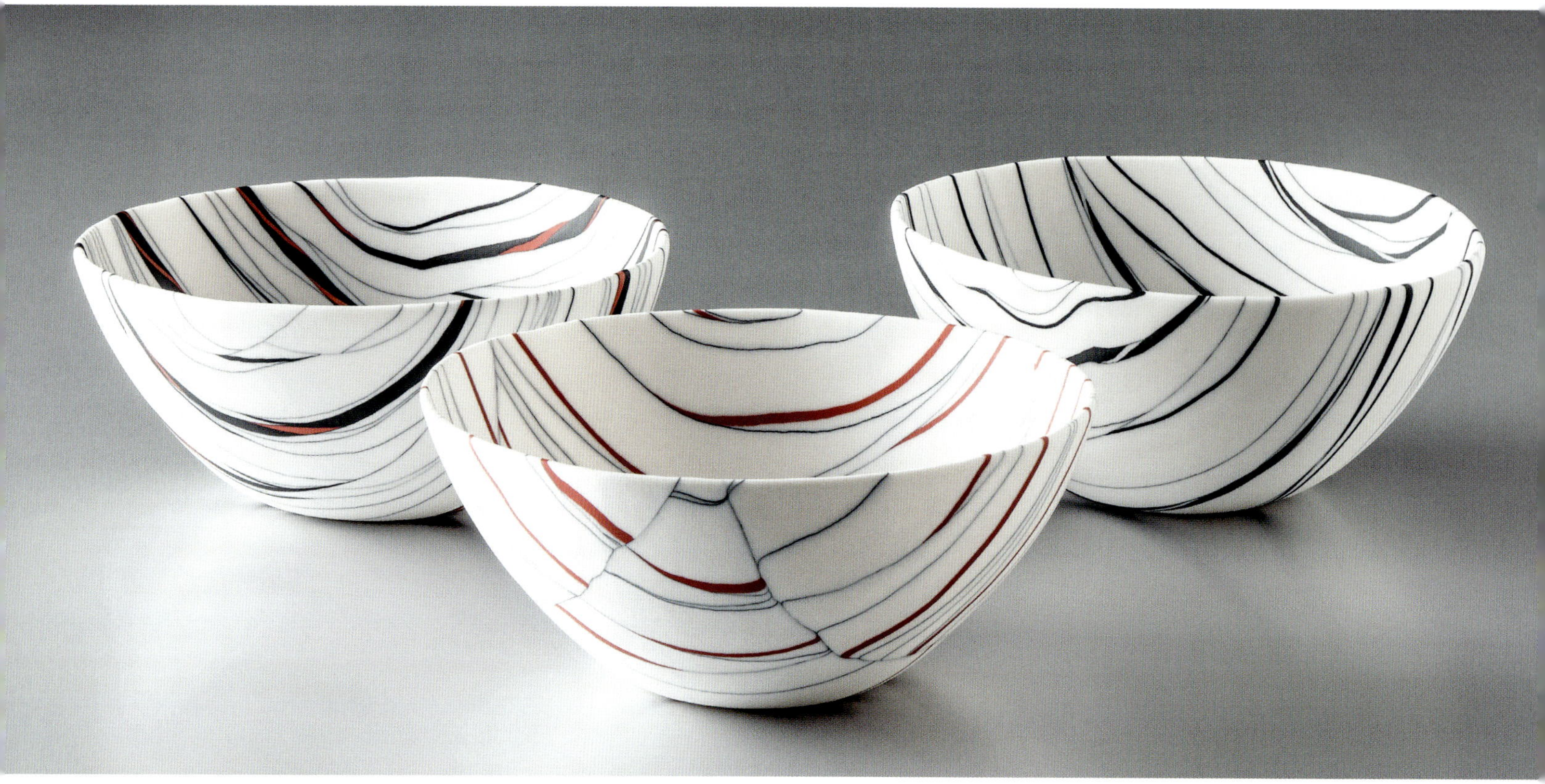

Line to Line #1, 2018. Height: 5⅛ inches, width: 10⅗ inches, depth: 10⅗ inches (13 x 27 x 27 cm)

These images all © Angela Burkhardt-Guallini

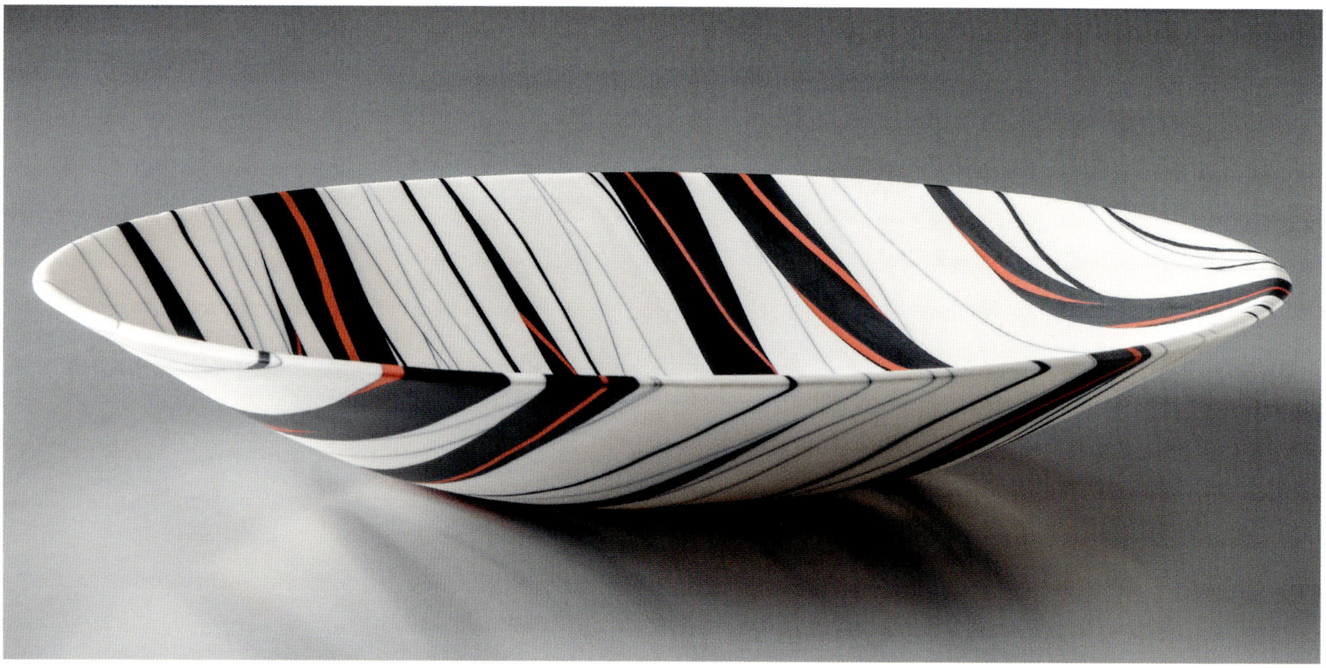

Flow, 2019. Height: 3⅛ inches, width: 15 inches, depth: 15 inches (8 x 38 x 38 cm)

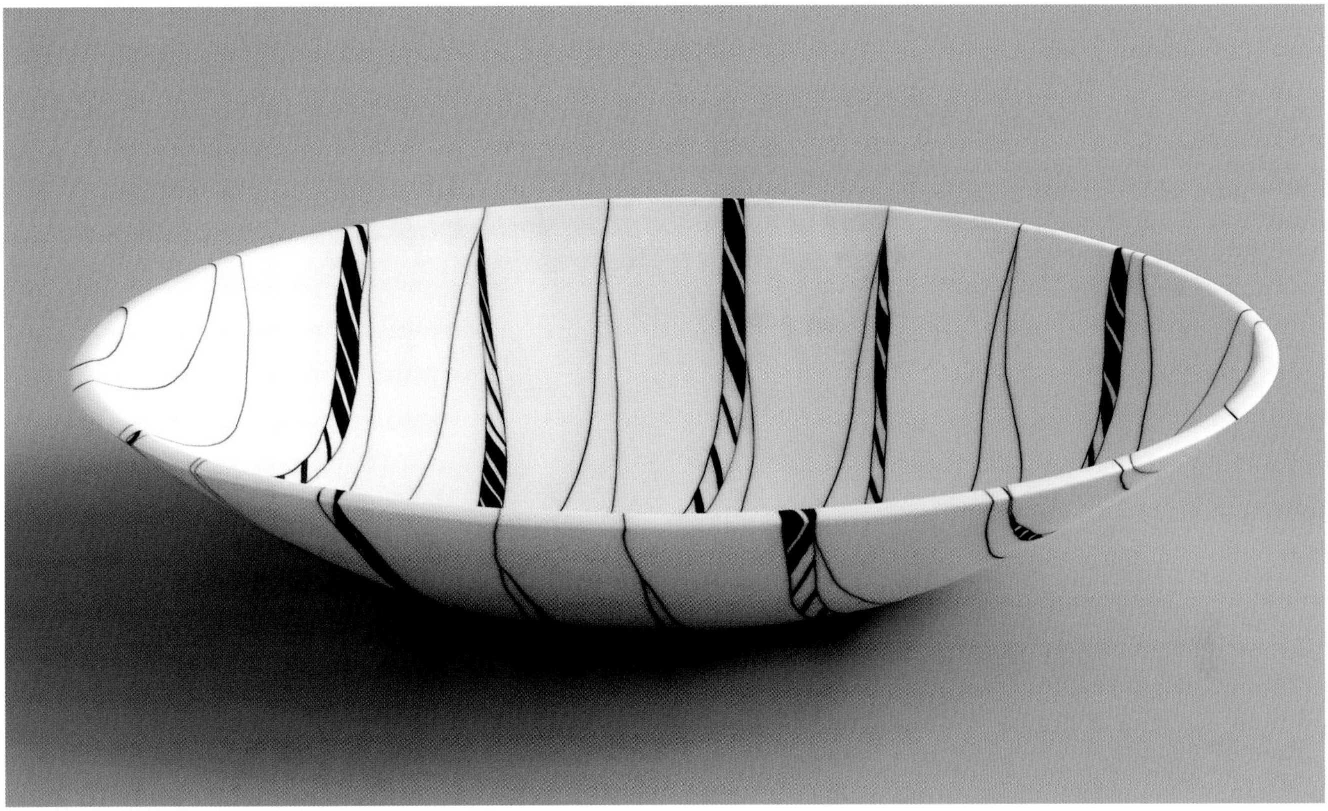

Flowing Lines #2, 2020. Height: 3⅛ inches, width:15 inches, depth: 15 inches (8 x 38 x 38 cm)

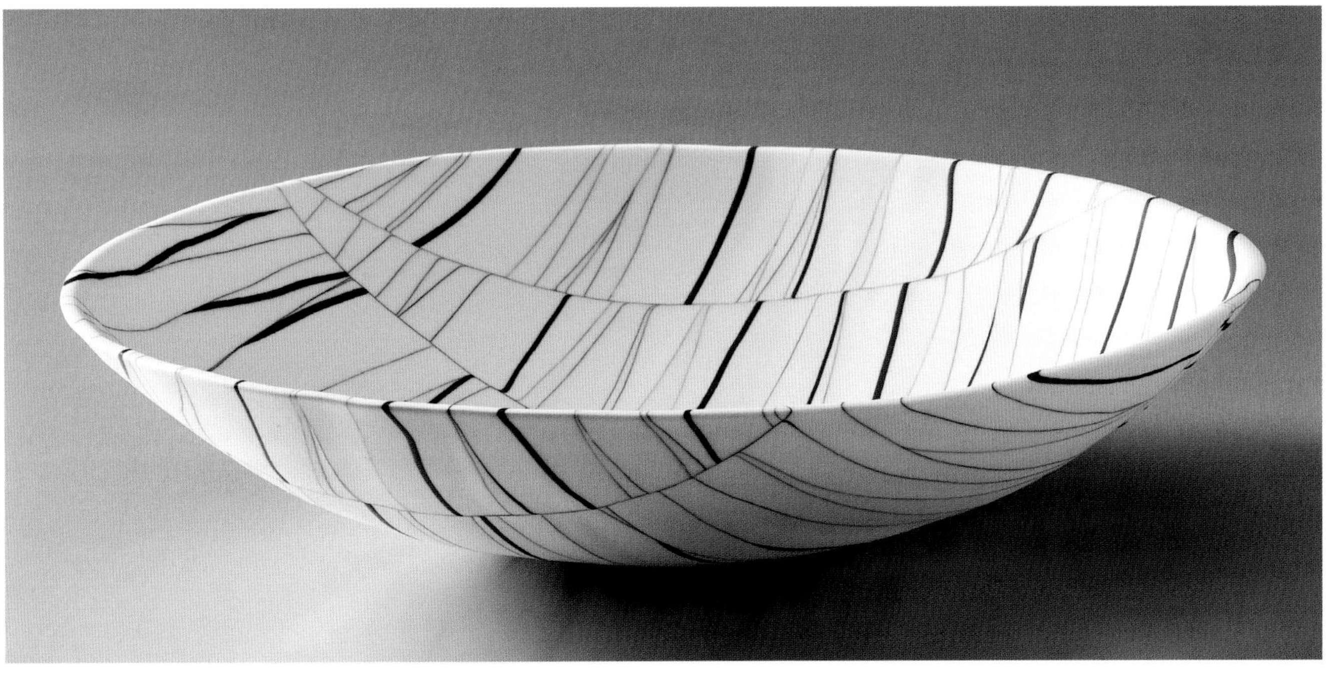

Bamboo, 2017.
Height: 3⅛ inches, width:
15 inches, depth: 15 inches (8 x 38 x 38 cm)

Angela Burkhardt-Guallini's studio process.

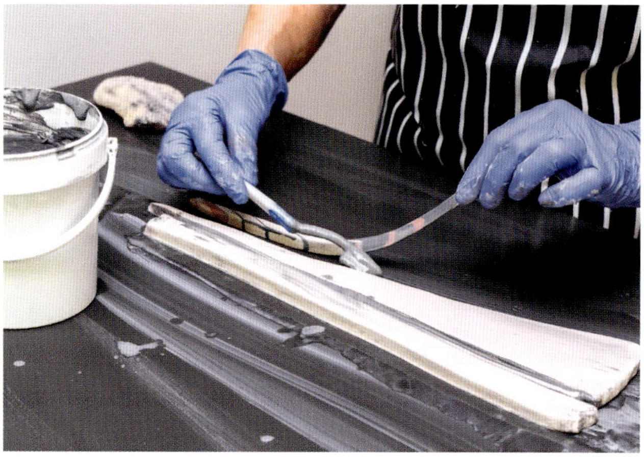

Different colored stripes are put together with dyed slip.

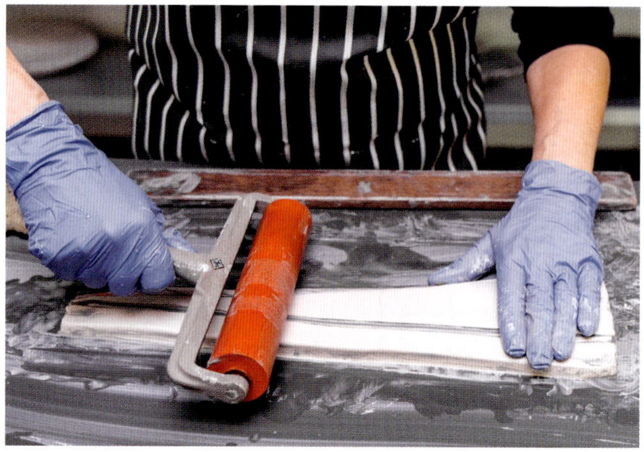

The assembled strips are rolled flat and more are added until the desired pattern is created.

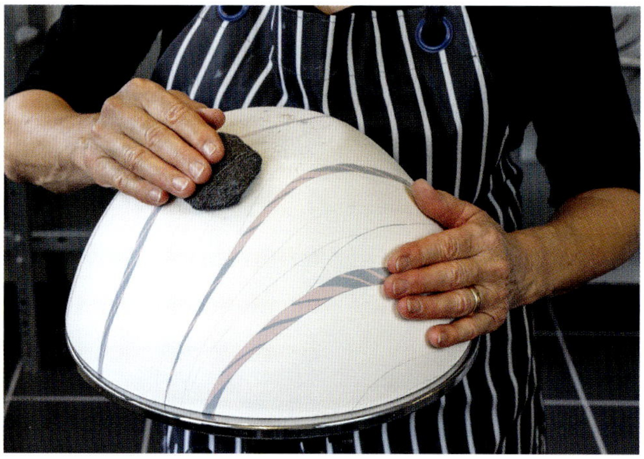

After four to eight weeks of drying, the object is sanded with steel wool until the pattern appears clear. Then the raw firing is carried out at 980°C (1796°F).

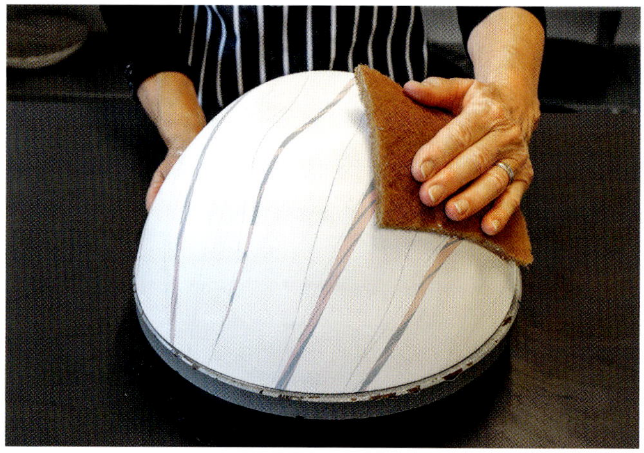

The object is polished several times with different pads and sandpapers.

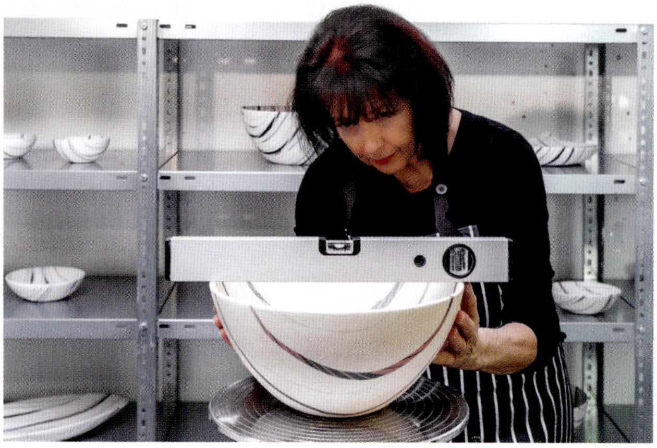

After the final inspection, the last firing takes place at 1250°C (2282°F).

AWARDS AND EXHIBITIONS

Angela Burkhardt-Guallini has exhibited her work in numerous exhibitions and competitions worldwide, including the International Ceramics Festival, Mino, Tajimi City, Japan where she received a Bronze Award in 2005 and Honorable Mentions in 2008, 2011, and 2014. She has also received awards in Switzerland, South Korea, Taiwan, and Croatia. Her work is found in the Sasama Museum, Japan; the Staufen Museum, Germany; the Icheon Museum, South Korea; and other museums in Switzerland, Japan, Taiwan, and Croatia.

Mandy Cheng, Ireland

The functional and creatively designed pottery of Mandy Cheng is a beautiful example of an ancient technique being used with a contemporary twist. Great care is taken to assemble her intricate pattern blocks. Thin slices taken from them are carefully hand-rolled to give an organic movement to the pattern. Her shapes and designs are decidedly contemporary, often using the boldness of black and white in partnership with the soft influence that the nerikomi technique can have over a geometric pattern.

This dichotomy between fragility and strength is an underlying theme of Cheng's work and is also a connection to the many forms of inspiration that influence her.

ARTIST'S STATEMENT

On the one hand the material is very fragile, on the other hand the patterns are lively and full of force.

This is like what I observe in nature, water flows smoothly around corners and yet has tremendous power, flowers are so delicate and come in millions of shapes and colors, birds are so fragile but fly swiftly and effortlessly everywhere.

I am inspired by what I observe, I design my works to capture the lightness, the movement and the delicacy in nature that is so beautiful.

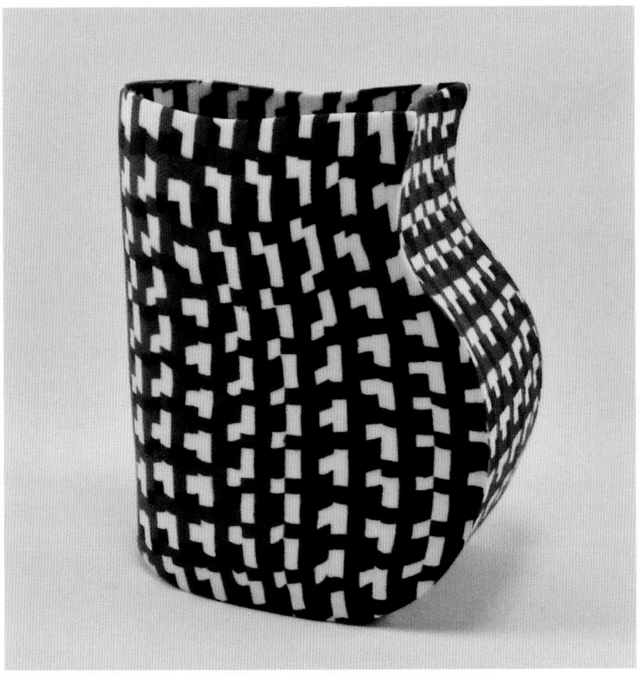

Bob, emboss, 2021. Height: 6⅓ inches, length: 5⅛ inches, depth: 2½ inches (16 x 13 x 6.5 cm)

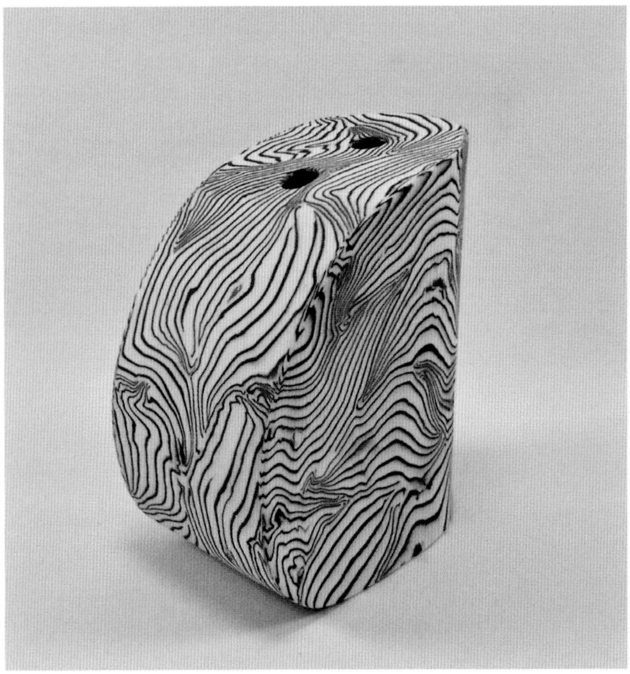

Half Moon, rainforest, 2020. Height: 5½ inches, length: 4¾ inches, depth: 2¾ inches (14 x 12 x 7 cm)

Cheng is dedicated to making forms that are minimalist yet ergonomically functional, an example being a line called "Collective Sense." She has deconstructed each of her products, from teapots and cups to milk jugs, plates, and bowls, to examine the function of every part of the form – body, handle, lid, saucer – to understand their essential functions. As architect Louis Sullivan famously stated, "form follows function," and Cheng's attention to this overriding principle has resulted in some unique details in her work that are both beautiful and ergonomic, such as the turned-up corner of the saucer and the cup handle in her *Wavy Cup and Saucer* from her *Zebra* three-piece espresso set. She explains:

> *Starting with the saucer, to actually lift or carry a cup, one needs to hold the saucer at one point – meaning that having only one area curved up to hold easily is sufficient. This is how the saucer with an upturned corner was created. Next was the cup handle. My thin porcelain cup is light, just needs a delicate handle to hold, and this concave shape fits snugly between the grip of thumb and index and middle fingers.*

A common theme among nerikomi artists is the combination of an artistic sensibility and a technical mind. Mandy Cheng is no exception: "My interest in the nerikomi technique reflects my dualistic character, being artistic enough to appreciate beauty in all forms of expression whilst also having a rational scientific mind, able to engineer complicated geometric forms."

Cheng's vessels are made from Porcelain Royale and incorporate Ceradel high-temperature coloring oxides. After a bisque firing, they are wet sanded and then glazed with a transparent glaze on the inside only, leaving the outside to be wet sanded to a smooth matt finish.

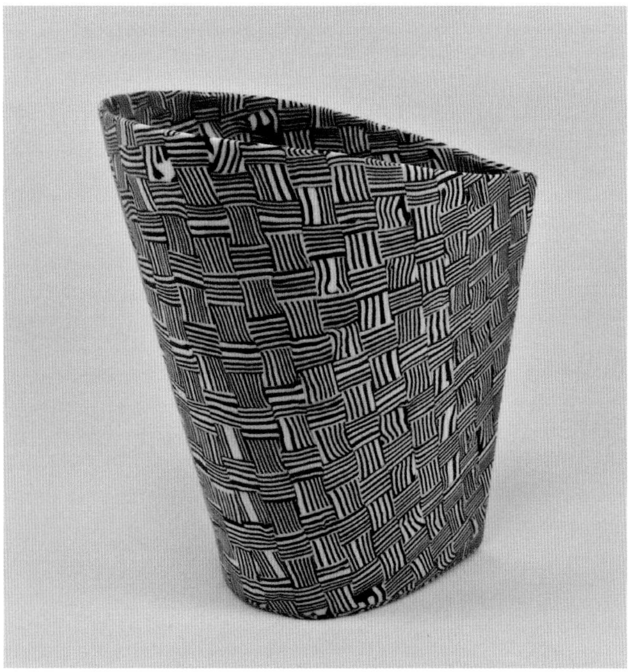

OM, basket, 2018. Height: 6 inches, length: 4⅓ inches, depth: 3⅛ inches (15 x 11 x 8 cm)

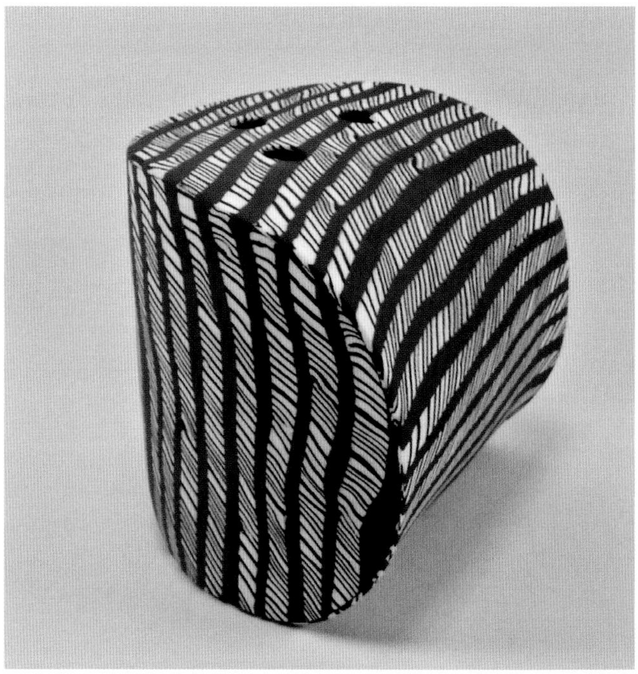

Semi-spheroid, twilight, 2019. Height: 8⅔ inches, length: 5⅛ inches, depth: 4⅓ inches (22 x 13 x 11 cm)

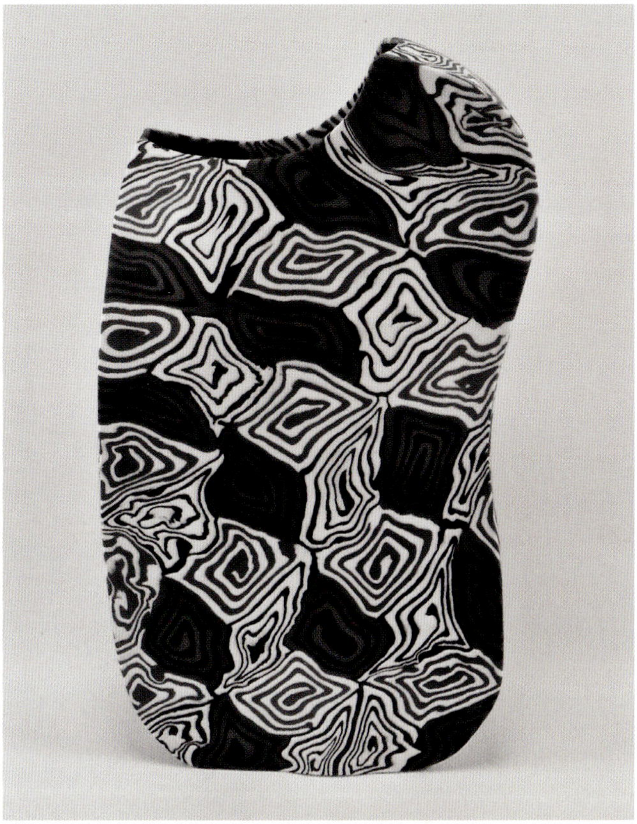

AWARDS AND EXHIBITIONS

Mandy Cheng's work is shown at the Contemporary Ceramics Center and the Ruthin Craft Centre in the UK; SO Fine Art Editions, Mill Cove Gallery, and Cahir Arts, all in Ireland; and Gallery 1608, Northern Ireland. In 2021 her work was included in the Ceramics Ireland Selected Members Exhibition, where she was the winner of the inaugural SO Fine Arts Editions award. Also in 2021, Cheng was represented in the Ceramic Expressions 4th Edition, Singapore. Her work was included in the 2017 British Ceramics Biennial and has been exhibited in Germany and the USA. In 2022 she received the Ceramics Ireland Creativity in Craftsmanship award and the Bronze Prize at the 2021 International Coffee Cup Competition, Taiwan. Cheng's ceramics are included in the collections of the OPW State Art Collection, Ireland; the New Taipei City Yingge Ceramics Museum, Taiwan; the Museum of Tea Ware, Hong Kong; and the Everson Museum, USA.

◄ *Roar,* cubic, 2021. Height: 11⅘ inches, length: 6⅓ inches, depth: 3⅛ inches (30 x 16 x 8 cm)

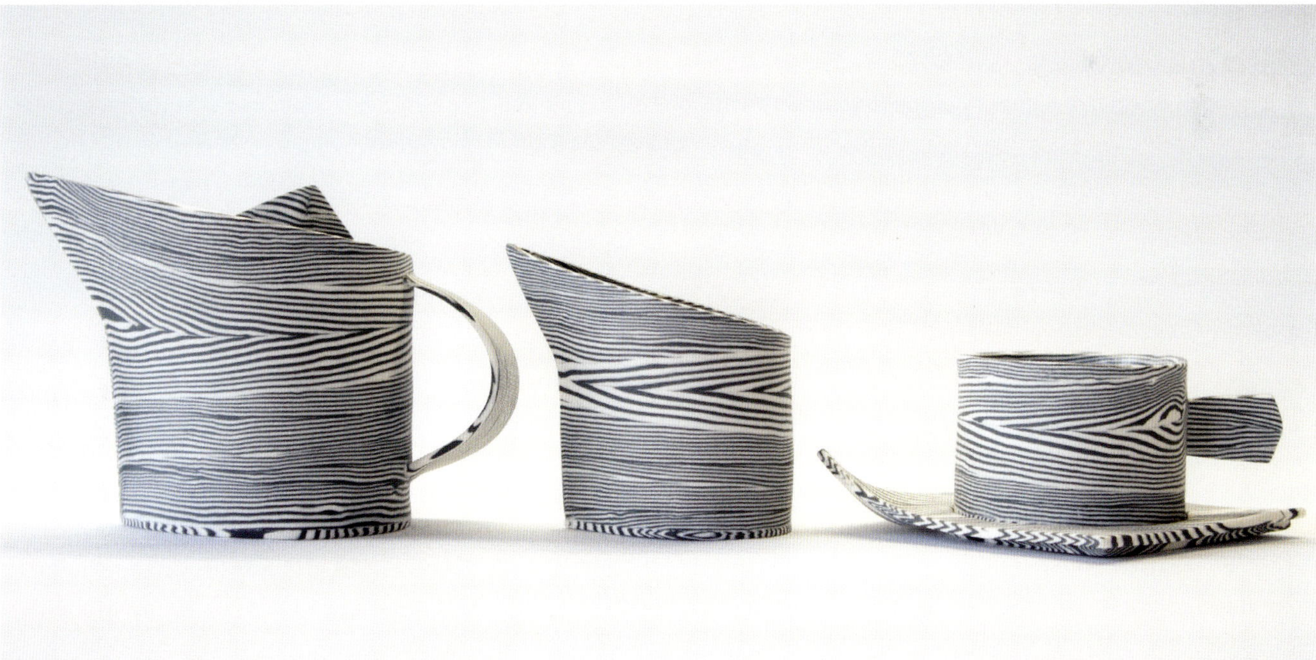

Zebra three-piece set, 2015. (Tea pot dimensions) Height: 4⅓ inches, length: 5 inches (11 x 13 cm)

Ben Davies, England

Ben Davies is a professional cellist and music teacher. He is also a talented potter. Inspirations, creative strategies, and teaching methods in these two disciplines often overlap and influence each other. From another angle, it turns out that after playing the cello for extended periods, working with soft clay is good for the hands.

ARTIST'S STATEMENT

The solitude of working alone is also good after the gregarious nature of working as part of an ensemble. A performance in music is ephemeral and often subject to the demands of a conductor, but the making of a pot is entirely my creation and results in a finished and permanent object which may exist for thousands of years. There are advantages and disadvantages to both! The rules of working with clay are defined by the material and feel much more free and intuitive than those linked with the discipline needed to arrive at the same freedom to make music a lifetime career.

As I child I was fascinated with nature and loved music. It was during this time that I was inspired by my pottery teacher at school and although I studied Art at Foundation Level, it wasn't until my thirties that I returned to ceramics. In between I studied Geography and Geology at university and trained as a professional cellist at the Royal Academy of Music.

I started exhibiting in my forties with work that developed as a result of having limited facilities. Initially I made unglazed coiled pots using colored and textured clays, employing burnishing techniques and occasionally smoke firing.

My current work explores textures and pattern inspired by geology and landscape. I use a variety of hand-building techniques to produce vessels which I then sand and polish to achieve smooth tactile surfaces.

My self-taught approach to making pots contrasts with the disciplined and formal training I received as a musician, but shares the same creative thinking. I wondered as a young adult whether it was possible to find a way of combining all my interests and take pleasure in reflecting on how all these strands seem to have come together.

The clays that Davies uses are predominantly earthenware (black, white, red, and buff – bought ready-made) with the addition of semi-porcelain. In addition to these colors, others are created by adding stains or oxides to one of the white bodies. Scrap clay is also collected and recycled into intermediate tones.

Davies' pottery doesn't fall into either of the traditional colored-clay techniques – nerikomi or neriage. From a basic hand-building pottery background, a variety of differently colored clays, and a mind that can critically imagine three-dimensional strategies for layering the clays, he has discovered methods that result in fascinating patterns within the clay that seem to emerge organically from his forms. After seeing some of Davies' pots at a ceramics fair, a museum specialist shared with him an image of a pot from an ancient era that was remarkably similar to his. An ancient mind had been thinking along the same lines (or swirls) and Davies felt a deep connection over the centuries.

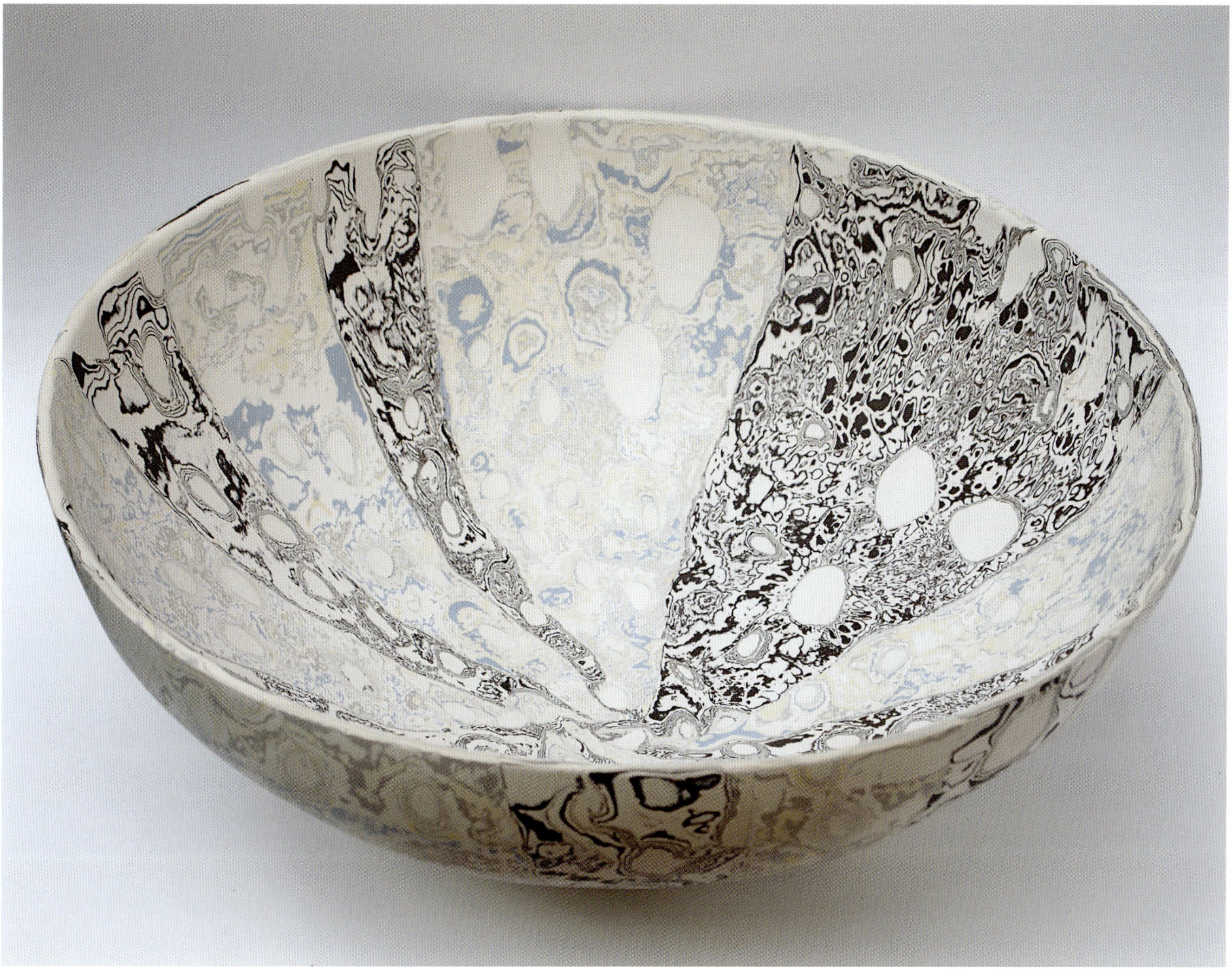

Large segmented "contour" bowl, 2022. Height: 5⅛ inches, width: 13 inches (13 x 33 cm). *These images all © Ben Davies*

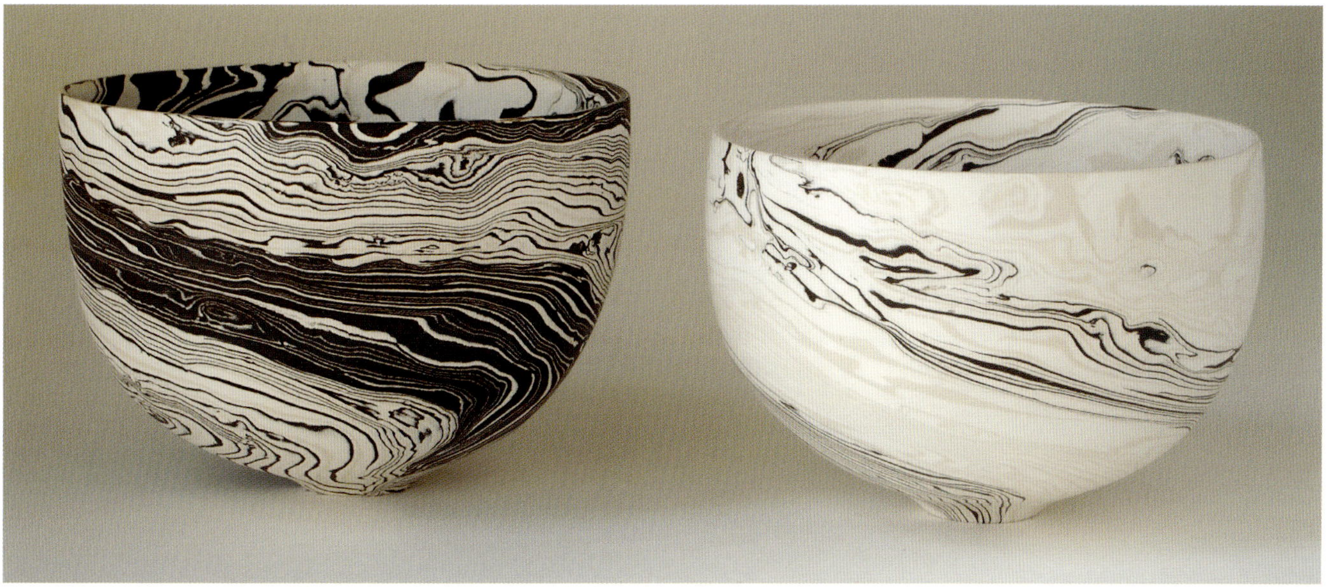

Black-and-white spiral pinch bowls, 2013. Left: height: 4¾ inches, width: 4 inches (12 x 10 cm);
Right: height: 4⅓ inches, width: 4 inches (11 x 10 cm)

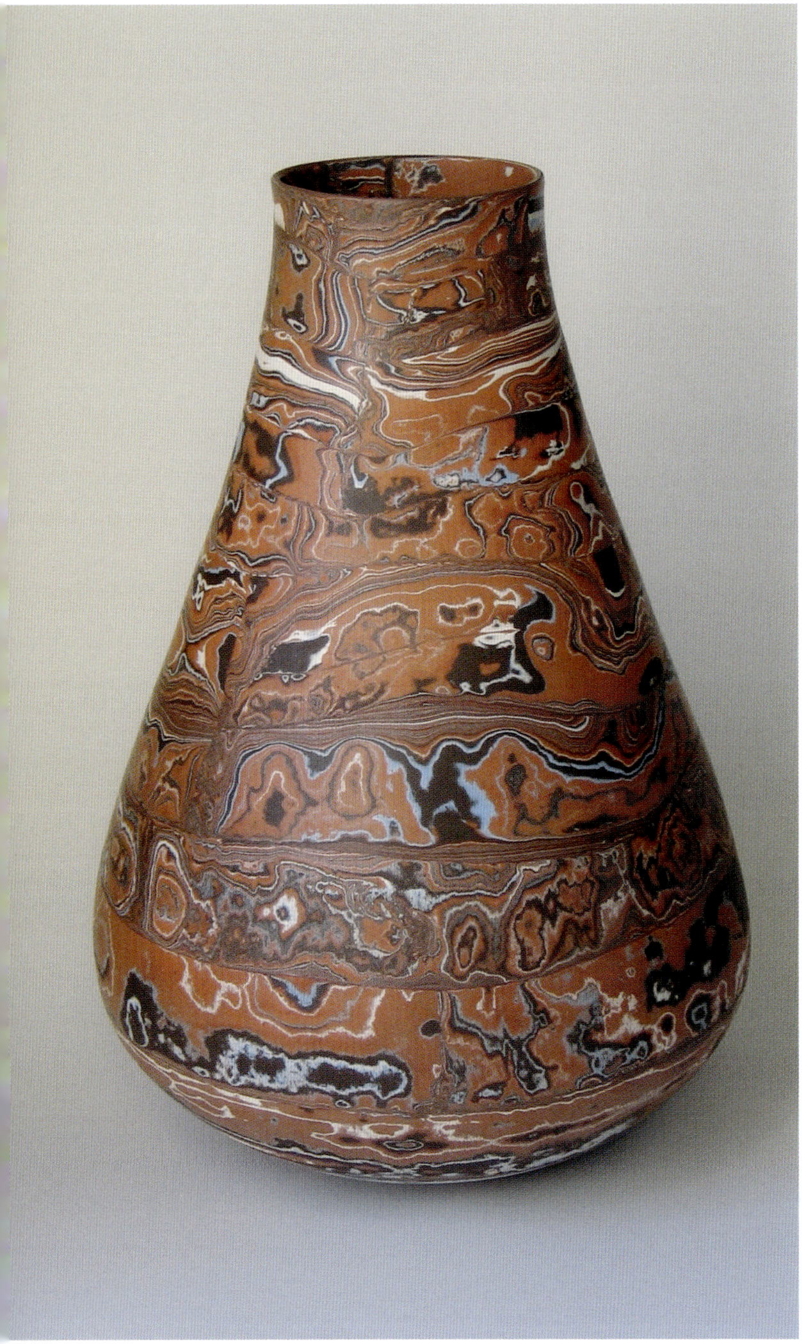

Large conical vase, 2008. Height: 17⅓ inches, width: 13 inches (44 x 33 cm)

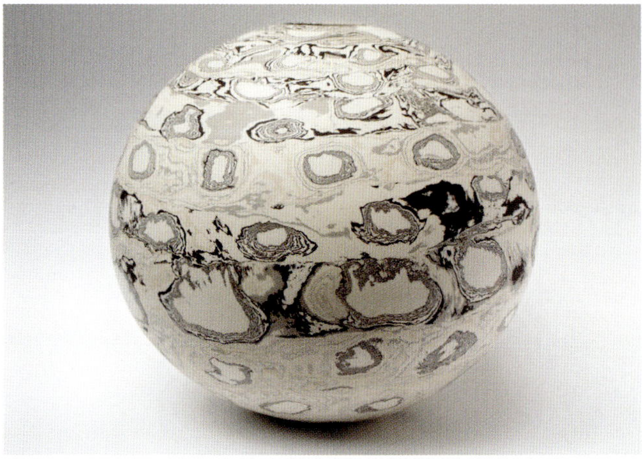

White, gray and black "contour" sphere, 2022. Height: 7⅘ inches, width: 8⅓ inches (20 x 21 cm)

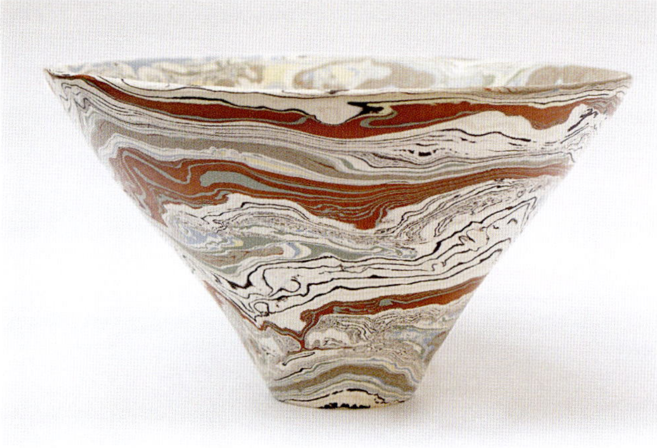

Layered conical pinch bowl, 2022. Height: 3⅗ inches, width: 5⅞ inches (9 x 15 cm)

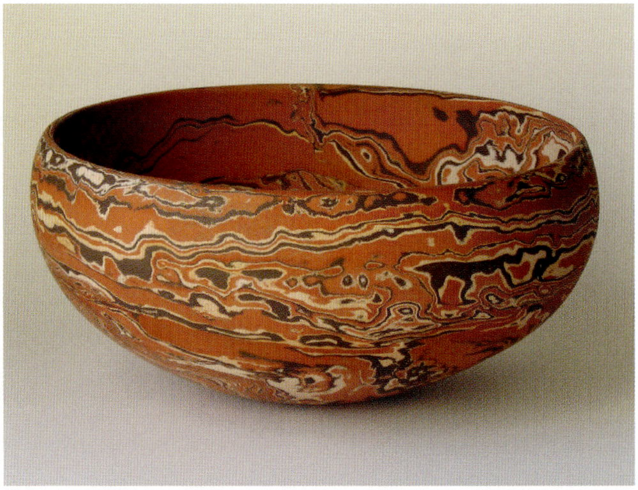

Large marble bowl, 2008. Height: 6 inches, width: 12½ inches (15 x 32 cm)

One of many forming methods by Ben Davies.

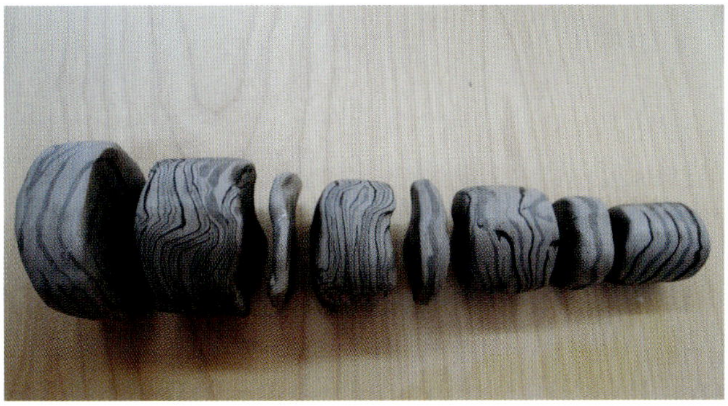

Circular sections cut from the varying "sandwiches" are organized side by side to create the pattern desired for the eventual bowl. *These images all © Ben Davies*

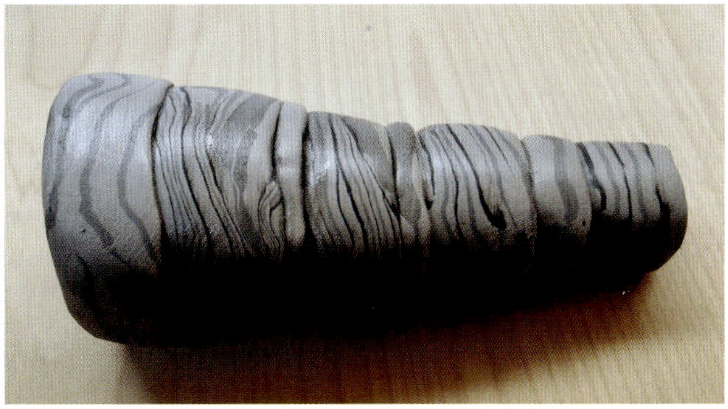

The circular sections are attached together using water and slip to create a cone of pattern.

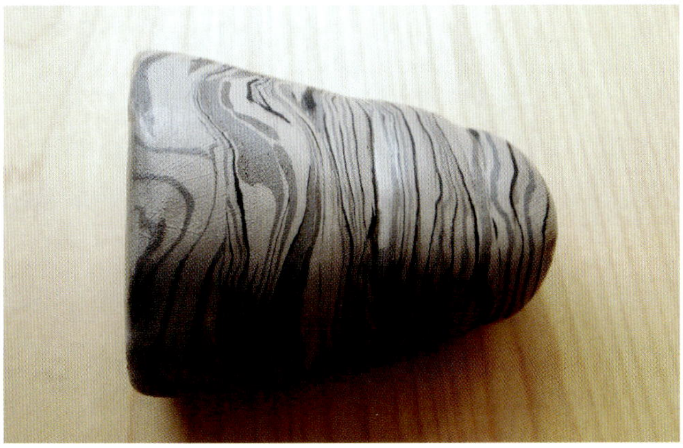

The cone is compressed to form a shorter fatter cone.

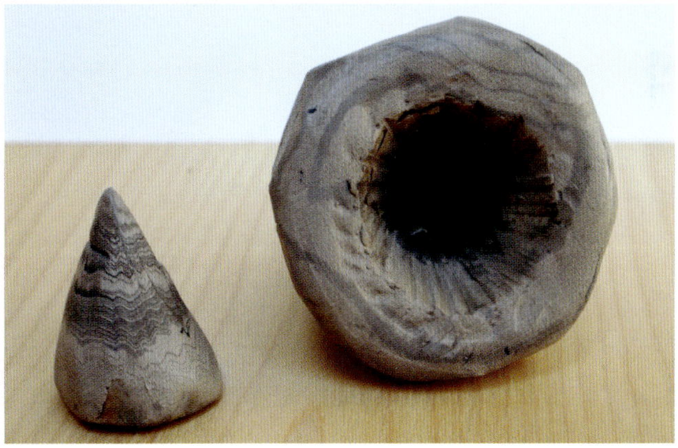

A knife is used to cut and extract a smaller cone from the larger cone.

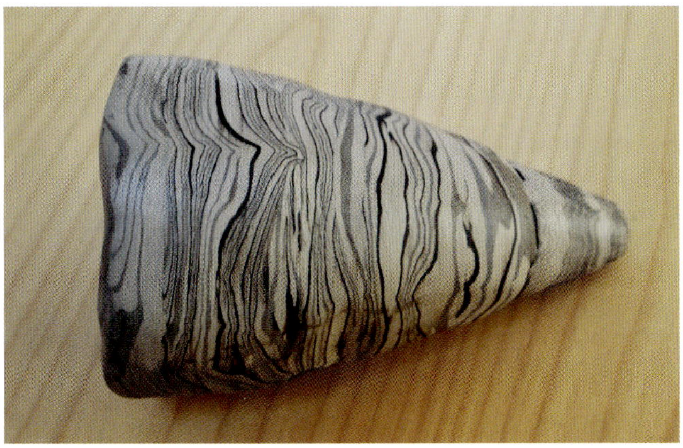

The smaller cone is attached to the tip of the larger cone.

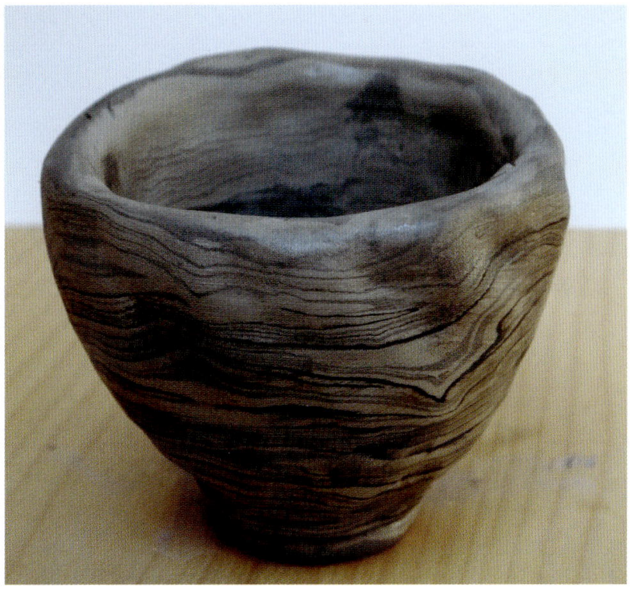

▶ The opening within the cone is gently expanded to create the early form of the vessel.

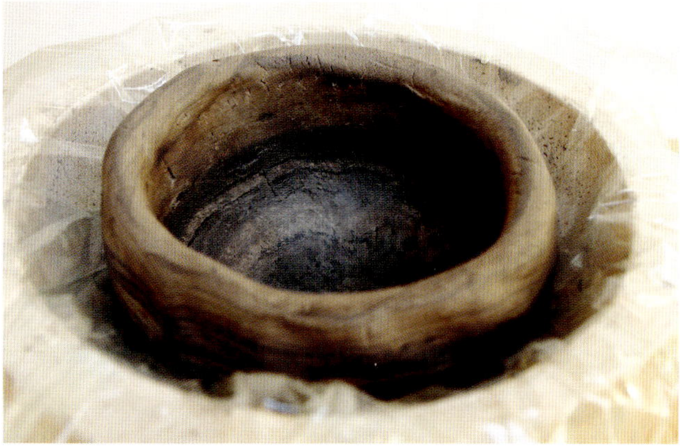

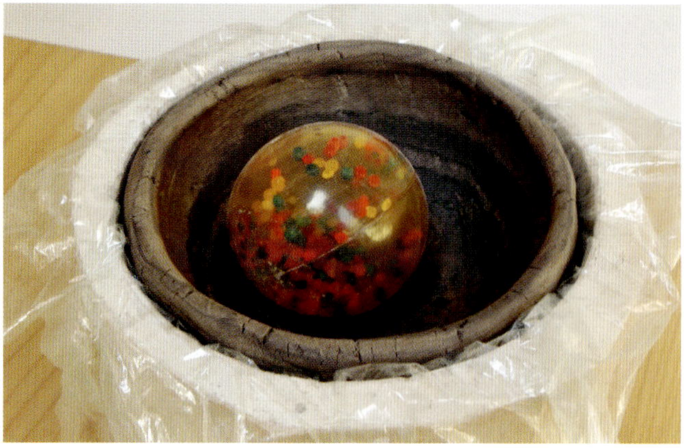

The vessel is placed in plaster mold and the process of opening the form continues using a combination of hands and various tools such as a plastic ball or spherical stone.

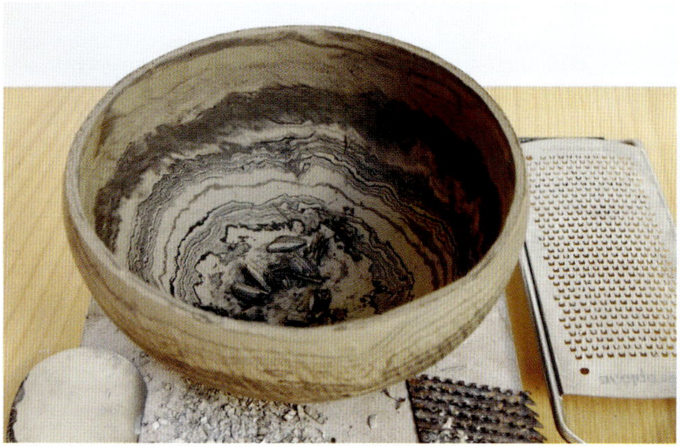

The form is impressed into the plaster mold and the walls are squeezed and drawn up by hand. When leather-hard the bowl is removed from the mold for further shaping using a range of tools – surform blades, microplane (from the kitchen!), sandpaper, damp sponges, and metal kidney scrapers.

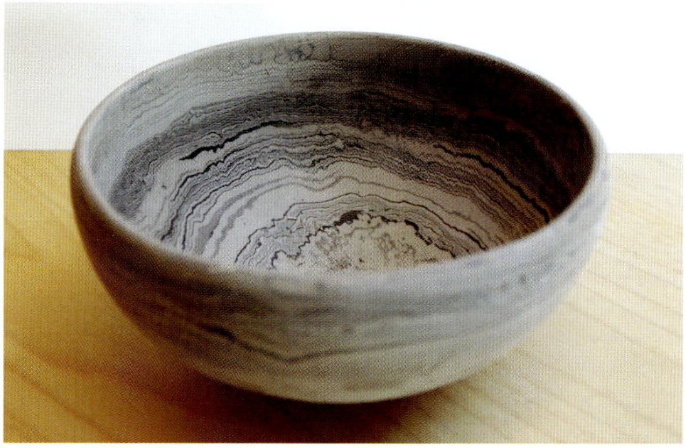

Finished bowl, drying slowly before first firing.

AWARDS AND EXHIBITIONS

Ben Davies has exhibited extensively throughout the UK in galleries, ceramics fairs, and exhibitions. In 2010 he was elected as a professional member of the Craft Potters Association.

Mieke Everaet, Belgium

The sculptural vessels of Mieke Everaet utilize colored clays, but in a manner that is quite different from traditional nerikomi. Rather than creating a pattern block from which slices are taken, she works with minute tiles of stiffening colored clay. She carefully inserts them directly into a plaster mold with brushed-on clay slip of a contrasting color as a bonding material. The slip becomes a lace-like network of color surrounding and connecting the individual tiles. This technique, similar to mosaic, is both the forming method as well as the designing method. It was developed over years of inspiration and experimentation.

Sources of inspiration have included plants, native art and craft, brick and stone architectural elements, as well as journeys to Assisi, Italy, China, and Burkina Faso. This last one was her twin brother's imaginary travel which for Mieke was inspirational. The vast historical ceramic heritage is also the bedrock of her influences. Exposure early in her career to the Chinese Tang Dynasty colored laminated dinnerware in the collections of the Musée Cernuschi and the Musée National des Arts Asiatiques – Guimet, both in Paris, was a dramatic turning point in her education.

ARTIST'S STATEMENT

I adore the repetitive element, the construction of the bricks in the cement, in the building of elements. It has a specific, tactile language, something very direct and also contemplative. It inspires me to adapt this visual element in my vessels and bowls. It is endless and shows eternity.

The relationship between shape, structure, and color within the formal space of the bowl fascinates me.

The constructed and designed walls of the bowl embrace or underline the content of the work. Inside and outside are connected to each other and together describe the surrounded emptiness. Every bowl sculpture has a three-dimensional experience. Each work is a little shrine and at the same time an experience.

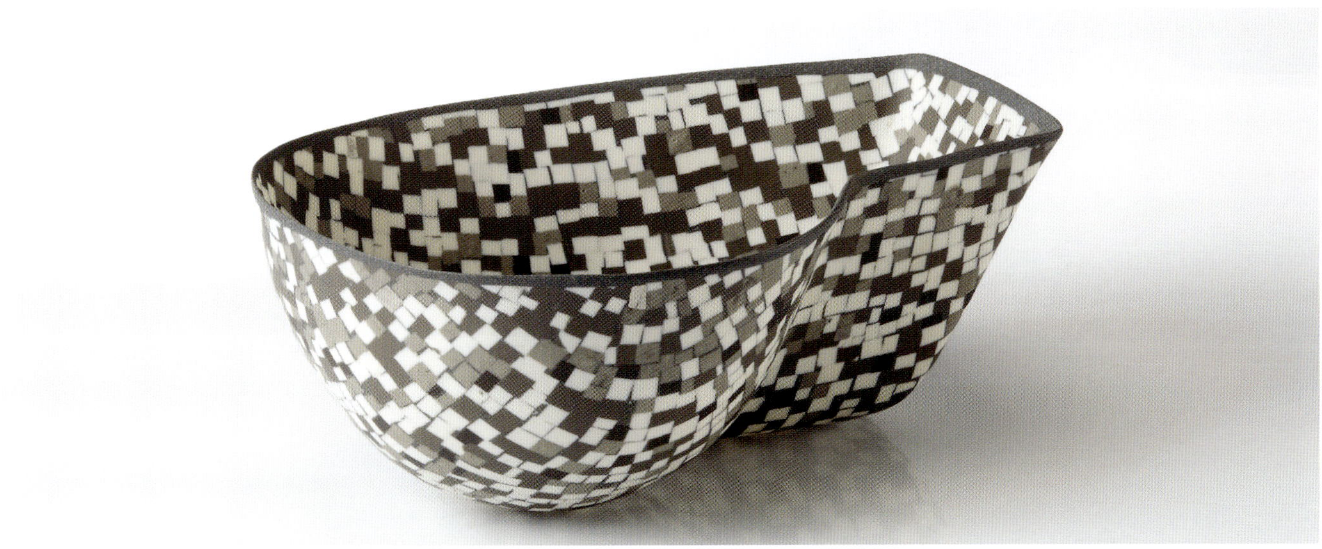

***Bowl with Rectangle,* 2018. Height: 10⅗ inches, width: 10⅗ inches, depth: 4⅓ inches (27 x 19.5 x 11 cm)**

Double-shaped colored porcelain bowl – inlay technique with white, gray, and black – electric fired at 1250°C (2282°F).

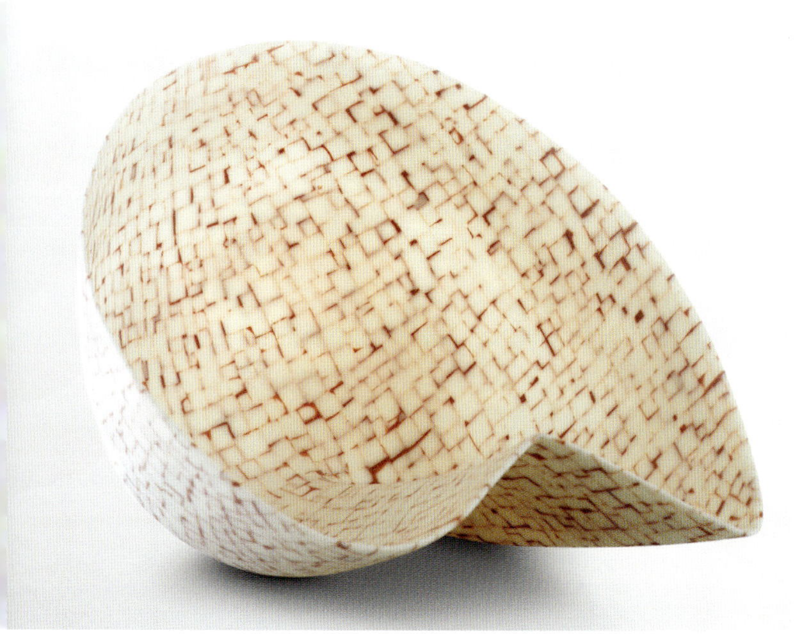

Pointed, **2020. Height: 9 inches, width: 8⅓ inches, depth: 4¾ inches (23 x 21 x 12 cm)**

Triangular-shaped and colored porcelain bowl – inlay technique with white porcelain in red – electric fired at 1260°C (2300°F).

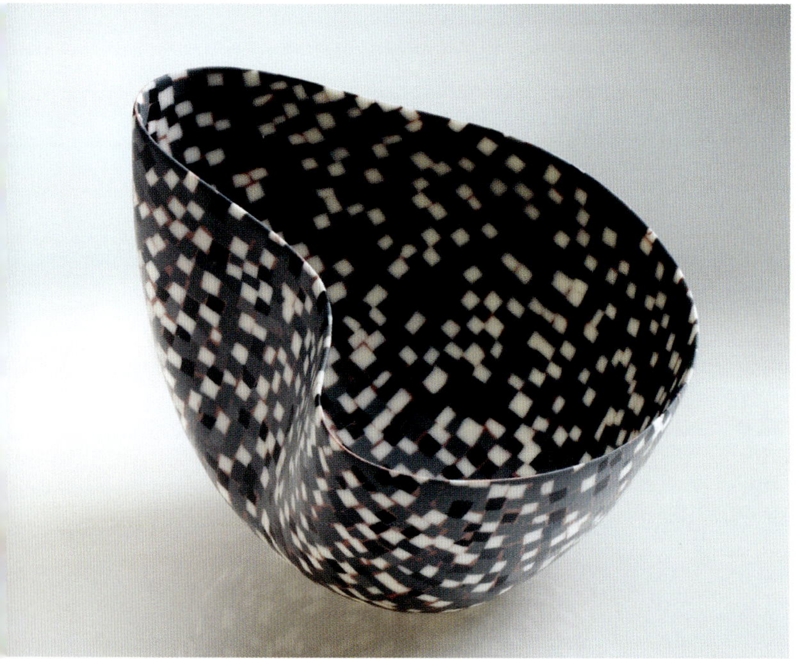

Kidney, **2017. Height: 9 inches, width: 9⅞ inches, depth: 6 inches (23 x 25 x 15 cm)**

Double-lobated porcelain bowl – inlay technique with gray, white, and black in pink – electric fired at 1260°C (2300°F).

Everaet's earlier work was influenced by Mexican and African woven baskets, composed in traditional circular bowl forms. More current work has broken away from those traditions and evolved into the realm of sculpture, incorporating complex asymmetrical vessel shapes which transcend the ordinary.

Nigel Atkins writes of Everaet's work:

> *The simplified and barely stable archetypal container is [...] no longer a bowl for use. It aspires to become art, to metamorphose into an object claiming ruminative contemplation. It has become a bowl-shaped canvas on which the artist can play out the spatial games she has in mind. These games clearly involve a sophisticated play between opacity and translucency, between virginal whiteness and seams of muted color, between seeming frailty and mechanical resistance, between pretended simplicity and evident complexity.*

Everaet was educated at the Royal Academy for Fine Art, Antwerp, and the National Higher Institute of Fine Arts, Antwerp, and teaches at the Art Academy for Visual Art in Halle, Belgium.

AWARDS AND EXHIBITIONS

Mieke Everaet has received numerous awards over her career from Germany, Belgium, Croatia, Japan, Switzerland, France, and South Korea. She has had solo exhibitions in Belgium and France and has participated in group exhibitions in France, Switzerland, Belgium, Germany, the UK, the USA, Greece, Taiwan, China, Sweden, the Netherlands, and Spain.

Her works are included in the Design Museum Ghent, Belgium; the Flicam Museum, Fuping, China; the Musée de la Porcelaine, Sèvres, Paris; and the Art & History Museum, Jubelpark, Brussels, Belgium; as well as other museums in Switzerland, Belgium, Germany, the Netherlands, France, and the USA.

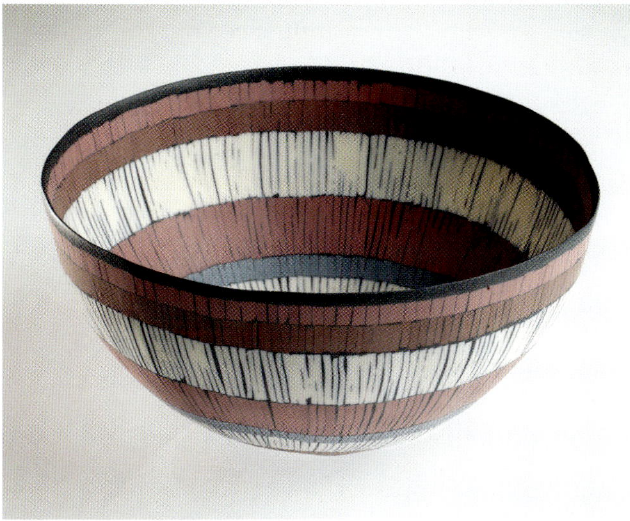

***Rhythm*, 2011. Height: 9⅞ inches, diameter: 5 inches (25 x 12.5 cm)**

Colored porcelain bowl – inlay technique with white, pink, claret-colored, and gray in black – electric fired at 1250°C (2282°F).

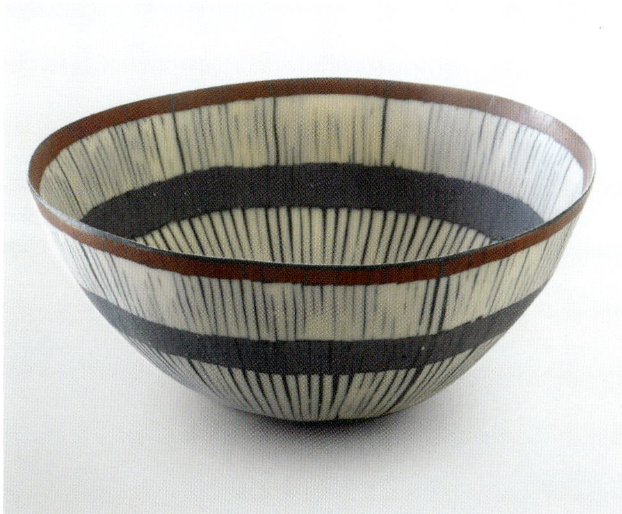

***Rhythm Assisi*, 2020. Height: 6⅓ inches, diameter: 3⅓ inches (16 x 8.5 cm)**

Colored porcelain bowl – inlay technique with white, gray, and brick-colored red porcelain in green – electric fired at 1260°C (2300°F).

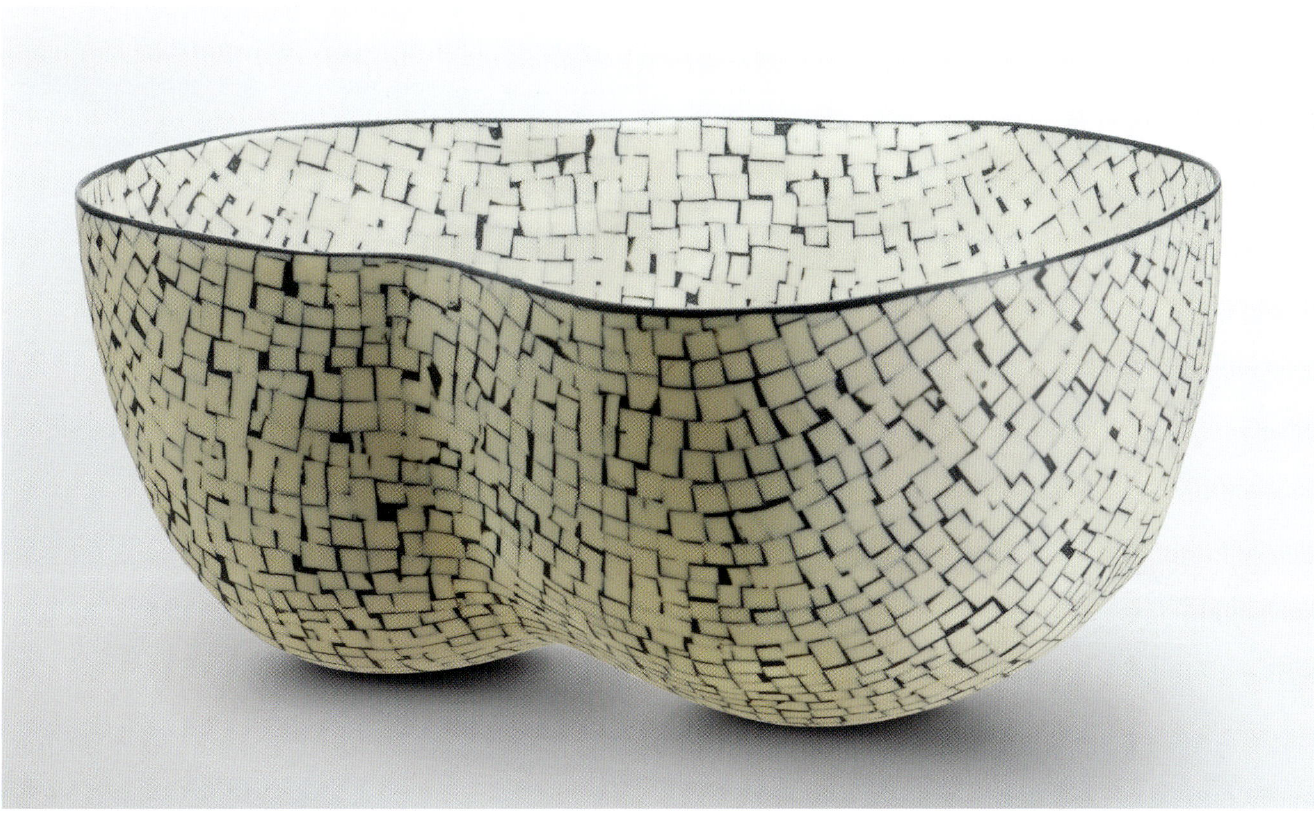

***One in another 3*, 2022. Height: 11 inches, width: 8⅔ inches, depth: 5⅛ inches (28 x 22 x 13 cm)**

Double-lobated porcelain bowl – inlay technique with white in black – electric fired at 1250°C (2282°F).

Mieke Everaet's forming and building process.

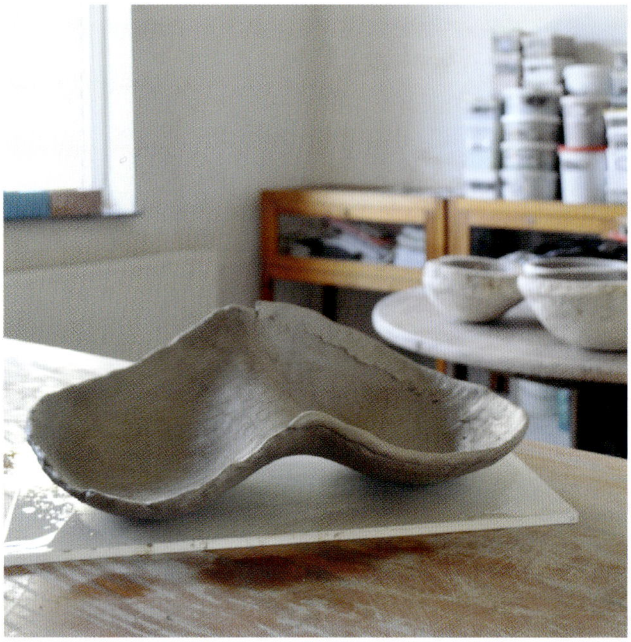

Coiling the shape.

I build a basic clay shape out of coils. Here it is a double bowl shape. I cast it in plaster and the resulting mold is used for the final construction. In the background, you see molds and prepared colored liquid porcelain slips in containers in the studio.

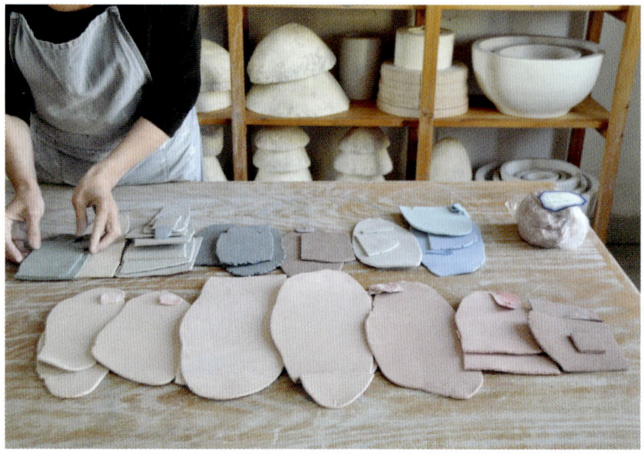

Colored porcelain sheets.

I prepare the inlay mass porcelain. I color the dry porcelain with body stain pigments. After wedging and kneading the porcelain I roll out the inlay mass into thin slabs of colored porcelain to cut later on.

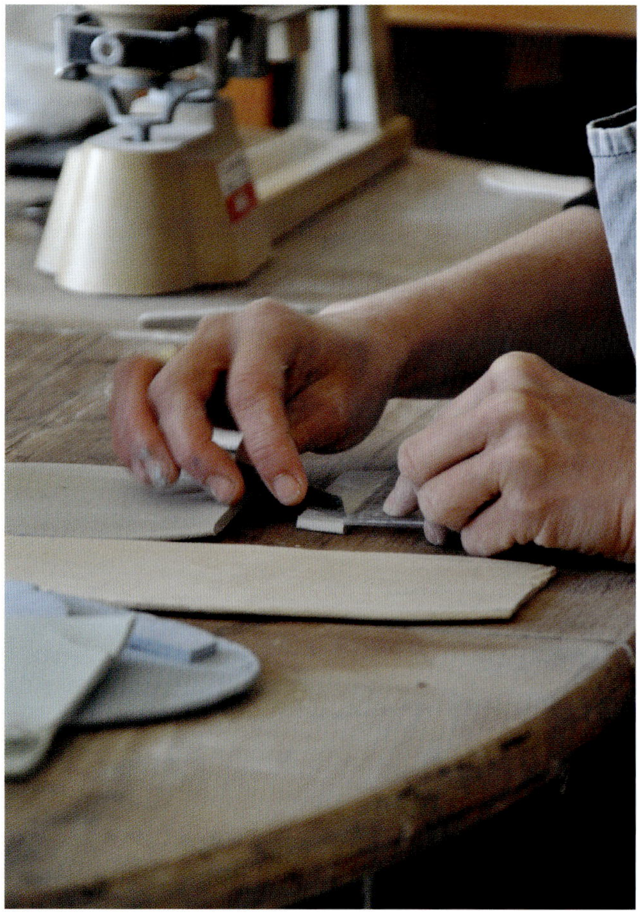

▶ I cut strips, blocks, bricks, lines, triangles, and ribbons in the porcelain sheets to use for building or inlaying on the mold.

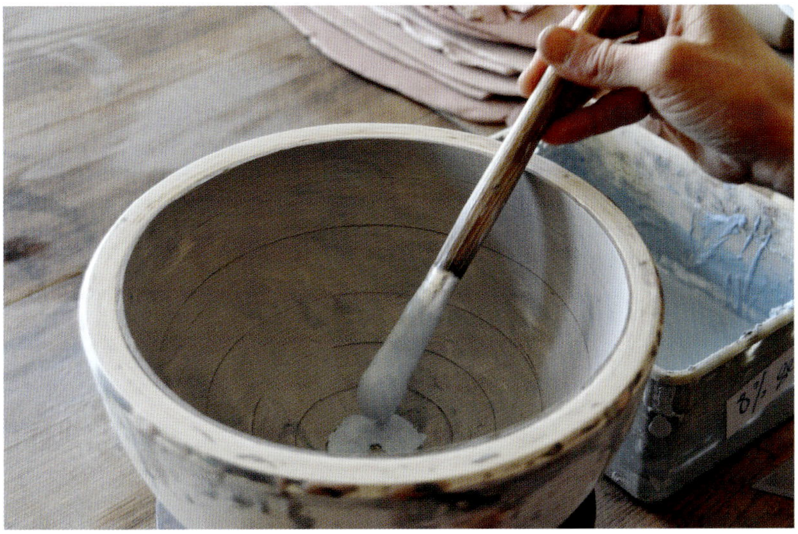

Dipping the colored porcelain slip with a brush into the mold.

The mold is prepared with a thin layer of porcelain where I marked the design of the pattern with lines to guide me during inlaying. I start with brushing the bottom with porcelain slip to start with the inlay. Each little piece of porcelain is covered with colored porcelain slip, following the lines of the pattern.

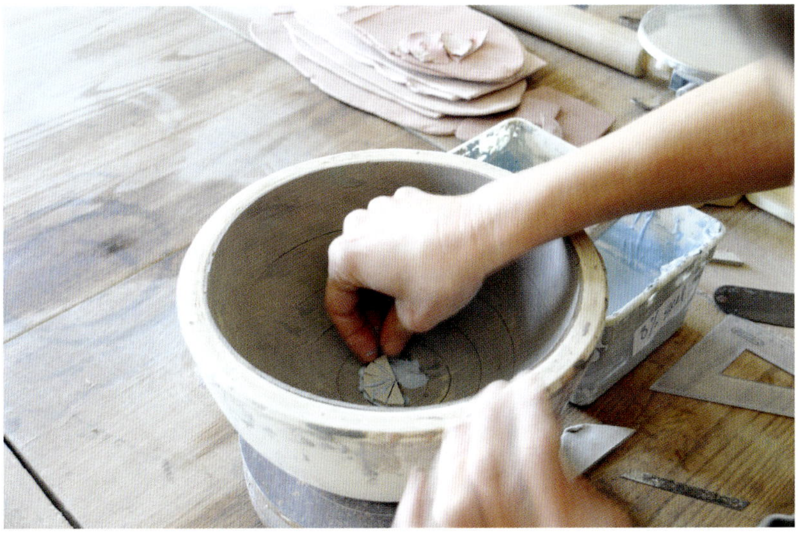

Assembling the design.

I start to build at the bottom of the shape on the intermediate layer. I work in a circular way, following the shape, row upon row till the rim of the border with the porcelain slip.

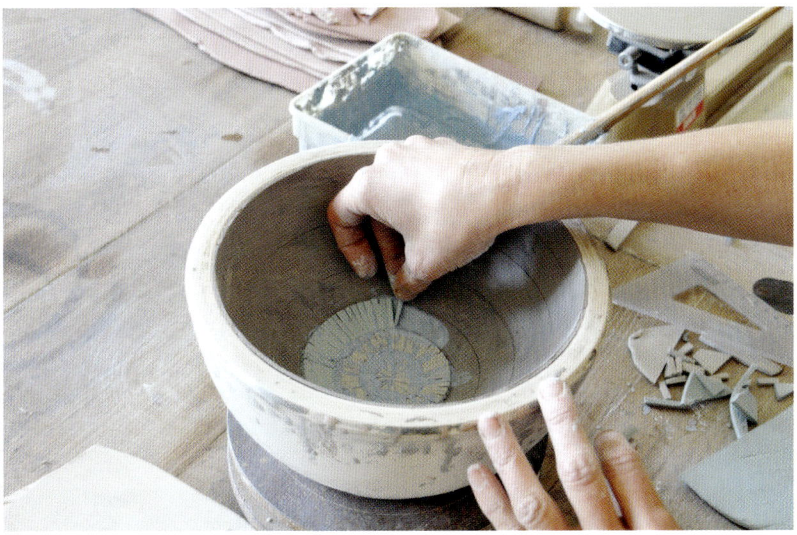

The design gradually takes shape.

When the addition of the pieces is complete, and after drying a little bit, before it is leather-hard, I scrape the inside and the image slowly appears. Later, I take the work out of the mold and remove the porcelain layer. I add slip again, shape, and clean the work by scraping.

Dorothy Feibleman, England, Japan

Born in the United States, Dorothy Feibleman spent more than two decades in the UK, followed by a move to Japan where she has worked in Tokoname, known for its many potteries, since 2000. Her early studies in ceramics, including her first work in nerikomi, were at the Rochester Institute of Technology where she received a BFA in 1973. During her studies there she first combined nerikomi with silver for jewelry and later explored vessels and installations. She then moved to England and, while studying Hungarian language at the University of London, set up her first ceramic studio. At about the same time, she rented a jewelry bench in another London studio where she later created earrings and necklaces from her own colored porcelain beads and gold. A beautiful example of this is a necklace in the collection of the Victoria and Albert Museum. Her early interest in jewelry came from a variety of sources. As a young girl she was fascinated by her grandmother's glass millefiori paperweights and the Egyptian faience and glass that she saw at a children's museum.

A born traveler, Feibleman has lectured around the world, with grants from the British Council and Craft Council to India, Turkey, Italy, Japan, and Finland between 1990 and 1998. During this period, in 1993 she received the INAX Design Prize, a yearly award especially given to Europeans working with colored clays by the INAX Corporation, a world leader in industrial ceramic production and research. She later worked for INAX developing clay bodies and nerikomi production methods for use in industry. Feibleman applied for and was awarded two tool and process patents to be used with extruders for making almost limitless nerikomi imaging for tiles, tableware, and objects. She currently collaborates with or has designed products for Noritake, Nikko, Art Clay (metals), Wedgwood, Herend, Amaco, and Kohler.

Although she received formal education in the field of ceramics, Feibleman is largely self-taught when it comes to nerikomi itself. Influenced by millefiori, she learned by "looking at the glass and figuring out how I could arrive at similar structures in porcelain."

In 1990 Feibleman became a member of the International Academy of Ceramics, Switzerland and by 1995 had developed her groundbreaking style of White/White Nerikomi which was shown in a solo exhibition at the Green Gallery in Tokyo in 1999. This style is similar to traditional nerikomi, which uses variously colored clays, but instead a variety of specialized white clay bodies are used which carry differing levels of translucency and texture. There can also be slight differences in the shades of white.

ARTIST'S STATEMEMT

As I developed as a potter, my working methods were refined and my tools changed. Through the years, these keep changing along with the clays, firing methods, and other elements. I am constantly making tests and find understanding my craft and materials exciting and necessary to allow myself mental freedom, similar to improvisation in music and dance. I feel that without good craft skills and material understanding, you cannot be free to create what you want to make.

I laminate either different colors of porcelain or different porcelain bodies together. Every change in color, texture, or translucency is structural. The forms are also dependent on the movement of the above elements in the construction, drying and especially, the firing, where the movement can be dramatic. All these elements and their relation to the atmosphere in the kiln and the temperature can be altered for very different images, using the same materials.

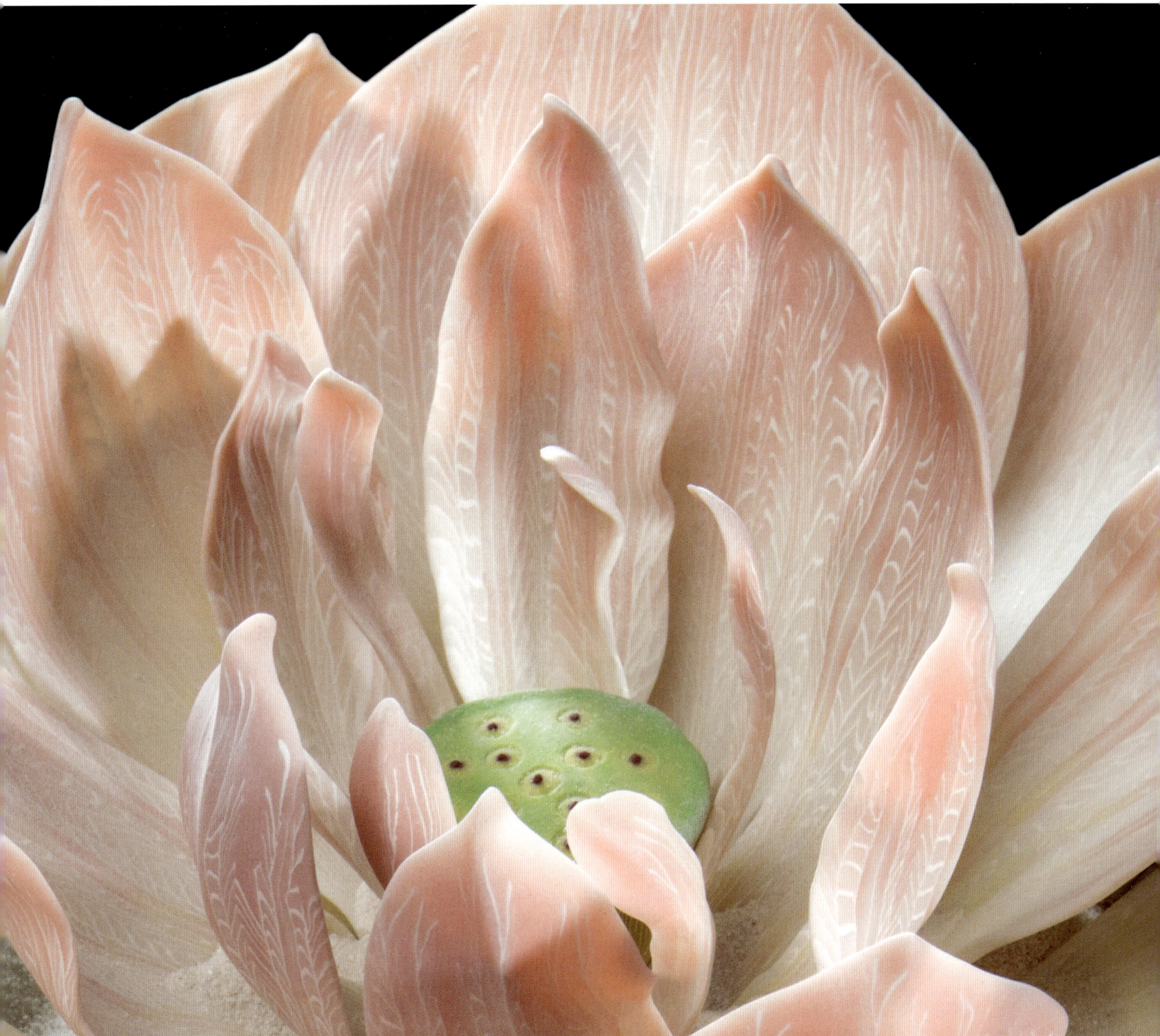

Detail of *Desert Lotus*, 2006. Total installation size: height: 23⅔ inches, diameter: 59 inches (60 x 150 cm)

Hand construction: laminated gradations of two different porcelains. *Image: © Dorothy Feibleman; photo by Ken Yanoviak*

The white/white pieces are made by laminating different white clays with different maturing temperatures. The white work is composed of up to five structurally very similar clays. The whitest high-fire porcelain acts as a structure for the more translucent bodies to almost melt within, resulting in textured structural elements with variations in translucency. The textured pieces are fired in oxidation.

Feibleman developed a porcelain clay body for Pottery Crafts in England which they made available in seven colors including white, from 1984 through 2002. Complex or more subtle colors could be created by mixing two or three of the base colors, or simply diluting a color with white. This body was fired to 1120–1180°C (2048–2156°F). She also developed a specialty clay for INAX in Japan to be fired in a tunnel kiln in one hour. It was fired to 1190°–1260°C (2174–2300°F).

AWARDS AND EXHIBITIONS

Dorothy Feibleman has lectured, taught, and demonstrated in numerous countries worldwide including Japan, China, the USA, South Korea, Canada, and the UK. She has exhibited her work worldwide in Japan, the USA, France, South Korea, Germany, China, Switzerland, Finland, Canada, Belgium, the UK, Hungary, Italy, Austria, and Holland. Over the past 50 years she has had over 20 solo shows in Europe, the USA, and Japan.

Feibleman has received many awards for her work including the 5th Mino International Ceramics Exhibition, Silver Award and three Honorable Mentions; 6th Mino International Ceramic Competition, Gold Prize; the Special Judges award from David Queensberry in the 7th Mino Competition; the 3rd Izushi Porcelain Triennale Competition, Second Prize; the 2nd Mashiko International Ceramics Contest, Judges Commendation Award; and the First Taiwan Ceramic Biennale, Bronze prize; as well as numerous others. She has been a guest artist at several residencies in both Europe and the USA.

Her work is in many international museum collections including the Metropolitan Museum, New York, and the Victoria and Albert Museum, London, as well as other American, British, Irish, German, Japanese, and Taiwanese museums.

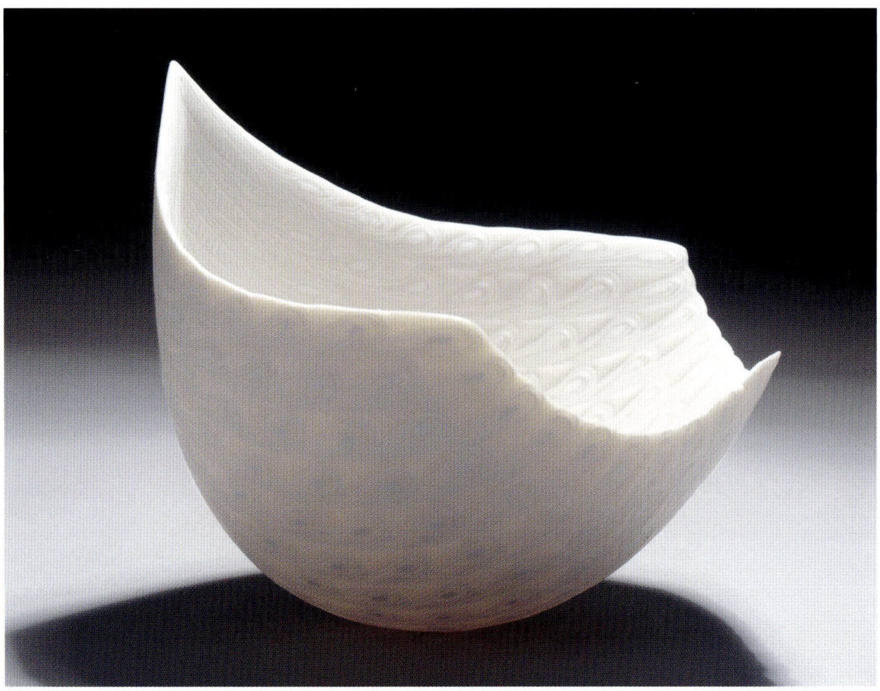

◂ ***White/White Asymmetrical* (textured vessel), *c.* 1997. Height: 6¾ inches, diameter: 5⅛ inches (17 x 13 cm)**

Hand construction of three porcelains, one with a lower melting temperature to acquire the texture. *Collection Indianapolis Museum of Art; image: © Dorothy Feibleman; photo by Ken Yanoviak*

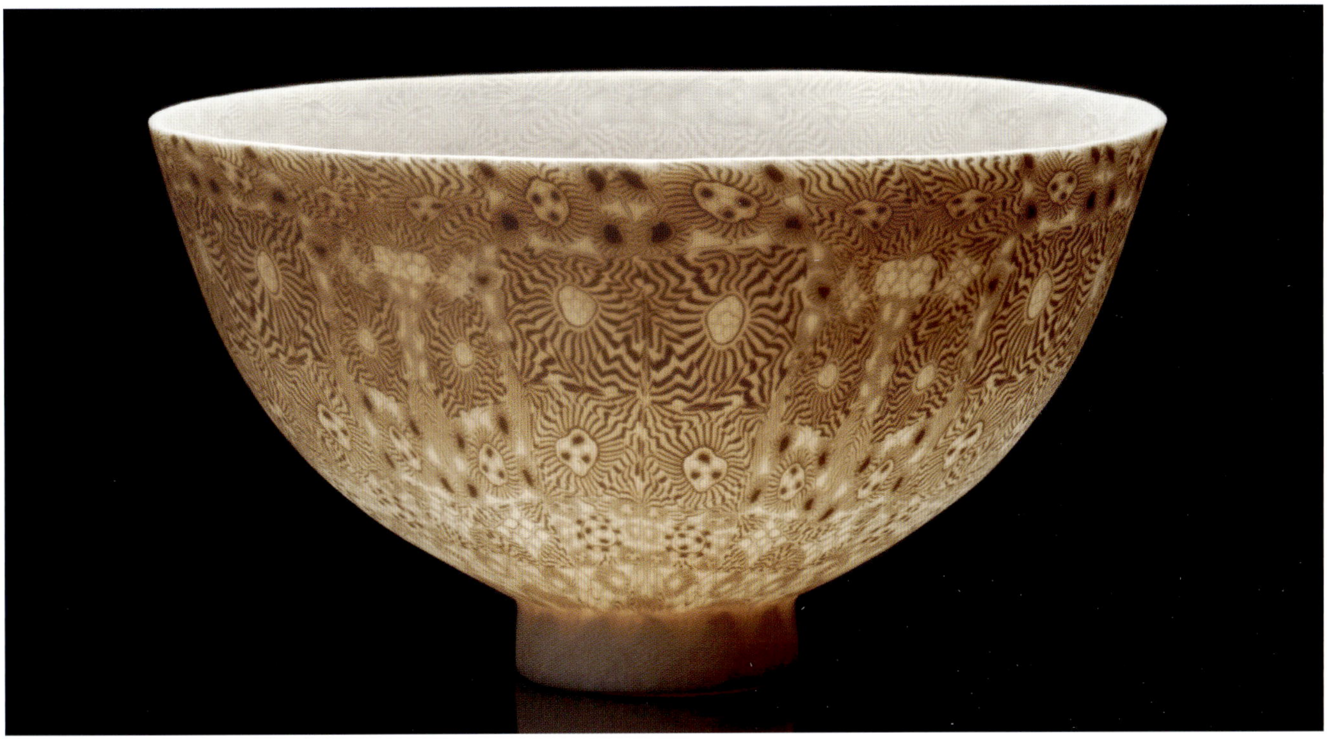

White/White Geometric Sake Cup **(unglazed), 1998. Height: 1¾ inches, diameter: 2½ inches (4.5 x 6.5 cm)**
Hand construction: pizza slice construction of structural imaging. *Image: © Dorothy Feibleman; photo by Ken Yanoviak*

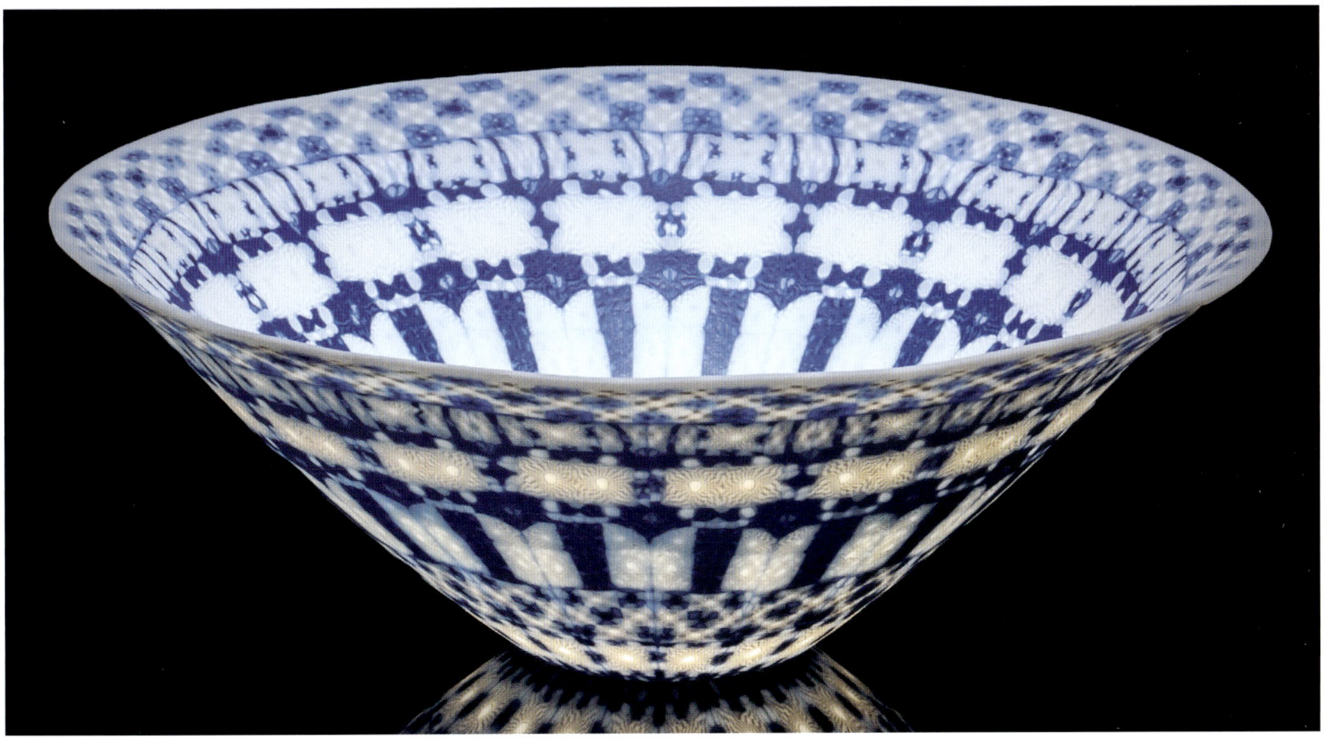

Blue White/White Tea Bowl**, 2003. Height: 2½ inches, diameter: 6¾ inches (6.4 x 17 cm)**
Hand construction: pizza slice construction of structural imaging. *Image: © Dorothy Feibleman; photo by Ken Yanoviak*

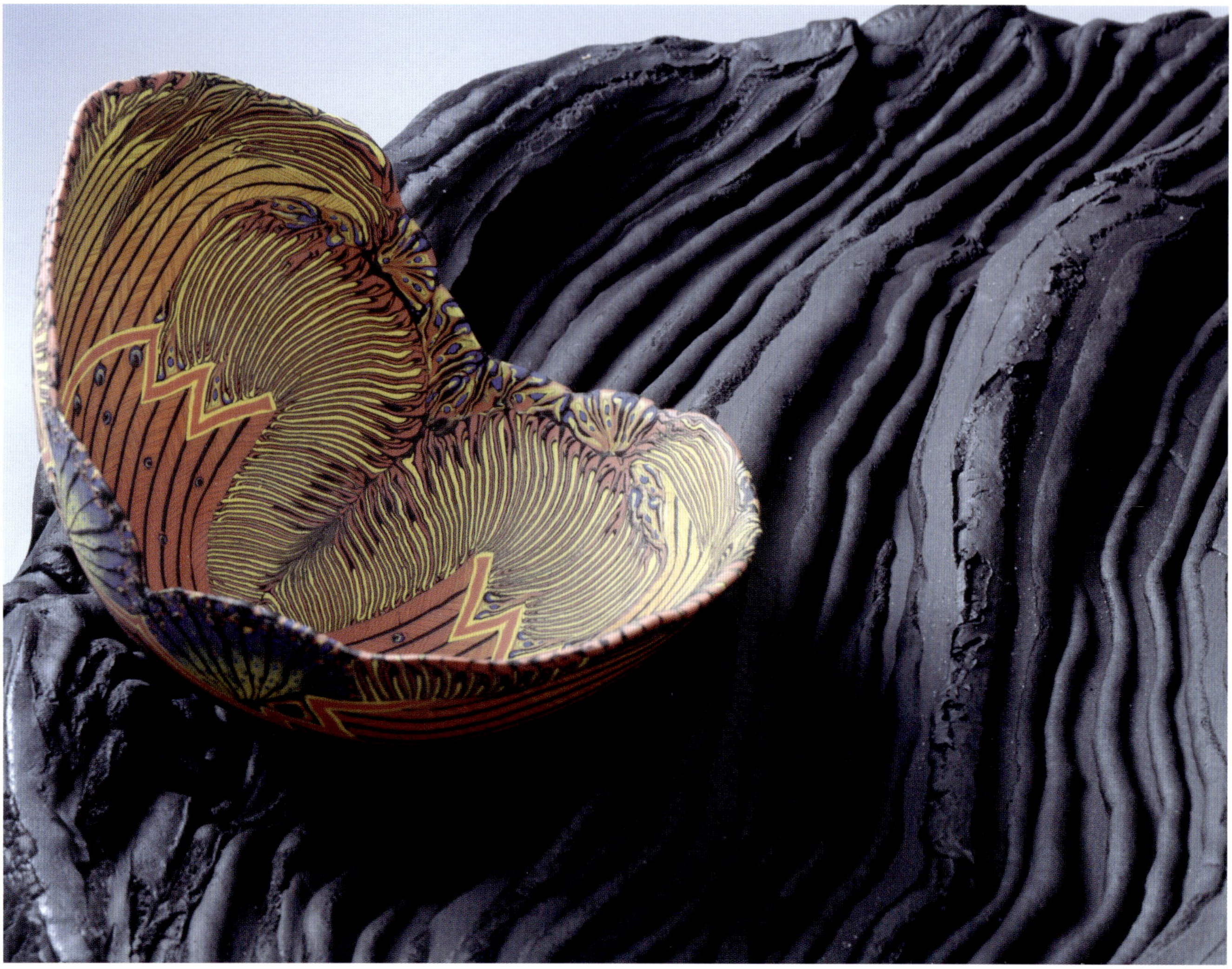

▲ ***Survival,*** **1980. Height: 2⅖ inches, diameter: 8⅓ inches, black base diameter: 27½ inches (6 x 21 x 70 cm)**

Multicolored bowl, colored porcelain construction which changes shape and "grows" in the kiln due to chemical additions. The black textured piece is made as a flat disk and then a structure is built around it to restrict its growth in the kiln during the firing so it expands in a restricted way. *Image: © Dorothy Feibleman*

Dorothy Feibleman's process for constructing a bowl.

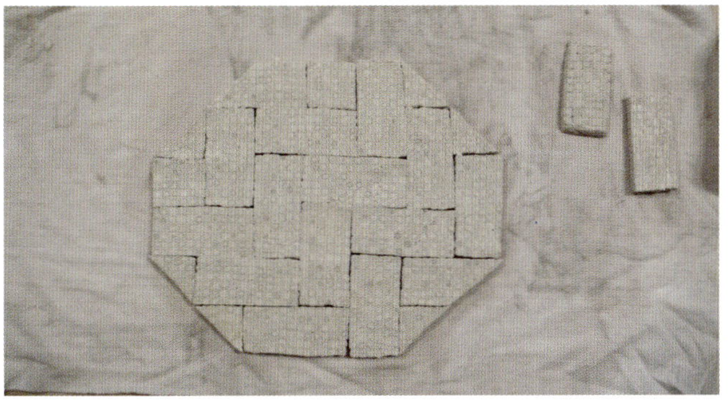

Slices from extruded pattern. *These Images: © Dorothy Feibleman; photo by her mother*

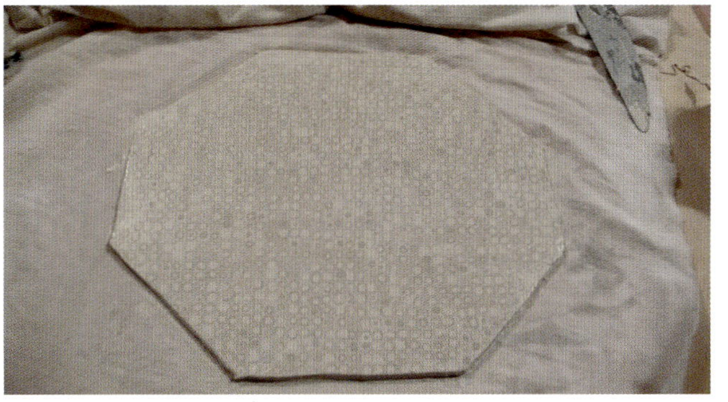

Slices after being compressed into a sheet.

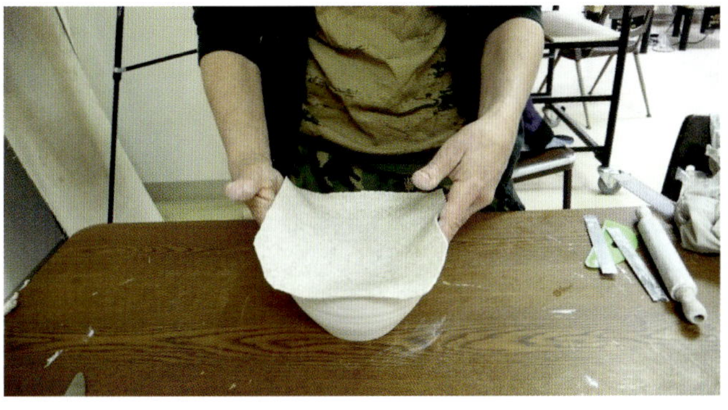

Sheet being lowered into a thrown form that fits interior dimensions. The difference in size of the thrown mold to the plaster mold allows for 3 mm of clay difference.

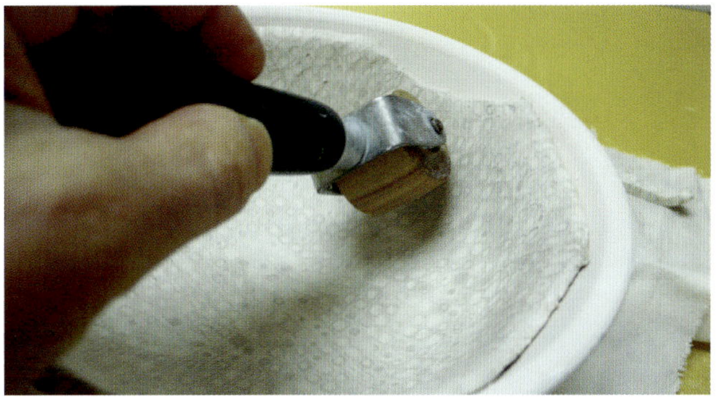

Inner compressing with a wooden wallpaper roller.

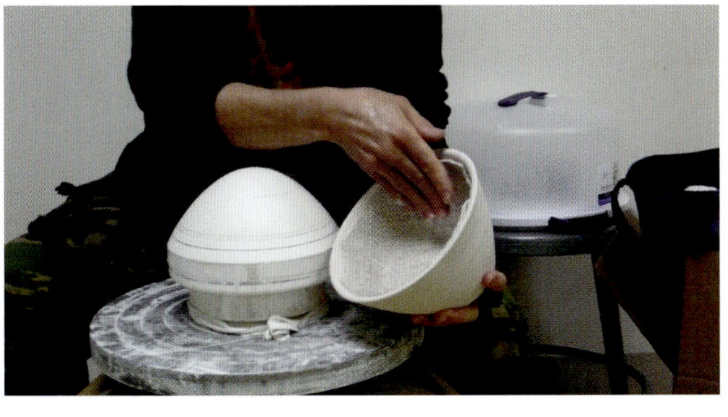

Transferring the clay bowl onto the plaster mold.

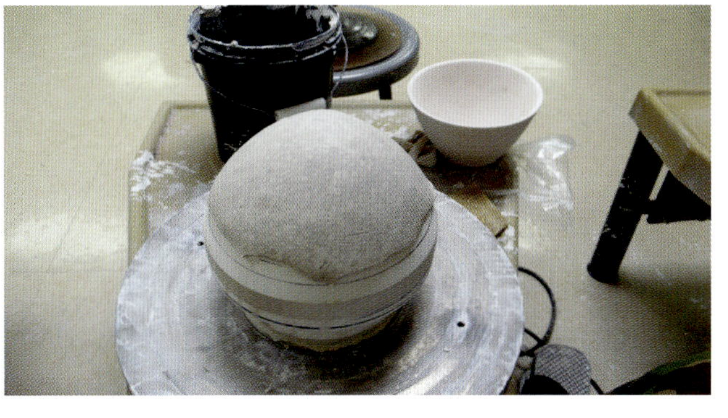

Transferred footless bowl on the plaster mold. Exterior compression to follow.

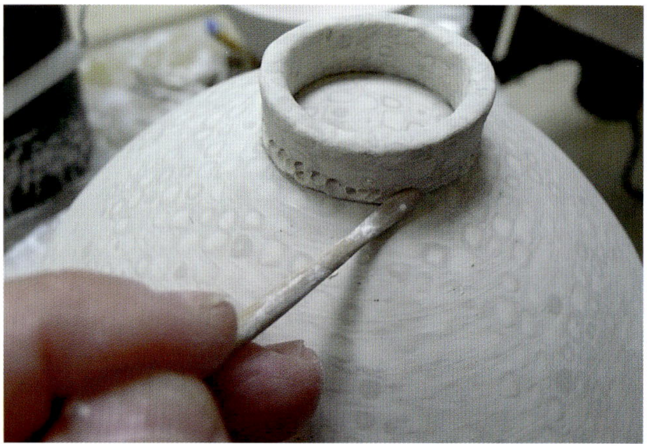

Application of the foot.

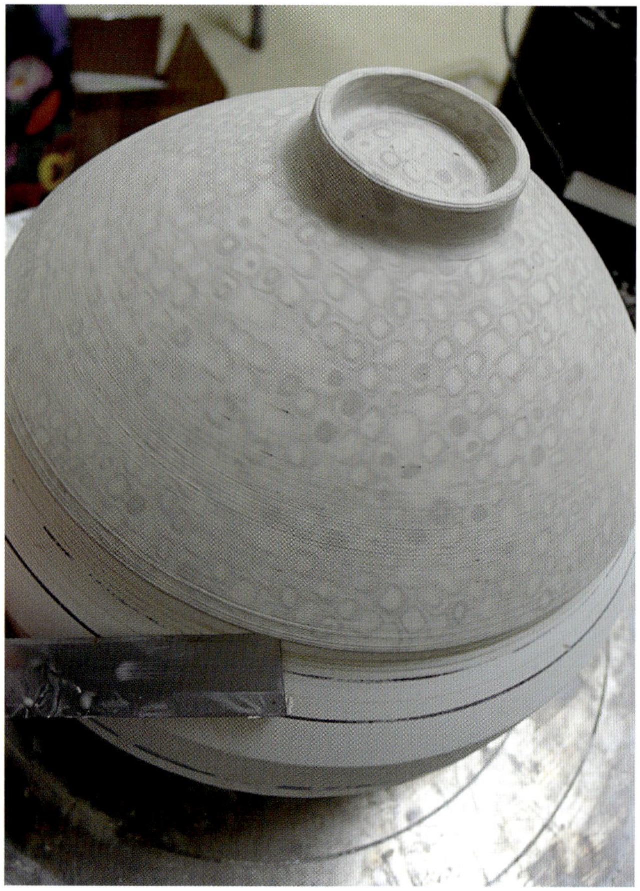

▶ Cleaning up the rim edge and the applied foot.

▼ Finished and fired piece.

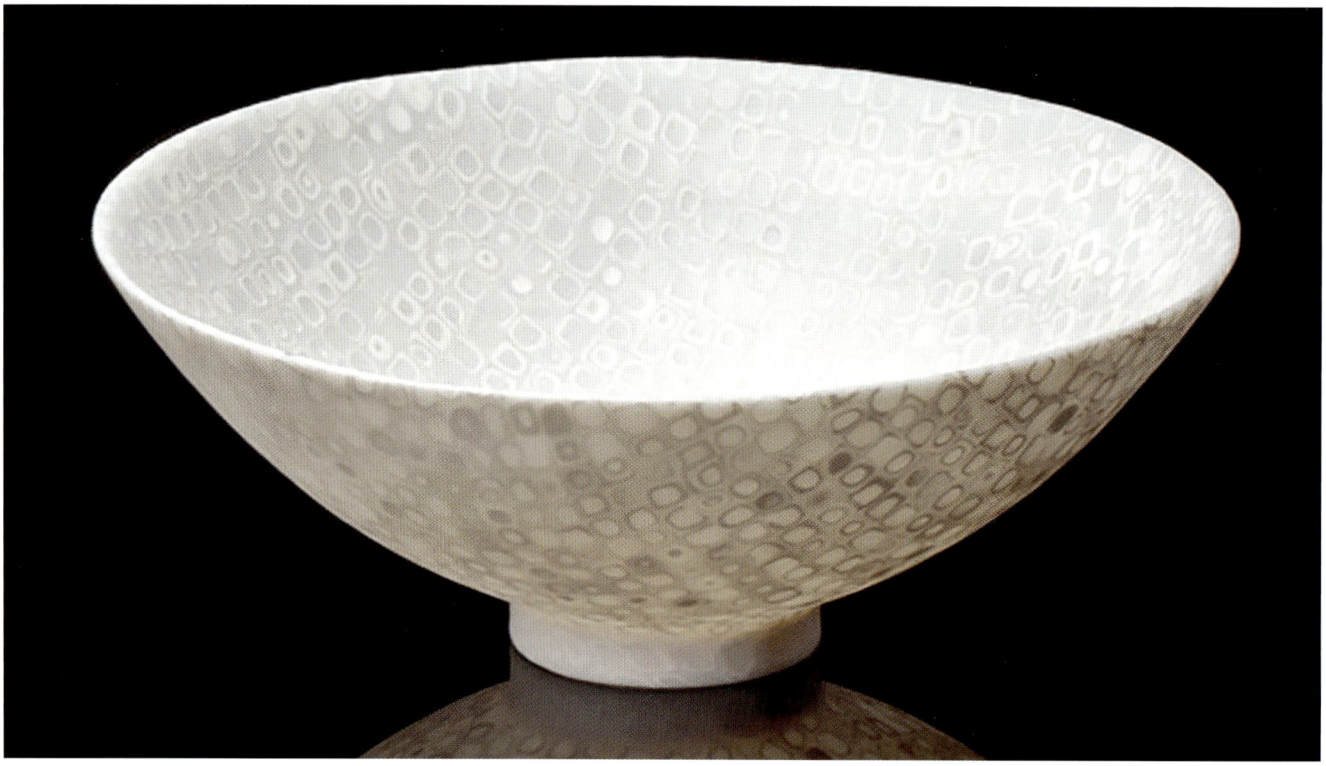

Barbara Gittings, England

Barbara Gittings spent her early years in South Africa, studied fashion and design in Johannesburg, and later moved to Cape Town where she worked in the fashion industry and created and sold her own designs with a friend. Wanderlust ensued and after European travel she settled in London, working as a designer and pattern cutter in the fashion world.

As it happens, this fashion design and pattern-making experience has been extremely beneficial in her now full-time occupation in ceramics. Gittings is largely self-taught and remembers her first discovery of a picture in a book of an ancient Chinese nerikomi bowl, and she has never looked back. Many years of learning – often by trial and error – followed, including a few adult education courses. One particular course was a week-long immersion in smoke firing with Jane Perryman, a firing technique she still uses with her porcelain nerikomi vessels. Inspiration came as well from established nerikomi artists like Dorothy Feibleman, Curtis Benzle, and others.

ARTIST'S STATEMENT

I now explore clay as an alternative medium to fabric. In fashion, the layering of textiles and the power of the cut merge to find new balances and forms, the biomorphic and geometric held in tension. My work in clay continues to explore this. I'm drawn to irregular repetition, primitive mark-making, and soft, earthy colors. I try to assimilate the poetry of things I see in the world and then allow them to rise up in my subconscious to inform the patterns I put through the clay.

The geometry of patterns in nature is a constant source of inspiration to me. Especially as random chaotic forces, such as growth, weathering and erosion push the initial perfect symmetry towards imperfection. I am always exploring this balance between symmetry and asymmetry in my work, trying to capture the imperfect perfection.

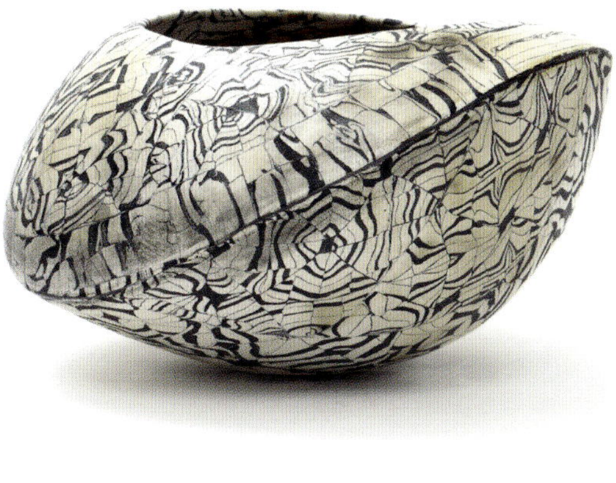

***Floating Bowl #6*, 2020. Height: 6⅓ inches, width: 11 inches (16 x 28 cm)** *These images all © Barbara Gittings; photo by Bipinchandra Mistry*

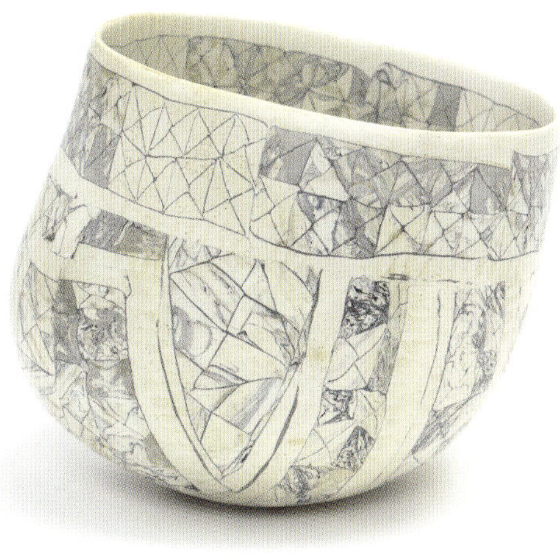

***Head Bowl #44*, 2020. Height: 5½ inches, width: 4¾ inches (14 x 12 cm)**

Gittings works with Earthstone finely grogged porcelain which can give strength to larger forms as well as help the pot survive the thermal shock of smoke firing. Another important device for improving strength and reducing cracking is the use of a paperclay slip, often colored black, to help join various pattern sections while building a large pattern block. Her bowl forms are made by pressing patterned nerikomi slices into a plaster mold, working her way from the bottom upward, and making design decisions as the piece progresses. The more geometric slab-built bottle forms are planned ahead in a manner harkening back to her garment-design years. After initial sketches a stiff paper pattern is taped together and adjusted as needed to give a realistic idea of how the form will work. When that is settled, a large slab of intricately patterned clay is then cut using the paper pattern and slab built using traditional joining methods.

Unique in the world of nerikomi is Gittings' use of smoke firing. After an initial low-temperature bisque firing in an electric kiln, wet sanding, refiring to a higher temperature (1046°C [1915°F]), and diamond pad sanding, the artist brushes on "crank slip," only in certain sections. The slip serves to mask areas which will avoid being "smoked" while leaving exposed the areas where extra smoke color is desired. After the smoke firing the crank slip is scrubbed off, leaving the pot ready for a final coat of beeswax furniture polish.

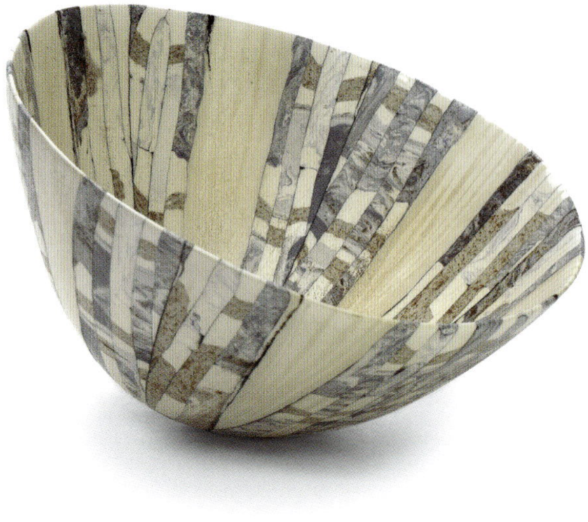

Tilted Bowl #34, 2022. Height: 7⅛ inches, width: 11 inches (18 x 28 cm)

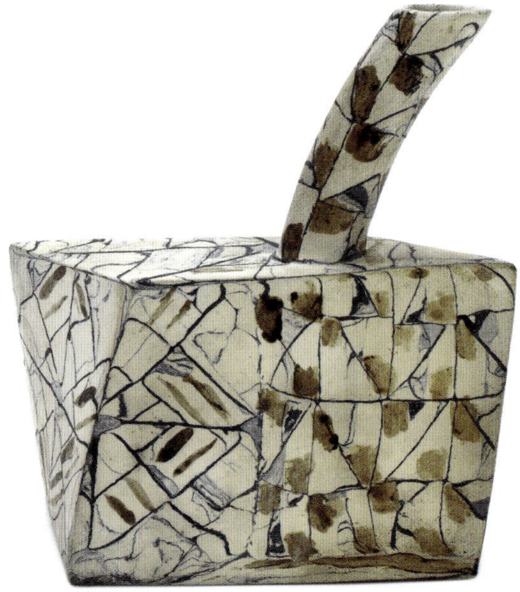

Bottle #65, 2021. Height: 4¾ inches, width: 3⅓ inches, depth: 2⅛ inches (12 x 8.5 x 5 cm)

AWARDS AND EXHIBITIONS

Barbara Gittings' work is widely recognized across the UK where she has had a large number of gallery exhibitions, as well as participating in many ceramics festivals.

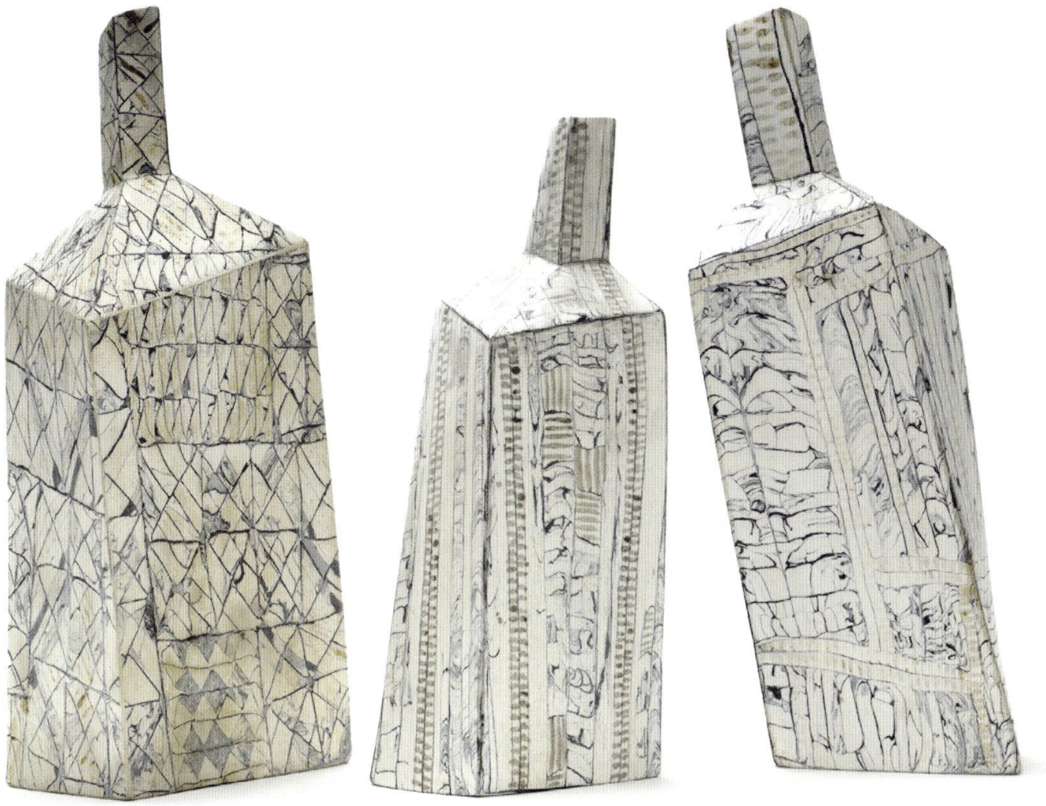

Three Bottles, 2021. (Left to right) #60: height: 13¾ inches, width: 5½ inches, depth: 4 inches (35 x 14 x 10 cm); #62: height: 11⅘ inches, width: 4⅓ inches, depth: 3⅛ inches (30 x 11 x 8 cm); #61: height: 13⅓ inches, width: 4⅓ inches, depth: 3⅛ inches (34 x 11 x 8 cm)

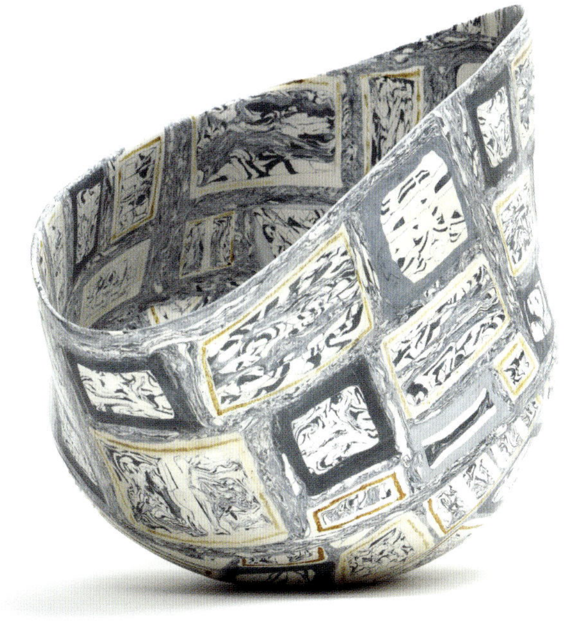

Large Sculptural Bowl #1, 2020. Height: 11⅘ inches, width: 9½ inches, depth: 13¾ inches (30 x 24 x 35 cm)

Barbara Gittings' studio process.

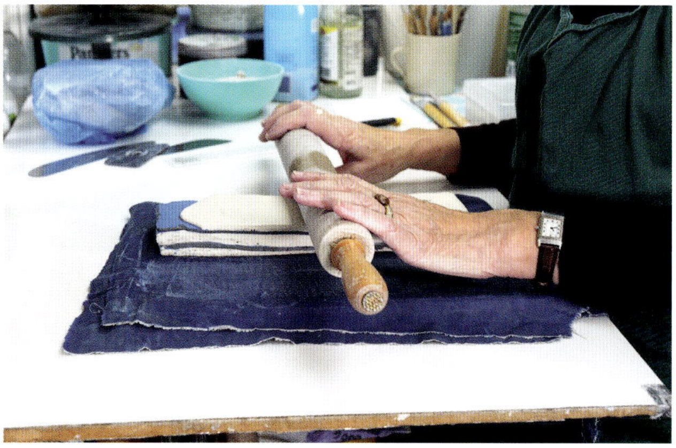

Having colored some clay black, I have rolled out slabs of various thicknesses of white and black clay, endeavoring to create rectangles of the same size. Using a different colored slip (blue) to join, I've alternated the black and white layers to form a block, gently rolling each additional layer to attain a good fit. *All method images: Barbara Gittings; photo: © Bipinchandra Mistry*

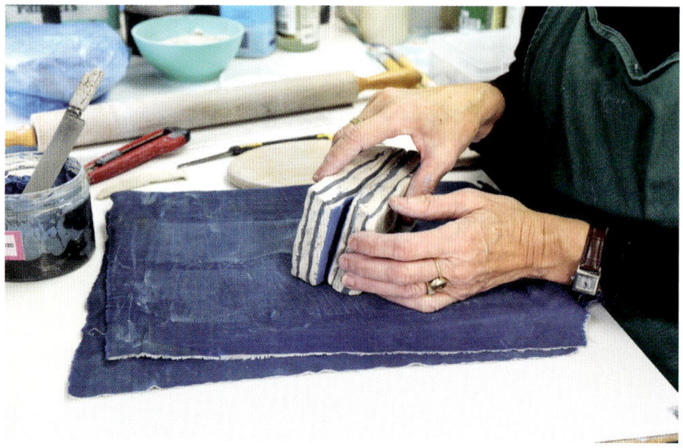

I've cut the block in half horizontally and painted an extra thick layer of the blue slip onto one face of each of the two halves. It's important to make sure that everything retains the same degree of moisture, so the slip must be allowed to go tacky before joining the layers. If too wet, the layers will slide against each other in subsequent stages and cracks will appear later. I've joined the two halves, blue to blue.

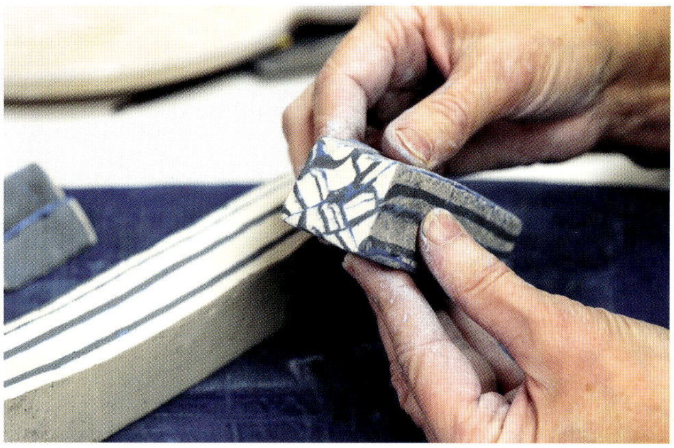

After whacking the block with a paddle to make sure joins are tight and the block has flattened, I've cut strips the same width as the depth. These I've cut through diagonally, so that I've ended up with large triangular strips. Some distortion will have occurred to the pattern by now from the whacking. I've made sure to join triangles that don't exactly mirror each other, forming patterned strips.

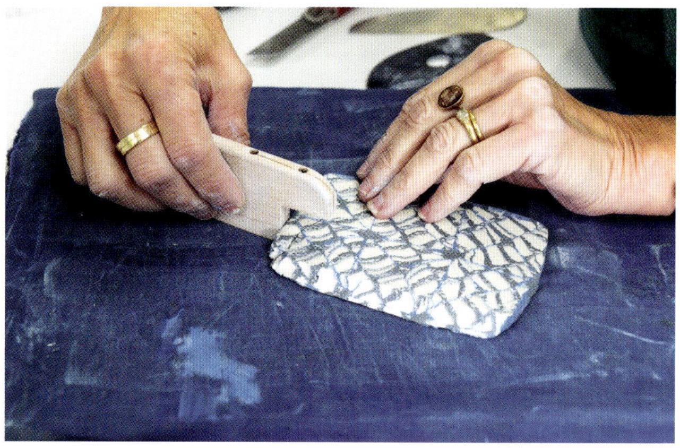

I've joined the smaller patterned strips to form a larger block, which I've then sliced to form slabs. These are then beveled on the edges prior to fitting together into the mold.

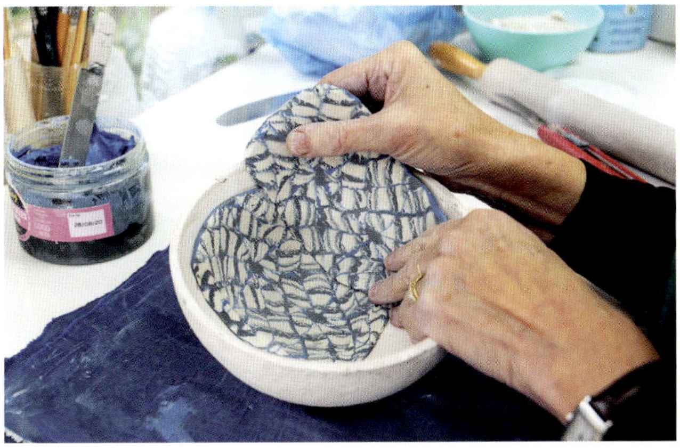

I place the slabs into the mold, joining the beveled edges with slip as I go, cutting away the excess. Sometimes I leave space to add a different patterned rim.

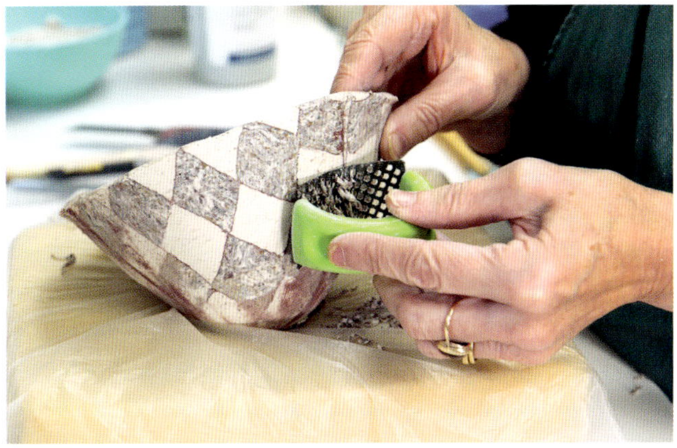

When the piece is firm enough to hold its shape, I scrape the surface with a Surform tool, then a very fine-toothed kidney. Finally, I scrape it with a smooth kidney and when it's completely dry I clean it with wire wool, wearing a mask for all the dust-creating processes. It's important to dry the piece very slowly.

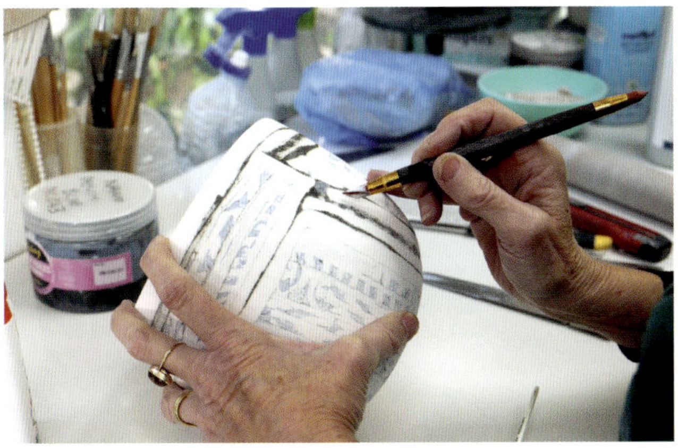

After firing to 1046°C (1915°F) in an electric kiln, I decorate the piece with various resists.

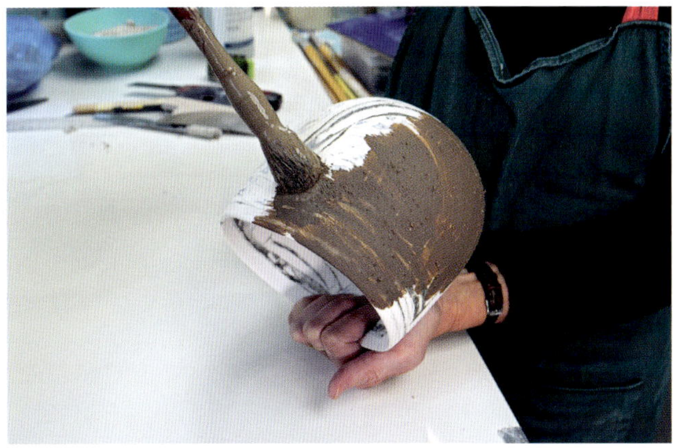

I then paint the piece with a heavily grogged crank slip (crank clay is a rough stoneware clay, the type used for outdoor pots. Because of its rough nature, it doesn't fuse onto the work during the smoke firing.)

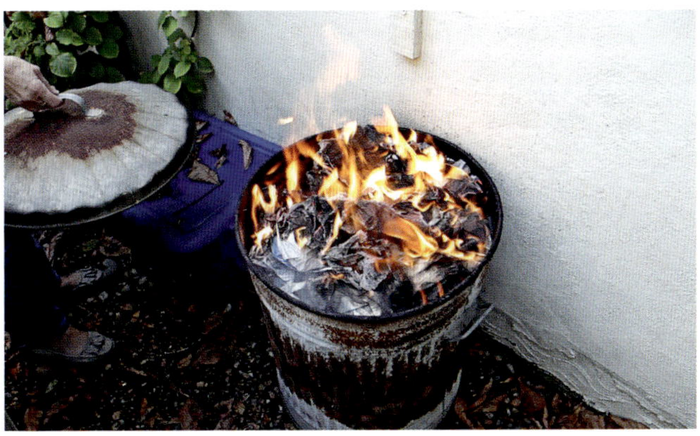

◀ The piece goes into a bin filled with crumpled newspaper and set alight. Once it is burning fiercely, I cover it with the lid. The bin has several holes punched through around the bottom. Once it's died down, I remove the pot from the ashes with fireproof gloves and, when it has cooled down, I scrub the slip off, revealing what the fire has "painted." I then wax the piece with furniture wax to give it a soft sheen.

Robert Hessler, USA

Robert Hessler's path toward a career in ceramics started at the School of Visual Arts in New York where he achieved a Bachelor's Degree in Photography and received the David Rhodes Award for Creative Excellence. There he concentrated on experimental approaches to the medium rather than traditional studio or commercial photography. Always questioning and reaching for something more, he experimented with printing manipulation, collage, and paint. Graduation can often be a rude awakening and Hessler struggled to discover his next creative chapter.

A few years later he enrolled in a ceramics class at The Potter's Wheel in Kew Gardens, Queens, New York and almost instantly felt a strong affinity with the materials and process of working with clay. Hessler explained: "I remember coming home that day feeling elated. It felt as if something internal shifted in me. I was excited to start creating again. Actively thinking about what I wanted to make and how I could make it charged my creative cells."

An intense period of study and experimentation began as he delved into the fascinating world of crystalline glazes. He was inspired by the work of early pioneers of this technique in the 1800s, and by the seminal research by Herbert Sanders starting in the 1930s, culminating in the book *Glazes for Special Effects* in 1974. Additional time was spent learning the ins and outs of the marketplace for ceramics which enabled Hessler to become a full-time potter specializing in crystalline glazes. As with his earlier photography experience he found himself again pushing the boundaries of the materials and processes to discover even more.

ARTIST'S STATEMENT

I've always been attracted to color and finding ceramic techniques that enable me to explore the use and creation of different colors and color combinations. This is what was most appealing about my glaze work and this is what drew me to explore the nerikomi technique. Once the pandemic hit and the art shows that I sold my work at were canceled, I found I had time and space to explore and learn something new after doing glaze work for 20 years.

Although this is a new technique and body of work, I still feel as though it's a continuation of my overall vision. The only thing that's changed is the process.

Hessler once again took matters into his own hands and figured out for himself how to use colored clay and the patterns that can be created with it to construct his vessels. His process is similar to nerikomi, and might be included in the broad definition of the term, but is unique among nerikomi practitioners. To be more precise, it might be thought of as a combination of both nerikomi and neriage, since the work is actually thrown on the potter's wheel part way through the forming process. He has often chosen vertical vase forms for the same reasons that he chose them for his crystalline glaze work. Hessler considers the surface of the pot to be his canvas on which he can compose a richly colored image. Small openings at the top tend to focus our attention on the surface which is rich with detail. Unlike his glazed work, the color and patterns of his nerikomi pieces show through to the inside as well, prompting Hessler to include more open forms where the inside surface is also visible.

Considering the relatively short period of time that Hessler has worked with colored clay he has created an impressive body of work. His pieces are made with a cone 6 porcelain body and fired in an electric kiln.

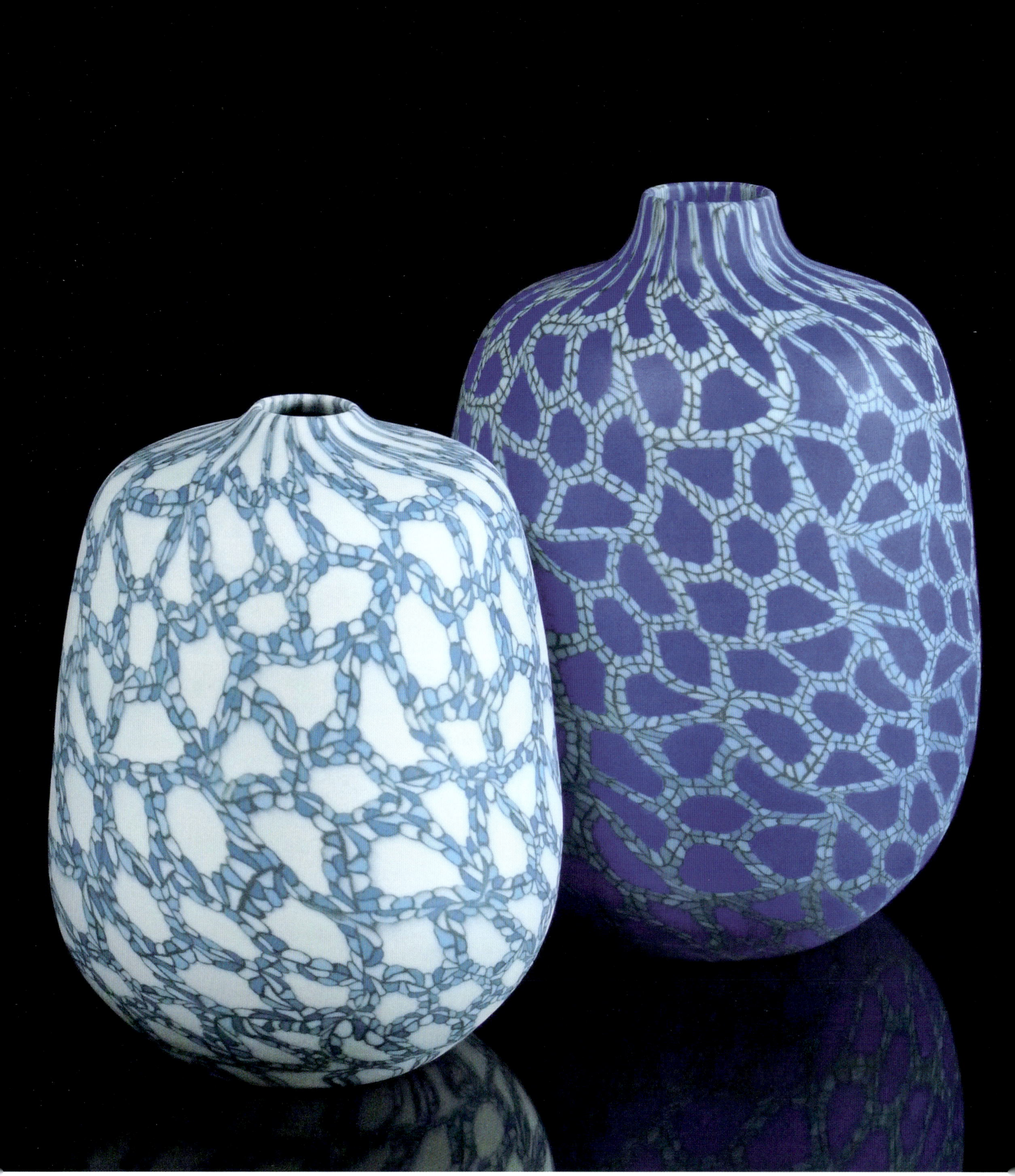

Nerikomi Vases #205–206, 2022. Left: height: 6½ inches, diameter: 4½ inches (16.5 x 11.4 cm); Right: height: 8 inches, diameter: 6 inches (20.3 x 15.5 cm) *These images all © Robert Hessler*

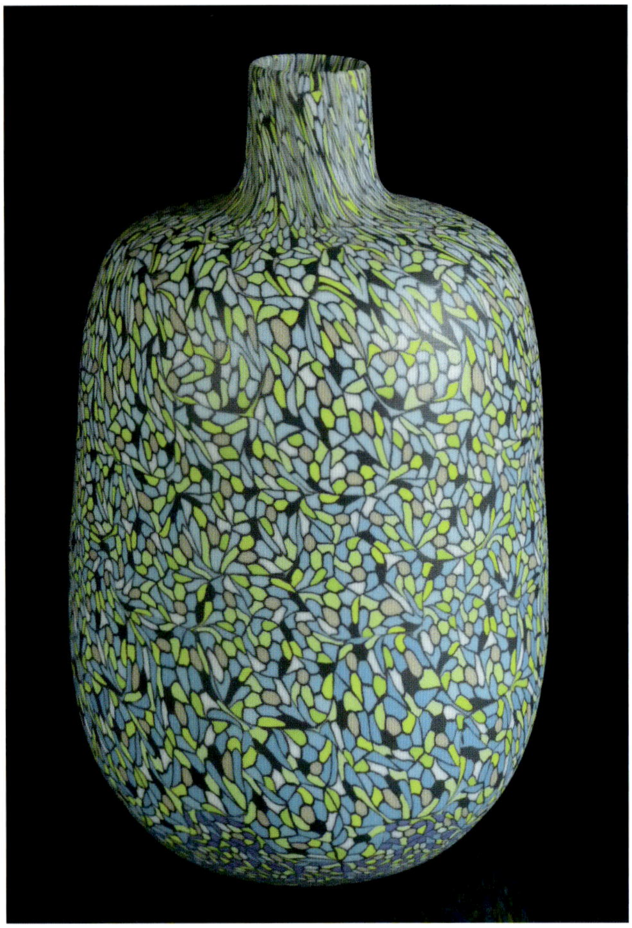

Nerikomi Vase #72, 2021. Height: 10 inches, diameter: 6 inches (25.4 x 15.5 cm)

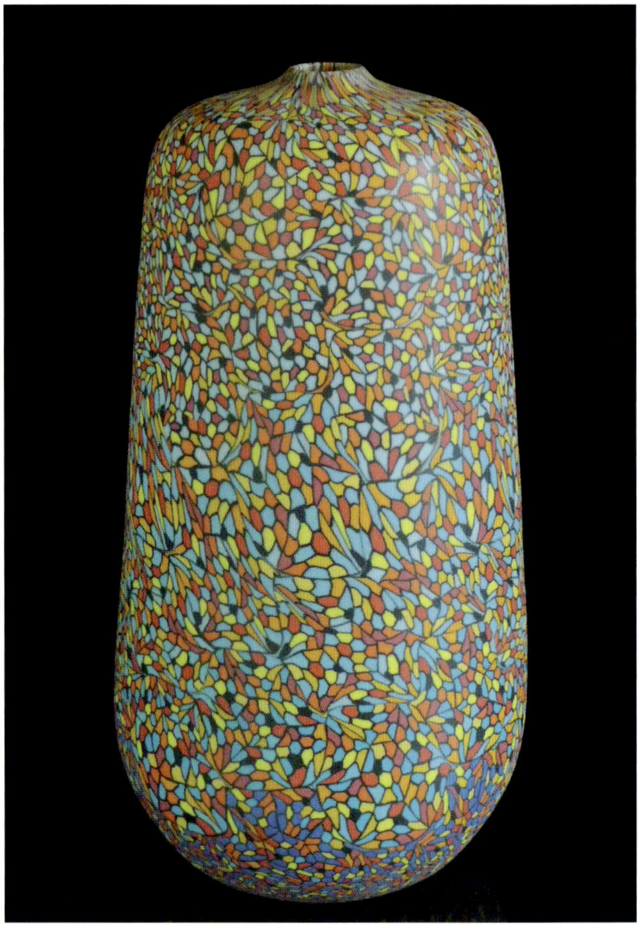

Nerikomi Vase #75, 2021. Height: 10½ inches, diameter: 5½ inches (26.6 x 14 cm)

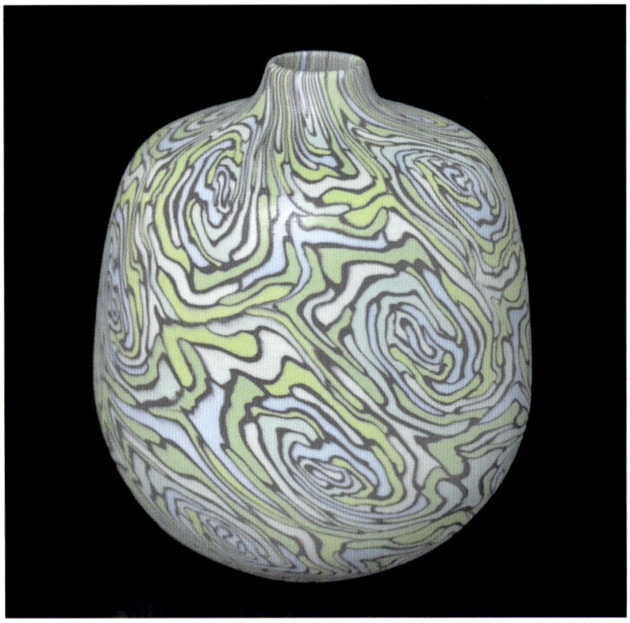

Nerikomi Vase #262, 2022. Height: 6 inches, diameter: 5 inches (15.5 x 12.7 cm)

AWARDS AND EXHIBITIONS

Since 2000 **Robert Hessler** has won numerous top awards at nationally recognized art shows around the US. These include awards at the Coconut Grove Arts Festival, the Westchester Fine Craft Show, CraftsBoston, the Cherry Creek Arts Festival, the Ann Arbor Original Art Festival, the Des Moines Art Festival, and the Armonk Outdoor Art Festival, to name a few. From 2014 to 2018 he was a ceramics instructor at Hudson Valley Pottery and has taught workshops in the art of crystalline glazes in New York and Florida. His work has been featured in magazines such as *Elle Decor*, *NY Spaces*, *GQ*, and *Harper's Bazaar*. Hessler's work is included in the permanent collection of the Crocker Museum in Sacramento, California.

Robert Hessler's studio process.

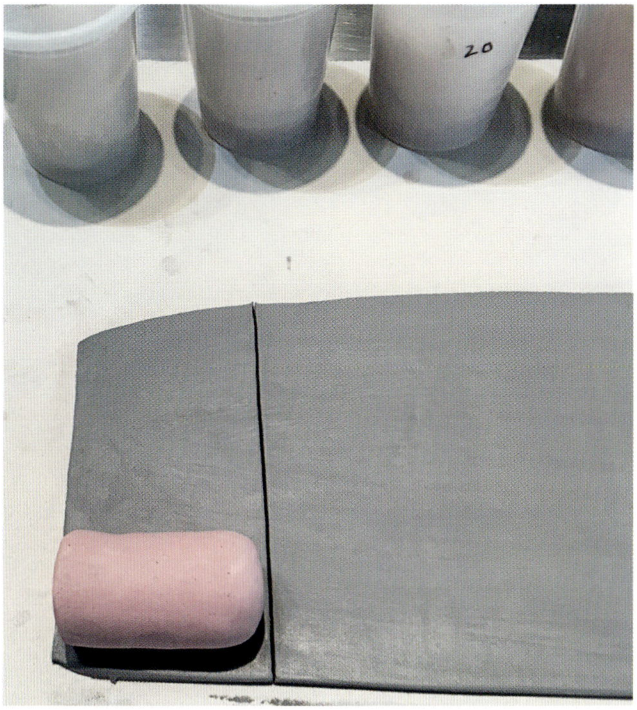

Preparing to enrobe the colored porcelain with black.

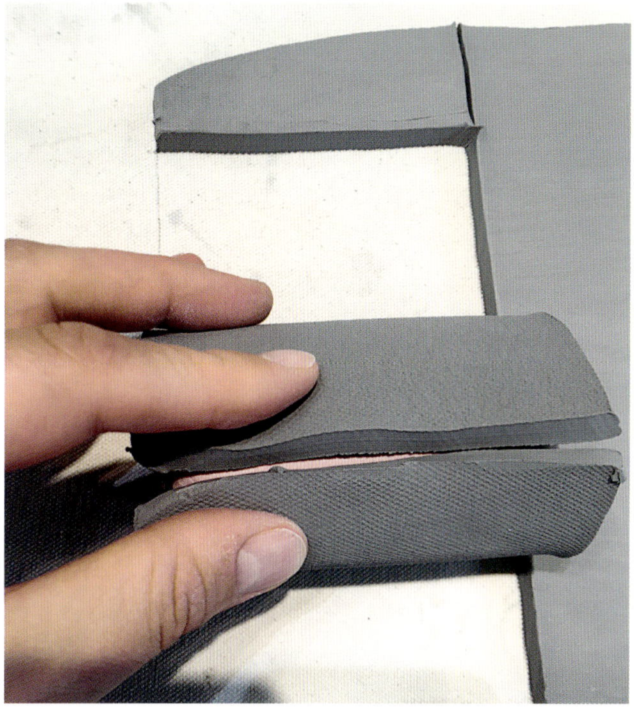

Pinching the seam closed.

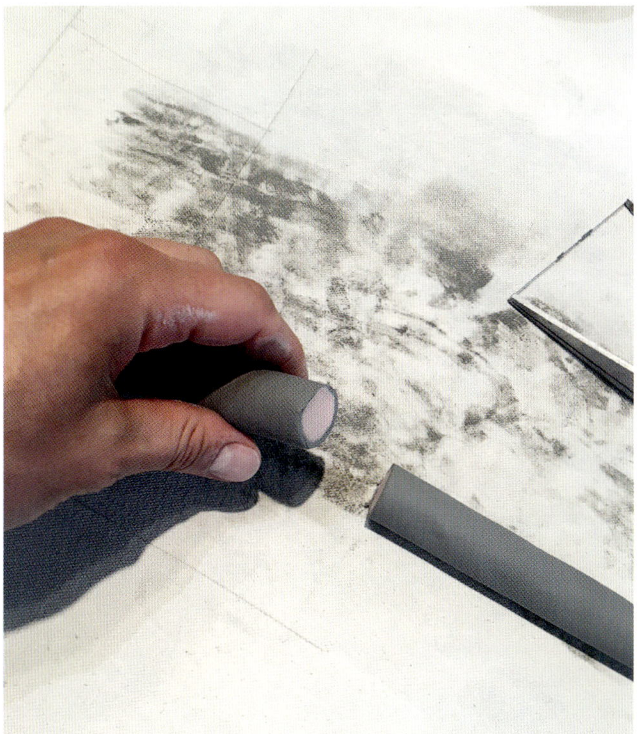

Cutting the rolled coil.

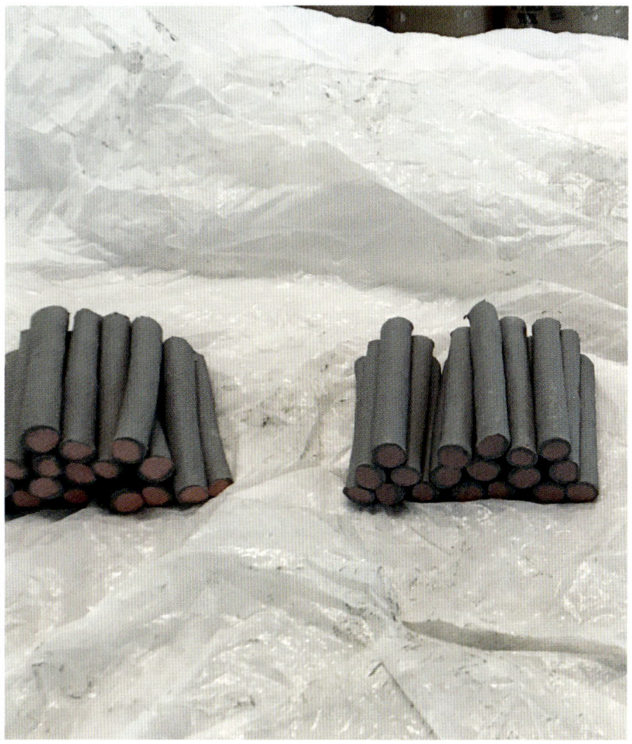

The rolled coils stored on plastic sheet ready to use.

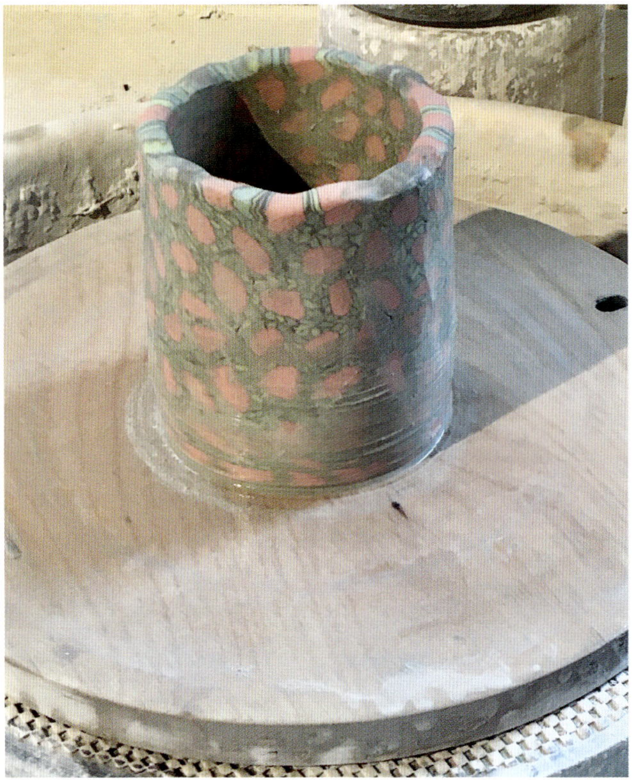

Coils hand-built into a cylinder and ready to finish on the wheel.

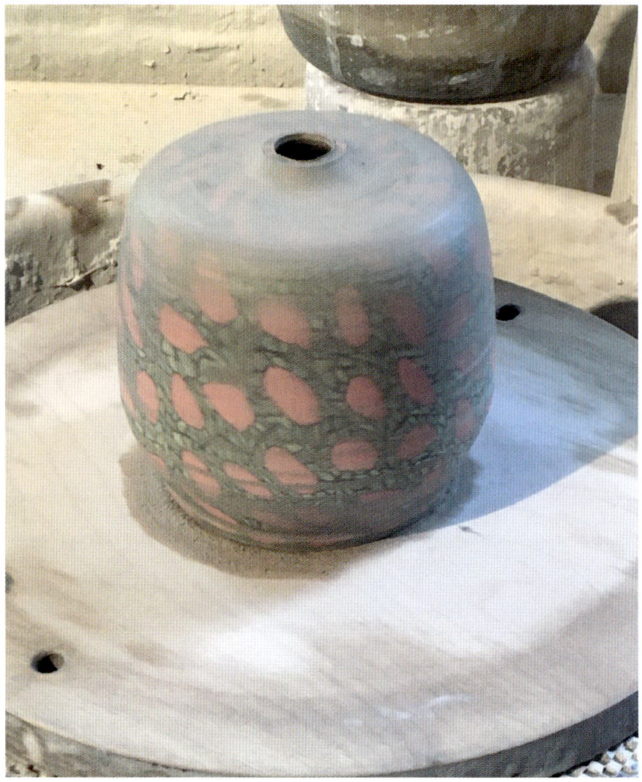

The formed piece after throwing.

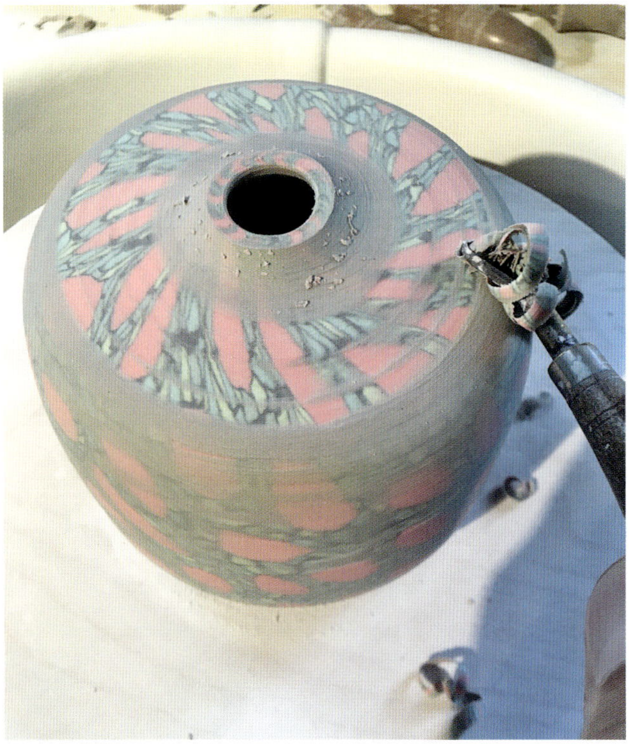

Trimming the leather-hard piece.

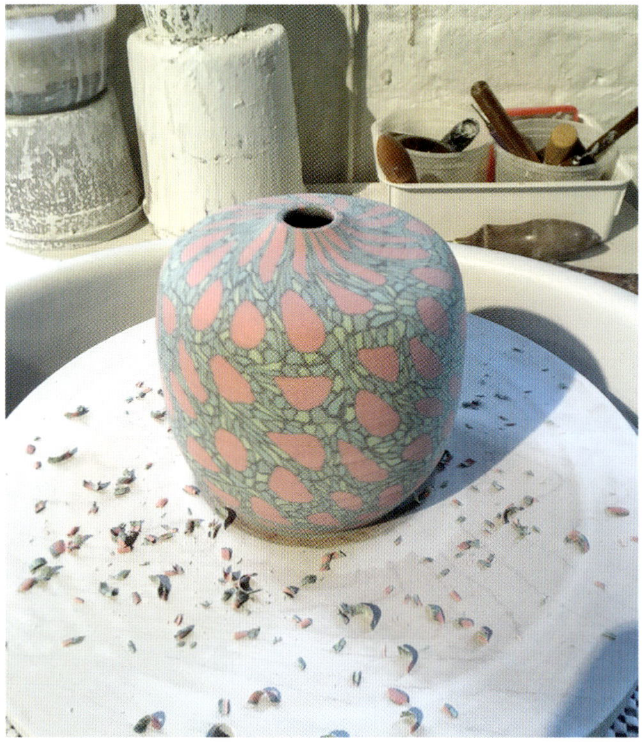

The trimmed piece.

Cody Hoyt, USA

Cody Hoyt is a unique artist in many ways. Unlike many of the artists in this book he started out with a BA in printmaking before his hands ever touched clay, but even his printmaking was unusual in that he would combine screen printing with lithography and finish off with etching, all on the same piece.

Working with clay came later, starting around 2012, and was mostly for a pleasant break from his artistic studio practice. Within two years, however, he had discovered techniques and a direction that led him to make clay his primary medium and focus.

ARTIST'S STATEMENT

I've always been interested in materials that have a utility and a tactility. I love color and texture. There's a warmth and vitality to ceramic that feels akin to a colored-pencil drawing or oil on linen. There's a definitive reality and finite-ness to the work. I think that's what drew me to nerikomi initially: building something felt more unambiguous than drawing it. Not necessarily the act of sculpting a form, but the act of constructing an image – breaking down the distinction between a thing and its surface. Questioning things about the usefulness of a drawing, or the superficiality of mark-making occupied me as someone who identified as an image maker. Nerikomi allows me to search a little deeper for meaning in my work.

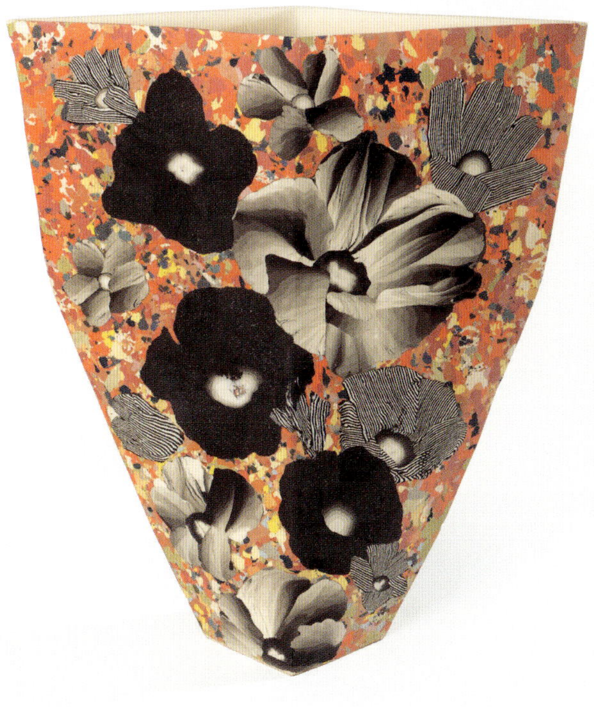

Landscape vessel, 2020. Height: 24 inches, width: 22 inches, depth: 12 inches (61 x 56 x 30.5 cm) *Image: © Cody Hoyt*

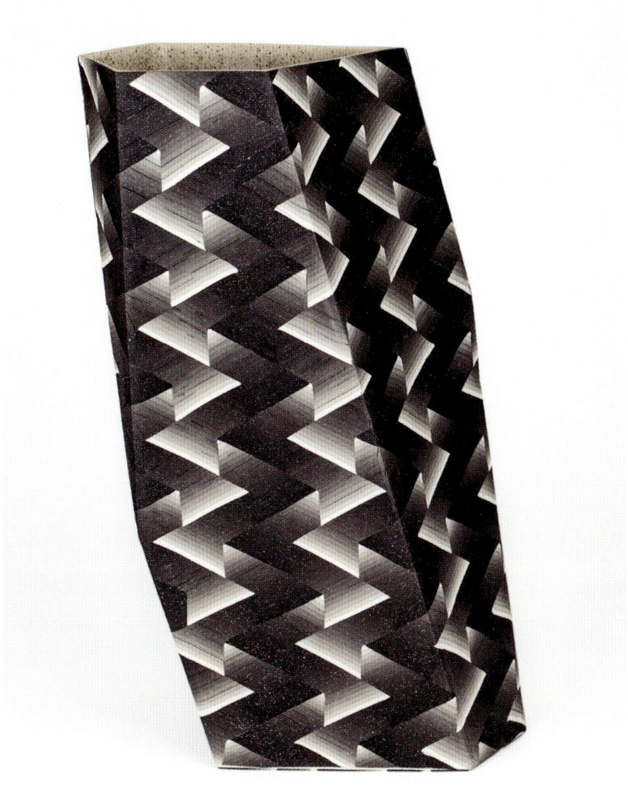

Oblique vessel, 2018. Height: 8 inches, width: 26 inches, depth: 14 inches (20.3 x 66 x 35.5 cm) *Image: © Cody Hoyt*

Hoyt finds inspiration in a variety of places, architecture being foremost. He is a fan of the Brutalist structures by Paul Rudolph and Louis Kahn, along with buildings by Eero Saarinen, Harry Wise, and I. M. Pei. It's easy to see how experiencing architecture has influenced him. The gradual switch from 2D work to 3D seems to be evidence of a growing interest and delight in complex sculptural ideas. What makes his work unique is that he has incorporated his visually vibrant and varied 2D designs onto the sculptural forms, which gain tremendously from the patterns. The inverse is true as well – the often multicolor designs take on added dimensionality by being placed on the forms. Hoyt often uses optical strategies in the designs which, in themselves, create illusions of three-dimensionality. The play between the illusions and the actual dimensional reality is brilliant.

After a period of experimentation through sketching, Hoyt will often work with cardboard to make mock-ups of his sculptural ideas, where adjustments are quick and easy. The cardboard templates are then used to cut slices of a grogged base clay onto which thin slices of the multicolored patterned clay are laminated as a veneer. With careful attention to the stiffness of the slabs they are then assembled to create the final form which is left to dry very slowly. Hoyt has devised a system of using pieces of sheetrock, which slowly absorb moisture from the clay, to surround the work and extend the drying cycle.

Hoyt's sculptures are generally made from a standard off white cone 6 clay body which he fires to cone 5 in an electric kiln. He uses a variety of Mason stains to create a wide range of colors and also enjoys using a manganese-rich dark-brown/black clay which he contrasts with much lighter shades of pastel colors.

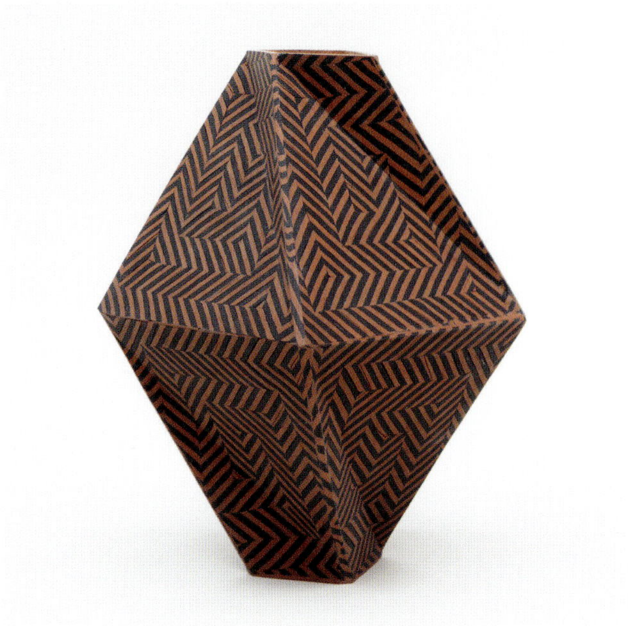

Twisted box vessel, 2014. Height: 30 inches, width: 24 inches, depth: 24 inches (76 x 61 x 61 cm) *Image: © Cody Hoyt; photo by Clemens Kols*

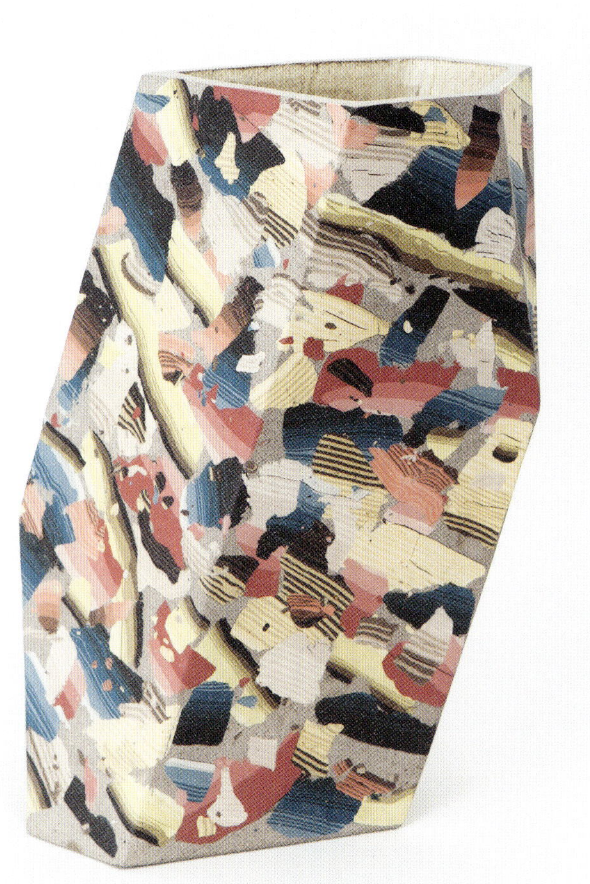

▶ **Oblique vessel, 2014. Height: 19 inches, width: 11 inches, depth: 7 inches (48 x 28 x 17.8 cm)** *Image: © Cody Hoyt; photo by Clemens Kols*

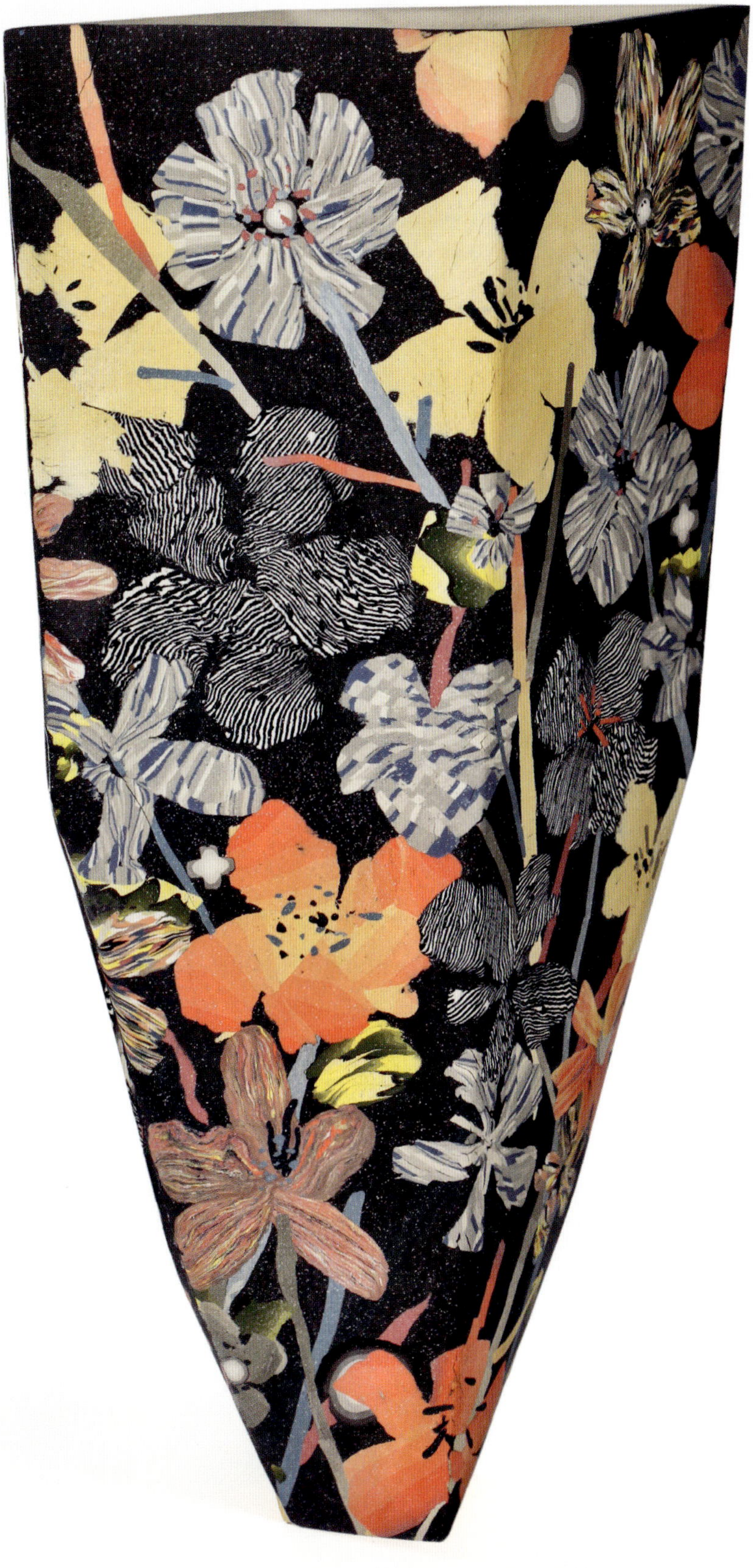

Landscape vessel, 2019. Height: 24 inches, width: 15 inches, depth: 8 inches (61 x 38 x 20.3 cm) *These images all © Cody Hoyt*

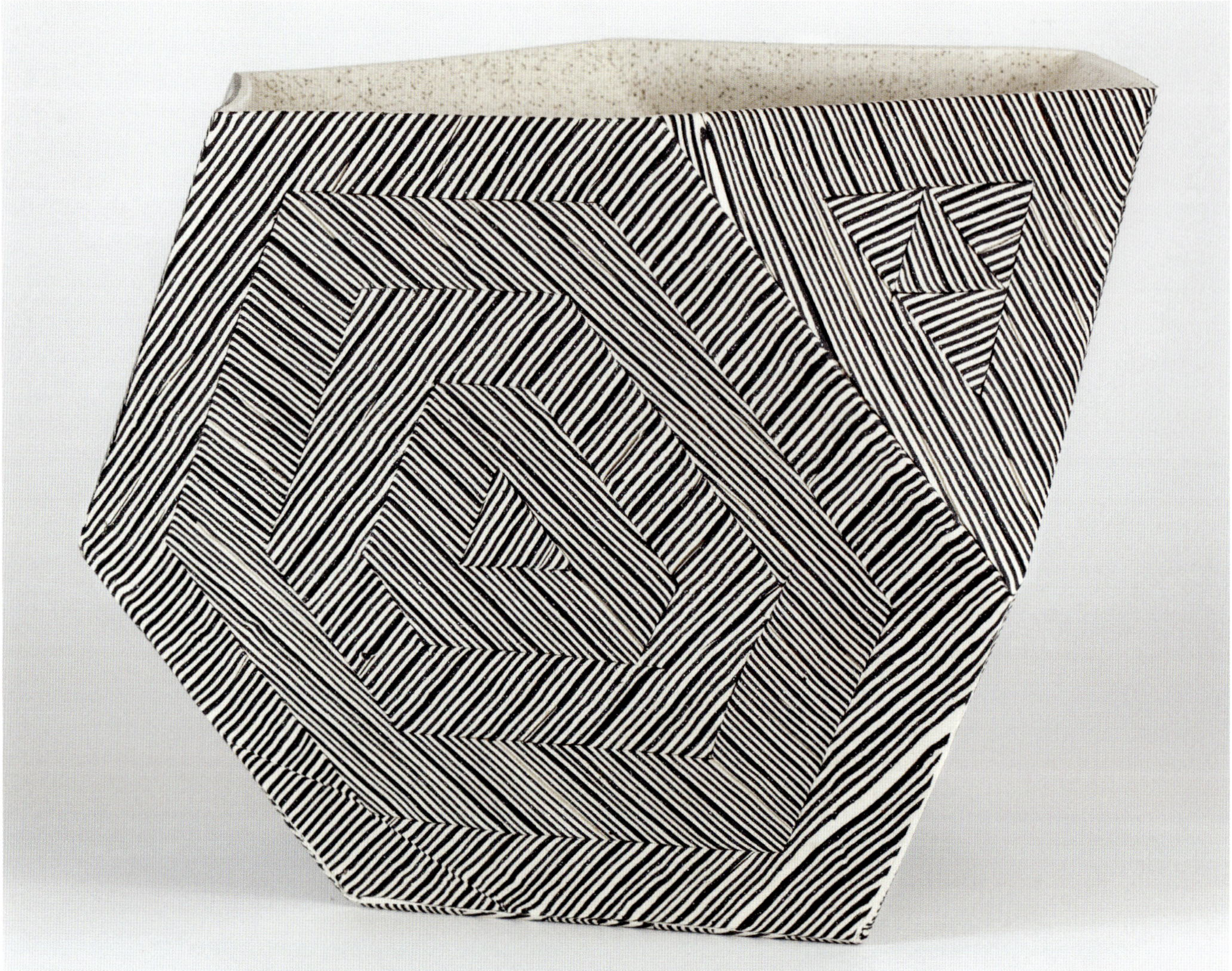

Truncated tetrahedron vessel, 2021. Height: 14 inches, width: 14 inches, depth: 14 inches (35.5 x 35.5 x 35.5 cm)

AWARDS AND EXHIBITIONS

Cody Hoyt has had five solo exhibitions at the Patrick Parrish Gallery in New York, solo or two-person exhibitions in other galleries in New York, Massachusetts, California, and Illinois, and group exhibitions in those states as well as in Florida and London, UK. He is the recipient of the Peter S. Reed Foundation Grant and has had his work published in *American Craft Magazine*, *Artsy*, *Elle Decor*, *Sight Unseen*, *Frame*, and *Juxtapoz Magazine*. Additionally, he is included in the books *Objects: USA 2020* by Glenn Adamson and James Zemaitis, Monacelli Press, 2020, and *The New Age of Ceramics* by Hannah Stouffer, Gingko Press, 2016.

Narumi Ii, Australia

In a few short years Narumi Ii has made an impressive contribution to the world of nerikomi. Art had always been part of her life since childhood, but mostly two-dimensional art. In 2016 a chance meeting with a friend spurred her on to enroll in a pottery class and almost immediately she knew she had found her medium. "The first time I touched clay, it felt very natural and familiar," she explains. "Clay has been with me ever since." Soon after, she chanced upon a YouTube video on agateware (the name for the English tradition of colored clay work – see page 18) and, intrigued, she set about learning more.

Ii is a self-proclaimed autodidact. Through reading and endless experimentation she has taught herself the basic techniques of nerikomi and, pressing forward, has searched for – and found – methods of taking these techniques to the next level. Understanding that multicolored designs could also be painted onto ceramic forms she has deliberately challenged herself to make such intricate designs with the clay that one couldn't possibly paint them. Creating an intricate design block of colored clay and then elongating the block so as to make the cross-section dimension significantly smaller is just one technique by which she has created such intricate and visually dazzling ceramics. Some of her most elaborate pieces include images of fish swimming along with their shadows which create an astounding sense of watery depth.

The experimental path to discovering these techniques invariably brings along with it a certain level of failure as well. Ii's philosophy luckily includes the valuable insight that any failure is also a learning experience and ultimately will enrich her understanding of both the process and physical dynamics of nerikomi as well as the art and skill of composition. In addition to this insight, Ii has also developed a strong sense of confidence to continue her quest for conquering new goals.

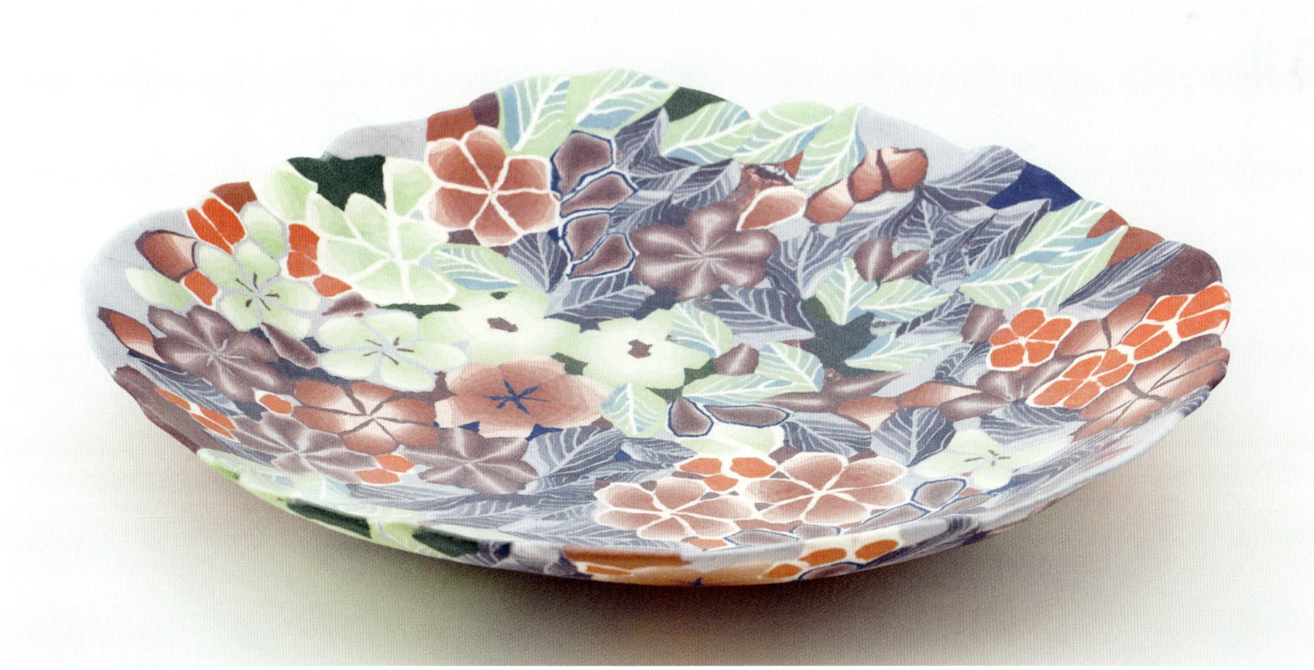

***A Morning in June*, 2017. Height: ¾ inch, diameter: 8⅔ inches (2 x 22 cm)** *These images all © Narumi Ii*

ARTIST'S STATEMENT

Getting rid of "attachment" is the key to enjoying what I do, there is no "failure" in anything you do, everything is a learning moment and part of growth, so I don't get upset or sad about things not going how I wanted or expected. This realization through work in ceramics made other aspects of my life so much different and enjoyable. Many artists have inspired me, but at the same time, looking at so many different and amazing works of others sometimes unsettled and confused me as well, to be honest. I have come to the conclusion that creating what I like and what comes to me using what I have mentally, spiritually, and physically is the best way to keep myself engaged and inspired.

What's possible and what's not is decided only by yourself. In other words, if you think it's possible, it's possible, if you think it's not, it's not. "Nobody has done it" doesn't mean it's impossible. I think I have managed to create something not many have done, and it's not because I am more skilled or talented than others, it's because I didn't think it was impossible. I had clay, stains, my hands, and time, all I needed was my own trust in myself.

In 2019 Ii started working with an additional technique called water etching. Shellac is painted in a variety of shapes onto greenware and, when dry, a wet sponge is used to wipe away clay from around the shapes leaving the shapes as raised areas surrounded by depressions in the clay. A bisque firing then burns off the shellac, and the design/texture then becomes highlighted through the use of semitransparent glazes. In the future she hopes to work with both techniques in a single piece, amplifying the sense of three dimensions.

Narumi Ii works with porcelain clay and a variety of stains, mixing in up to 200 different colors (often blending shades intuitively without test tiles), and fires her work to 1220°C (222°F) – cone 6.

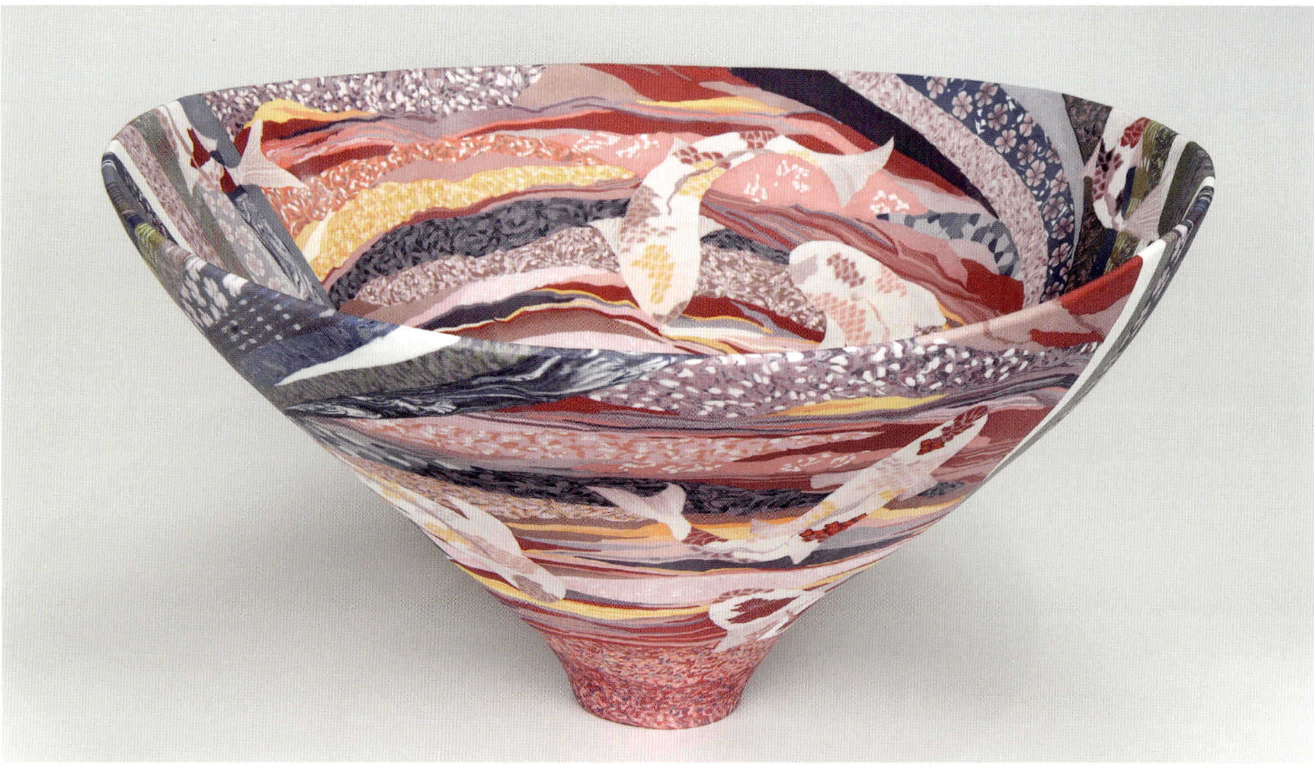

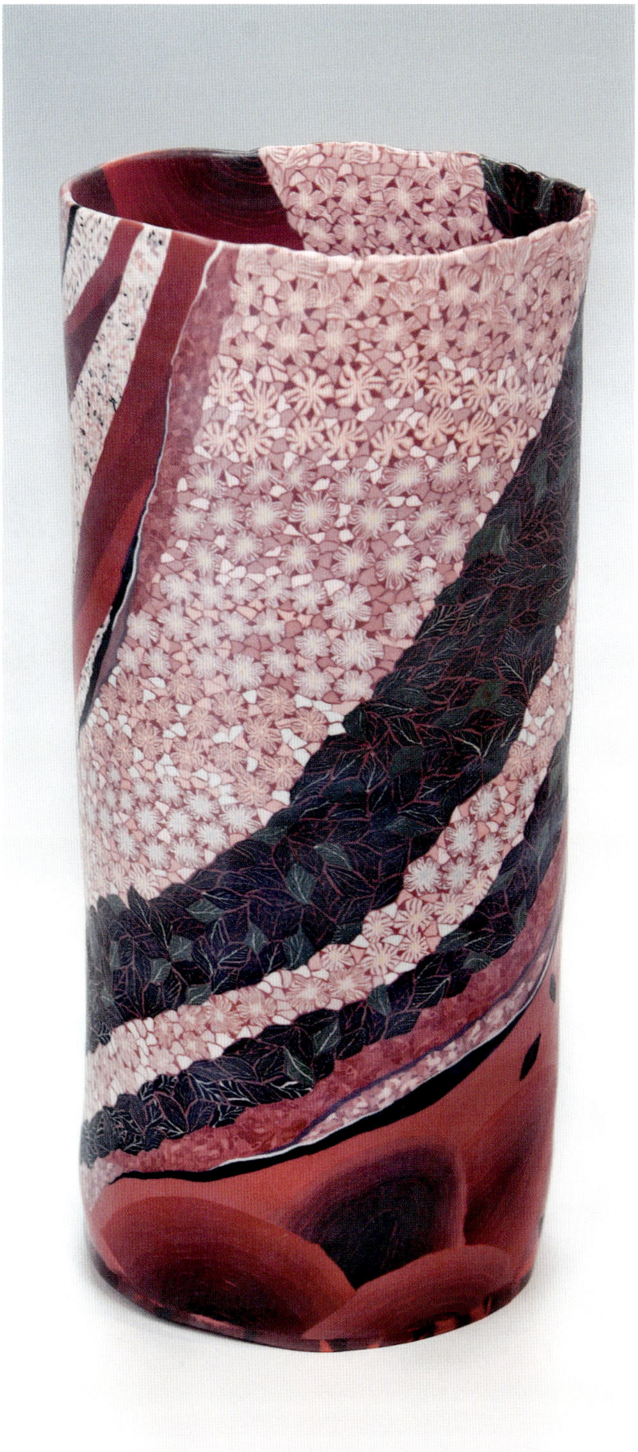

Somewhere in my Dream, 2018. Height: 12½ inches, diameter: 5¾ inches (32 x 14.5 cm)

AWARDS AND EXHIBITIONS

By the fall of 2017 **Narumi Ii** was already winning awards – finalist in the Clunes Ceramic Awards 2017, Victoria, Australia. The year 2018 brought her a Mark Brabham Emerging Artist Award, Klytie Pate Ceramic Award, Victoria, Australia; the position of finalist in the Stanthorpe Art Prize 2018, Queensland, Australia; finalist in the Biennial North Queensland Ceramic Award, Queensland, Australia; the Acquisition Prize in Toowoomba Biennial Emerging Artist Award, Queensland, Australia; and an Honorable Mention in Apero October Catalogue, California.

Narumi Ii has also been published in the November 2018 issue of the *Journal of Australian Ceramics* and the July/August 2020 issue of *Ceramic Review* magazine, UK.

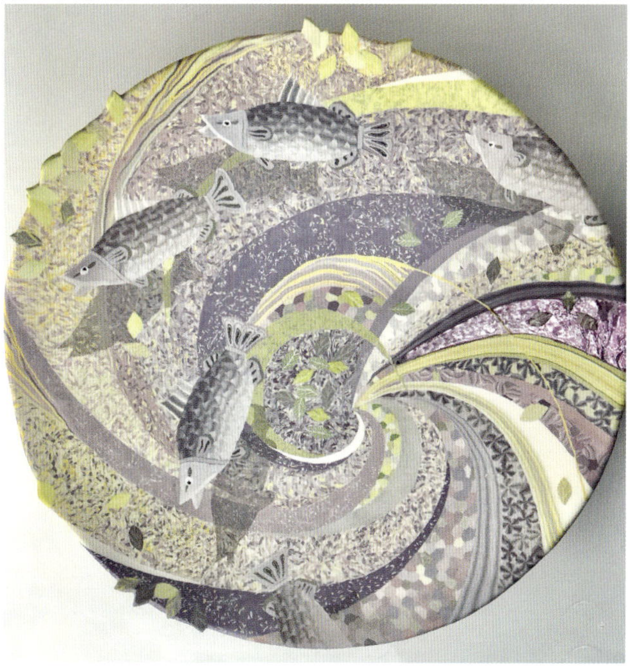

Coming Home, 2020. Height: 3⅛ inches, diameter: 9½ inches (8 x 24 cm)

◄ *Dancing in the Dusk*, 2019 Height: 6¾ inches, diameter: 13 inches (17 x 33 cm)

Dancing in the Dusk – Narumi Ii's making process.

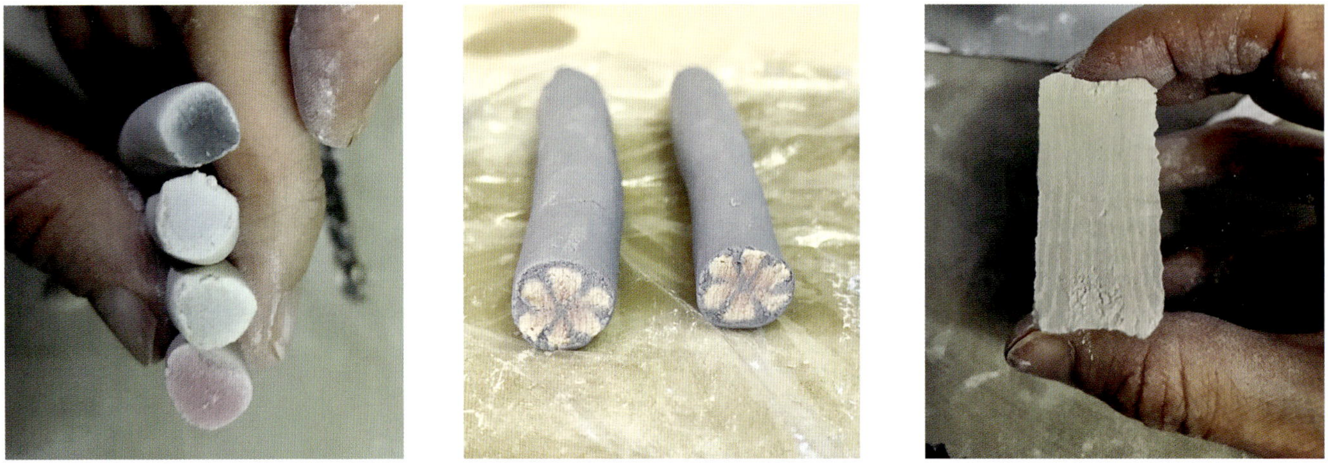

The component-making process.

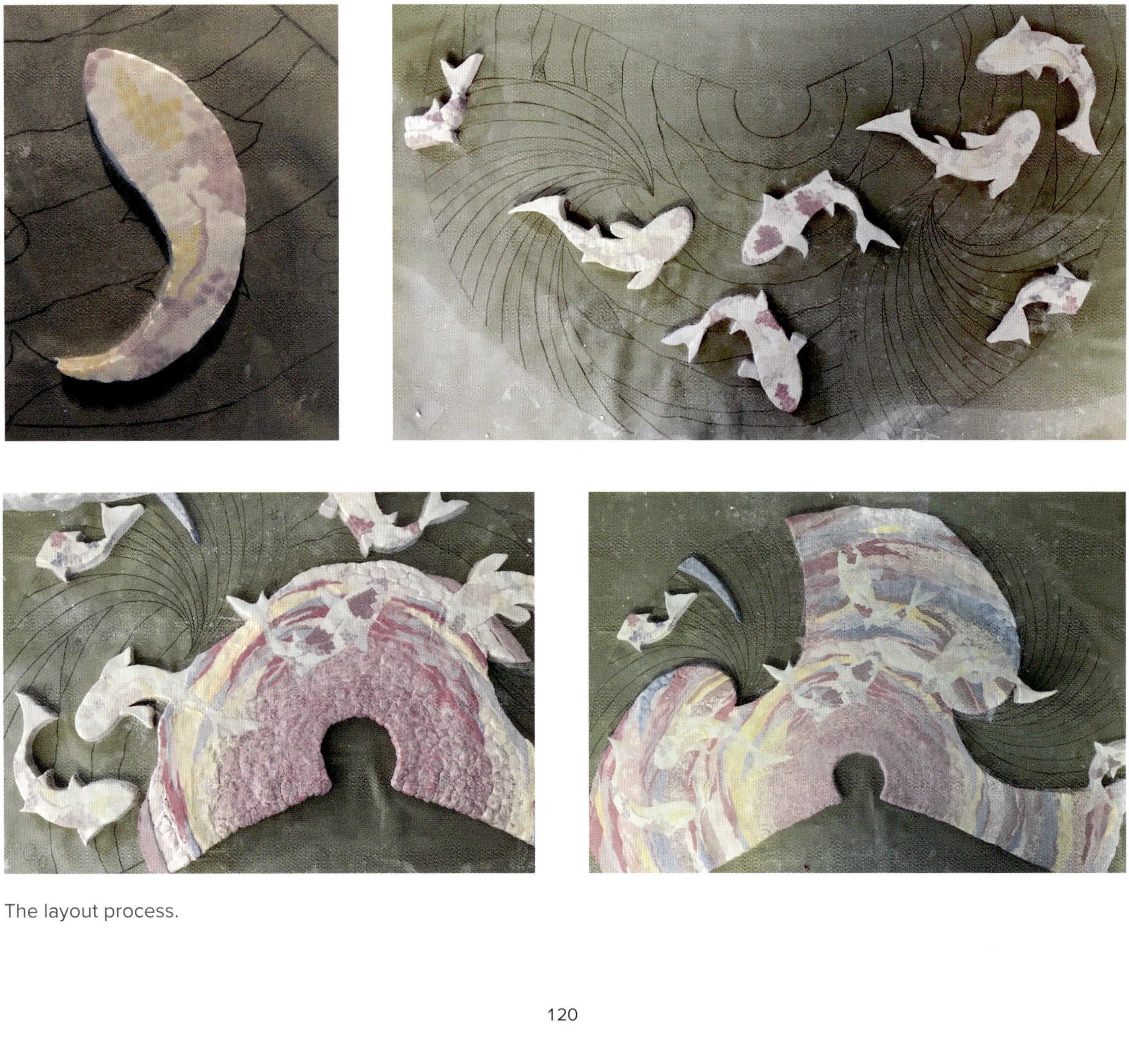

The layout process.

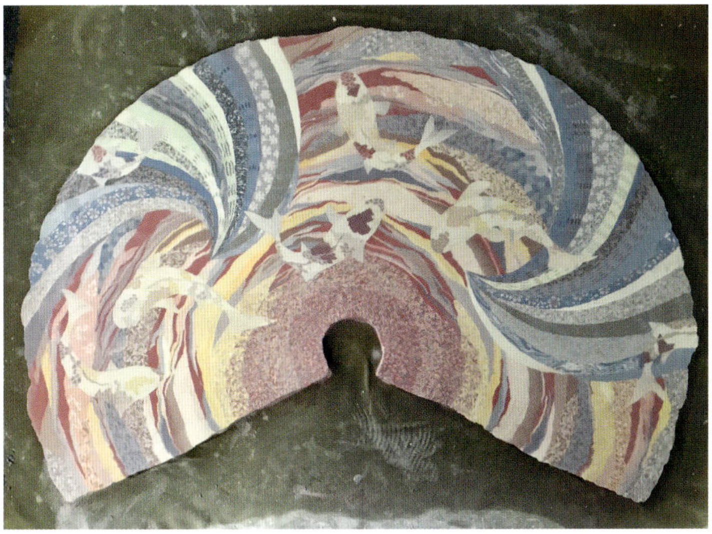

The finished slab.

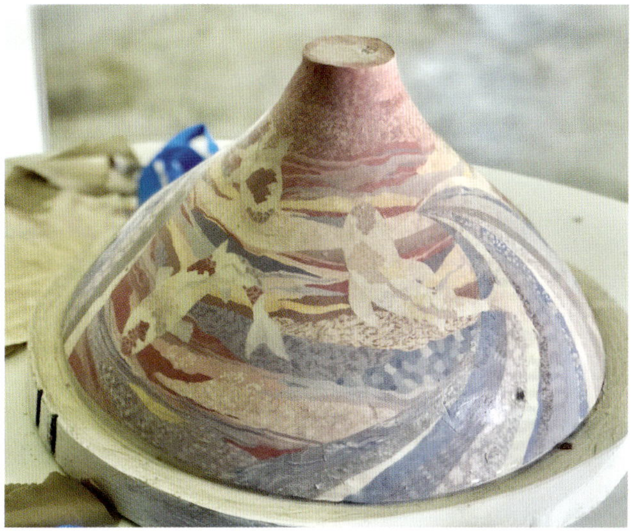

Forming on the hump mold.

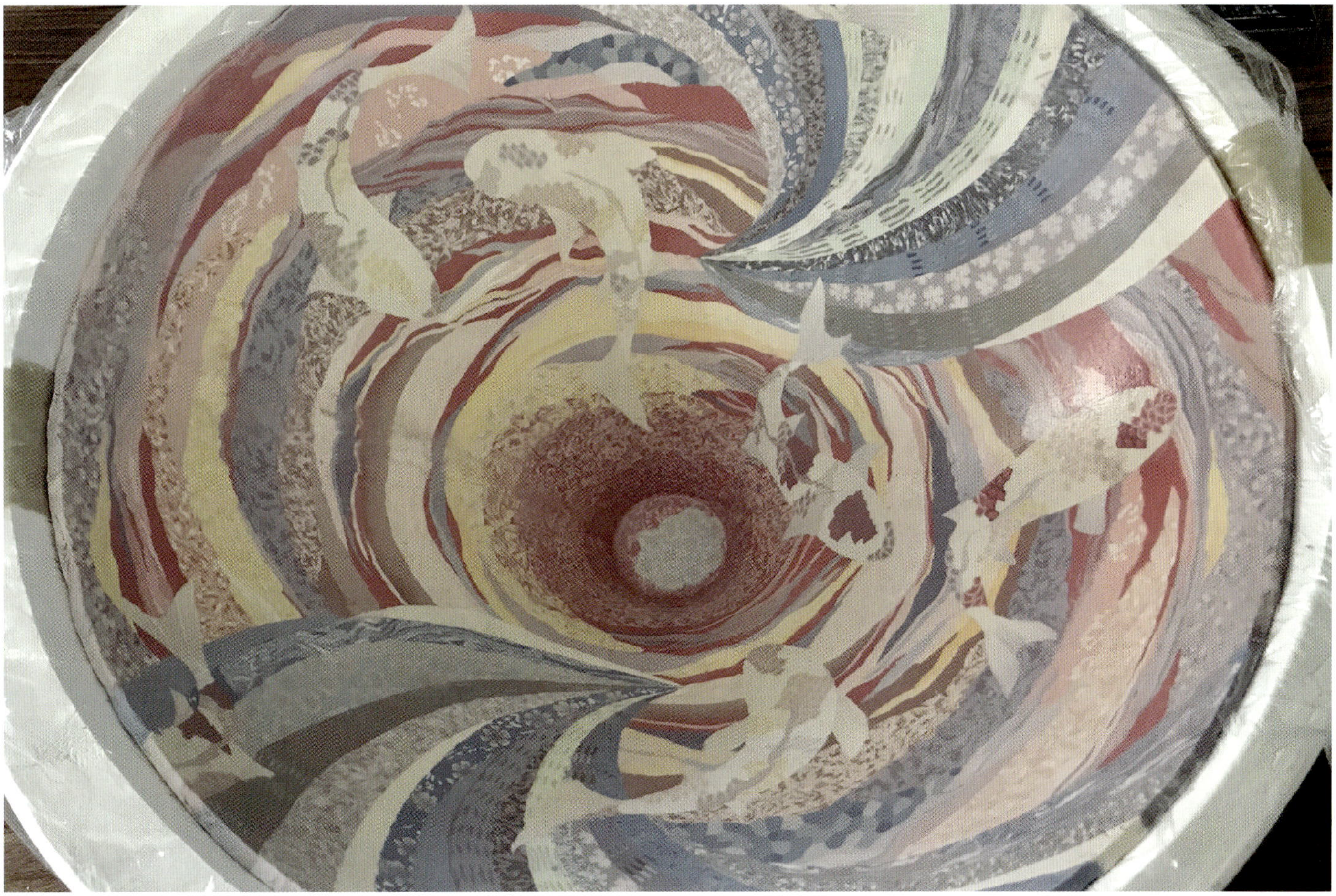

Inside the slump mold: it slowly dries over a few weeks inside the mold and is covered up with plastic.

Françoise Joris, Belgium

The delicate, translucent, and almost impossibly thin vessels and sculptures of Françoise Joris take the viewer into another world. Her works are reminiscent of the fascinating beauty of undersea creatures or fragile blossoms and seed pods. Under her hands thin sheets of porcelain take on a sense of life, where growth patterns become both structure and embedded design.

Joris remembers her childhood years living in Africa (Democratic Republic of Congo) where she enjoyed the feel of clay in her hands and had fun making small pots. She writes, "my childhood was influenced by the rhythm of the seasons and the surrounding nature. I have always felt in harmony with my environment very close to land and water. So, yes, we can say that the land-based and aquatic nature influences my current work. Even today throughout my travels, nature continues to feed my imagination."

Joris began her ceramics studies at the Academy of Beaux Arts Charleroi in Belgium in 2003, including a wide range of techniques: turning, stamping, plate work, creating molds, studying high-temperature Shino enamels, and working with polished and sigillated clays.

ARTIST'S STATEMENT

After more than ten years of experimenting, I made very thin and translucent bowls where the thickness of the porcelain and the enamel was almost identical.

I then decided to do personal research to open porcelain to almost all possible fields, by associating it with fibers and/or cellulose. At the beginning my pieces were white because the porcelain is sufficient on its own. I then introduced color through nerikomi techniques and made personal adaptations to these techniques.

My porcelain is colored in the mass, there is no transfer, or coloring by hand.

Her use of fiber inclusions in the clay body, often referred to as "paper clay" has enabled Joris to work with extremely thin slices of patterned clay while also creating forms that, themselves, are light and airy, seeming to defy the pull of gravity.

When asked what she loves most about being a ceramics artist she said: "I love the moment when I start a new artwork, because I do not have a preconceived plan. It is a journey, led by my imagination. A piece may look simple in appearance, but in fact requires a wide range of techniques and expertise, and time."

Joris's works, made from Audrey porcelain from the UK, are fired in an electric kiln at 1250°C (2282°F).

AWARDS AND EXHIBITIONS

The recipient of numerous awards and prizes (in Japan, Belgium, and Germany), **Françoise Joris's** pieces have been exhibited nationally and internationally at the International Ceramics Competition, Mino, Japan, where she was awarded Honorable Mentions in both the 9th and 11th exhibits; the Biennale de la Céramique Andenne, Belgium; Keramikmuseum Westerwald, Germany; Ceramica Multiplex Varazdin, Croatia; Terralha St. Quentin la Poterie, France; the Cluj Ceramics Biennale, Romania; the Musée Ianchelevici in La Louvière, Belgium; the Biennale Internationale Vallauris, France; and the Taiwan Ceramics Biennale, as well as other exhibitions in Belgium, France, Switzerland, Japan, Romania, Spain, the Netherlands, Germany, Austria, Croatia, Scotland, China, Slovenia, Italy, Luxembourg, and the USA. Joris's works are included in the collections of the New Taipei City Yingge Ceramics Museum, Taiwan, and the Musée des Beaux Arts de Lyon, France.

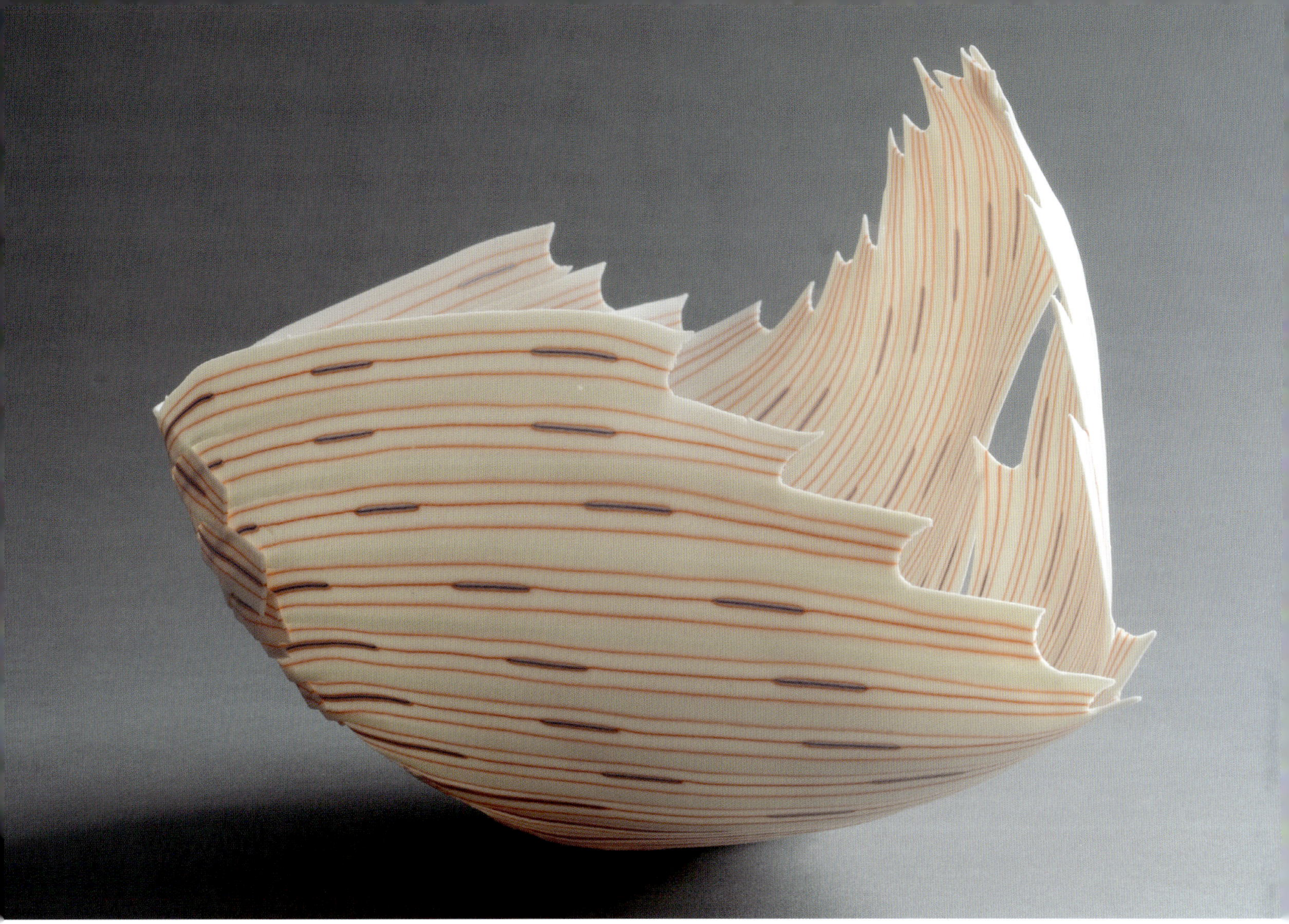

Flight, 2018. Length: 7⅛ inches, width: 8⅔ inches (18 x 22 cm) *These images all © Françoise Joris*

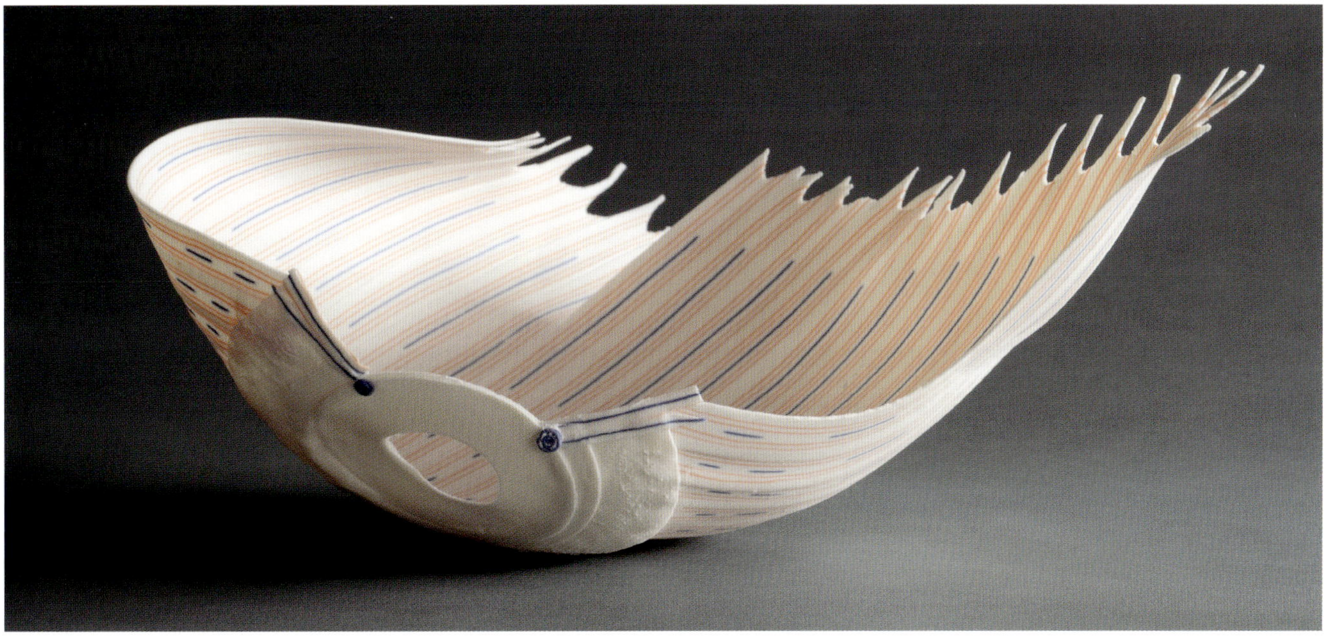

Undulation, 2019. Length: 9⅞ inches, width: 8⅔ inches (25 x 22 cm)

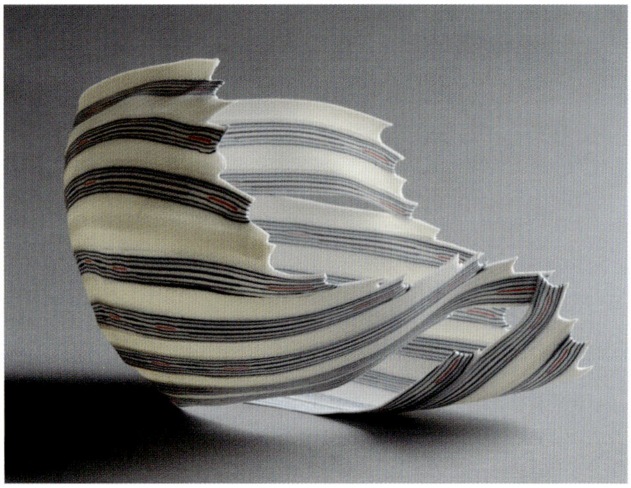

Untitled#1, 2019. Length: 8⅔ inches, width: 6 inches (22 x 15 cm)

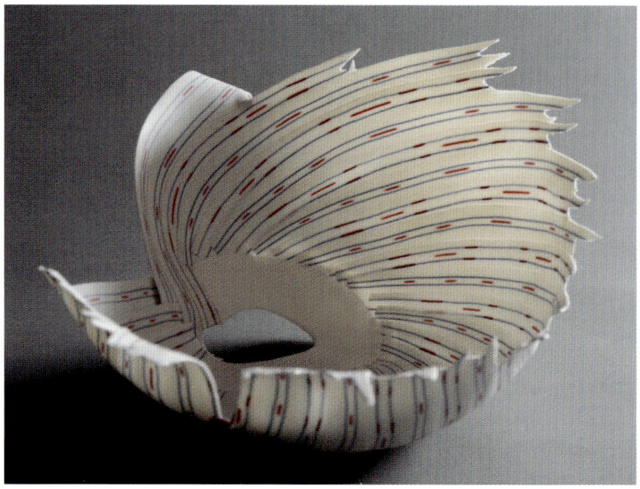

Untitled#2, 2019. Length: 11⅖ inches, width: 6¾ inches (29 x 17 cm)

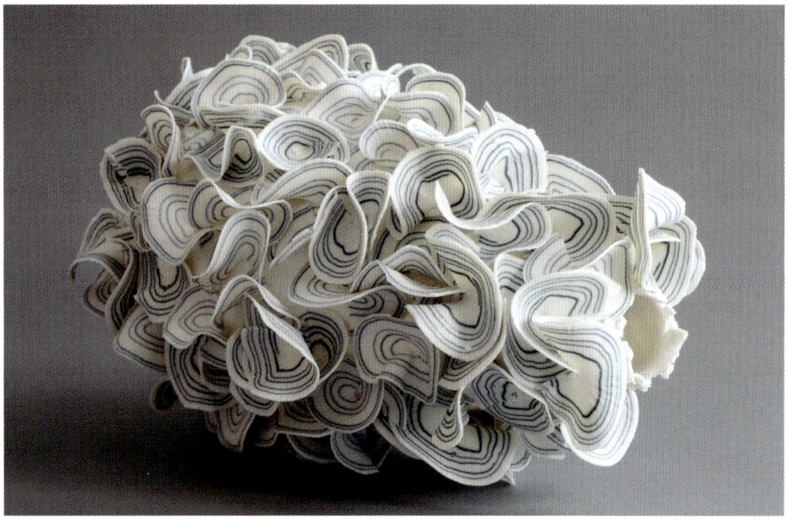

◀ *Black Waves 2*, 2022. Length: 8⅓ inches, width: 4⅓ inches (21 x 11 cm)

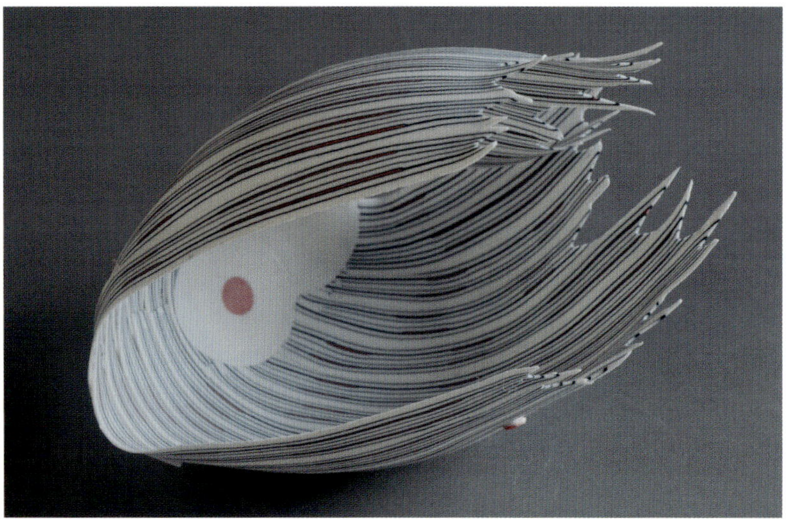

◀ *Untitled#3*, 2020. Length: 6 inches, width: 4¾ inches (15 x 12 cm)

Françoise Joris's paper clay technique for pattern making.

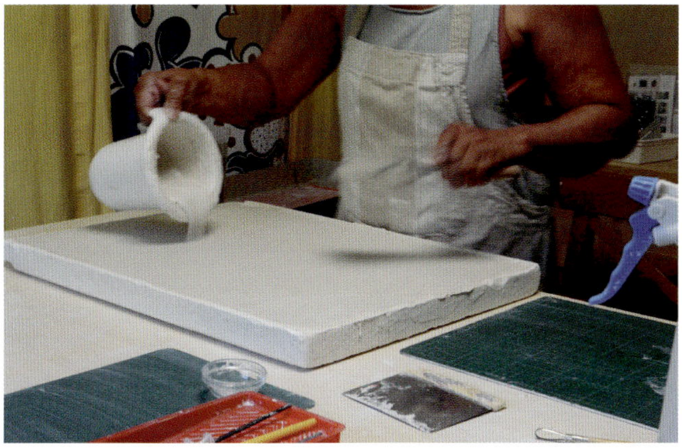

Pour the liquid porcelain on a plaster slab and spread it well to obtain a very thin sheet.

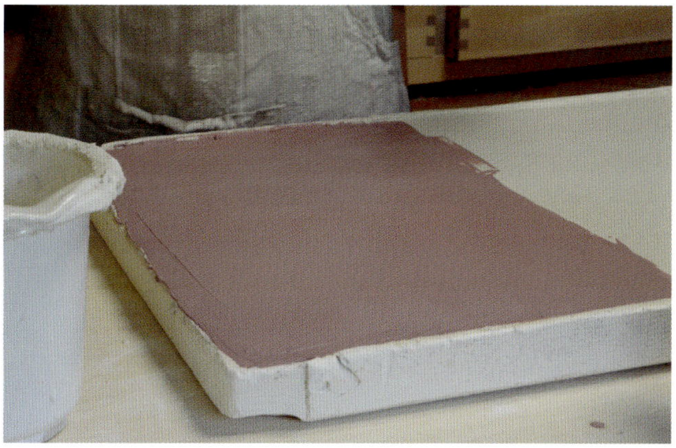

Pour the colored porcelain on one half of the sheet and proceed in the same way.

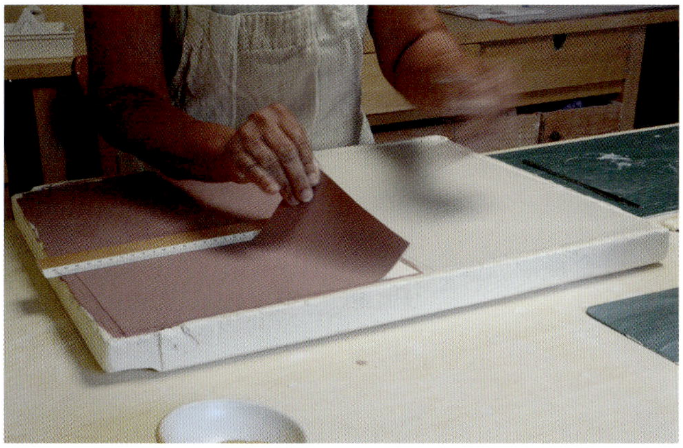

Cut two rectangles out of each half and peel each rectangle off the plaster.

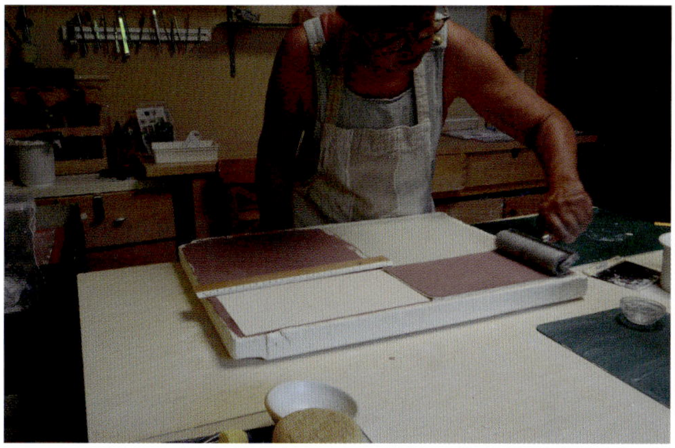

Put a colored rectangle on top of a white rectangle and roll to remove air bubbles.

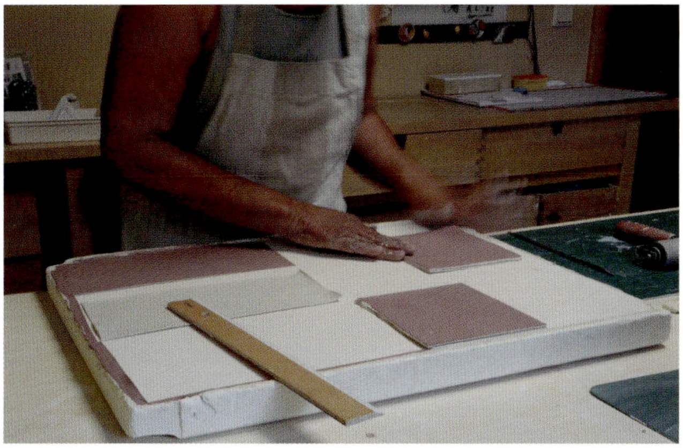

Proceed the same way with the two remaining rectangles and cut layer into two squares.

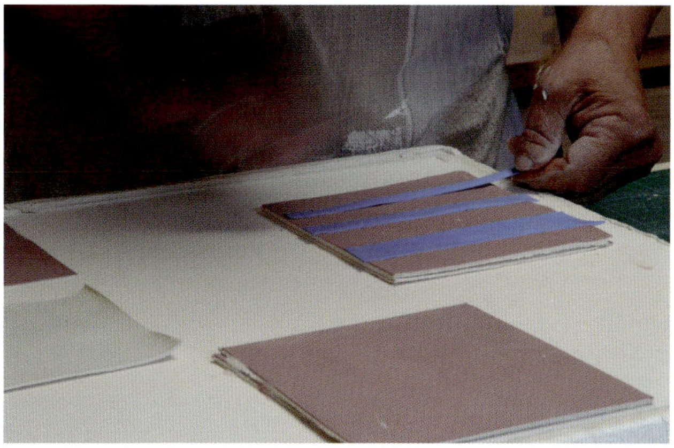

Place previously prepared colored strips on top of a square to obtain a checkerboard pattern.

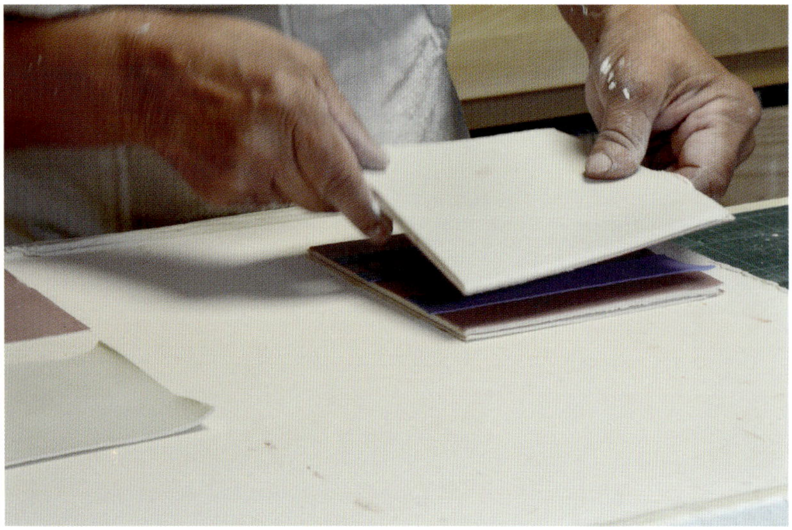

Put the colored side of the second square on top of the checkerboard pattern to obtain a layer of superimposed sheets.

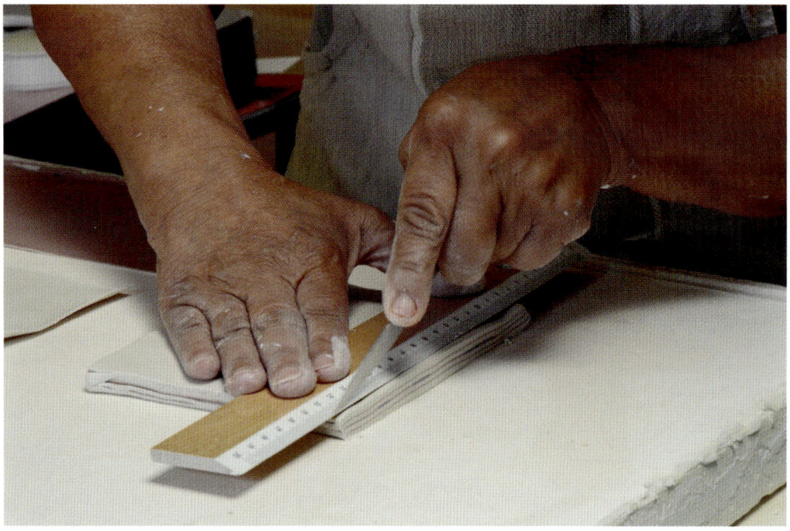

Cut thin strips vertically.

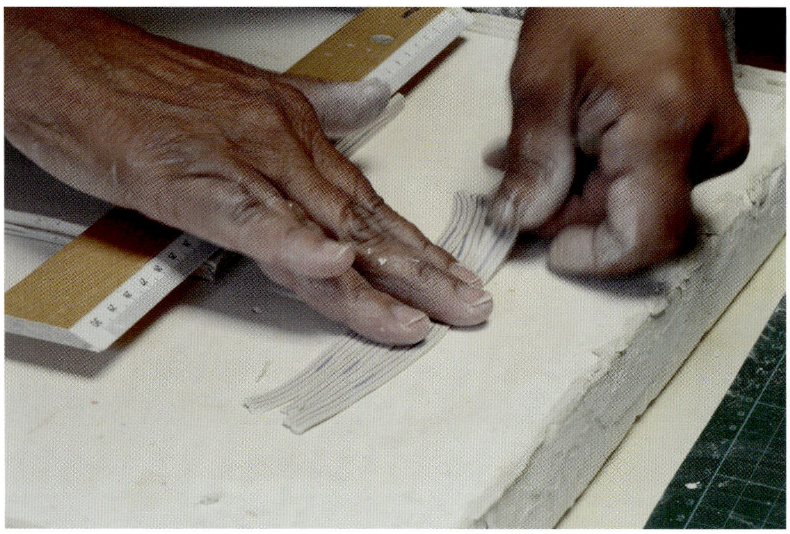

Assemble cut strips to obtain wider strips.

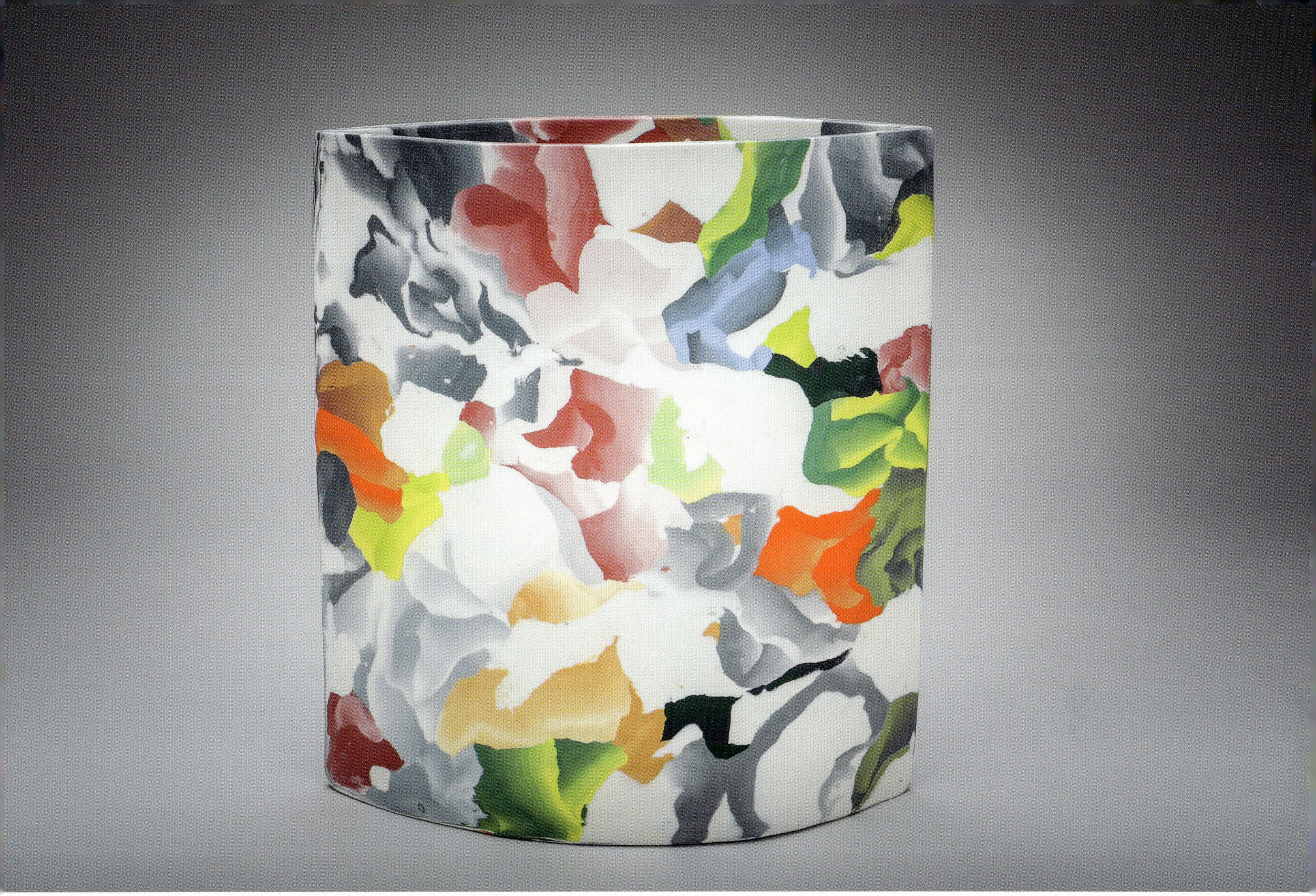

Earthly Joys, 2022. Height: 9½ inches, width: 8⅓ inches, depth: 4¾ inches (24 x 21 x 12 cm)

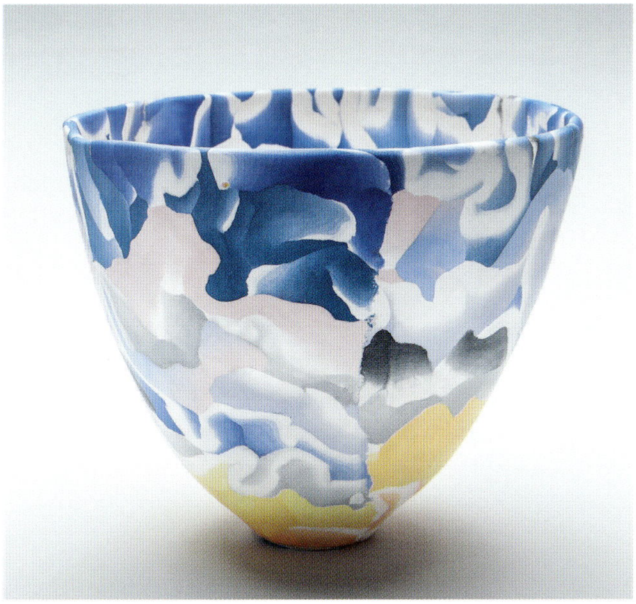

Pink Cloud, 2020. Height: 6⅓ inches, diameter: 4¾ inches (16 x 12 cm)

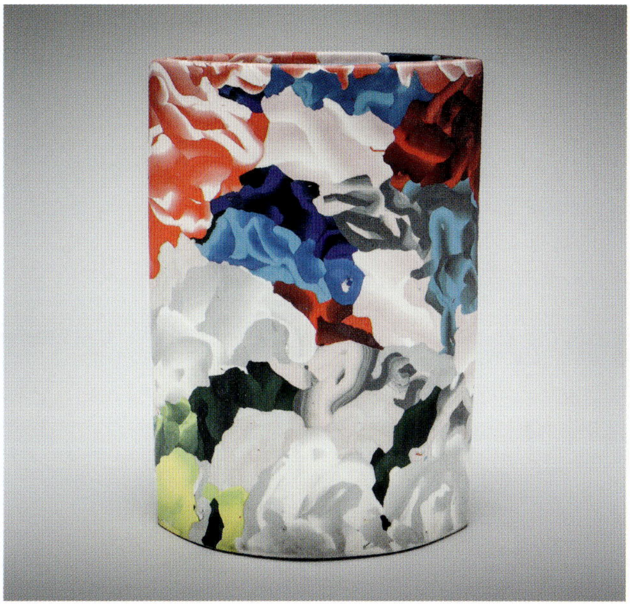

Lake Natron (silver kintsugi detail), 2022. Height: 7½ inches, width: 5⅛ inches, depth: 3 inches (19 x 13 x 7.5 cm)

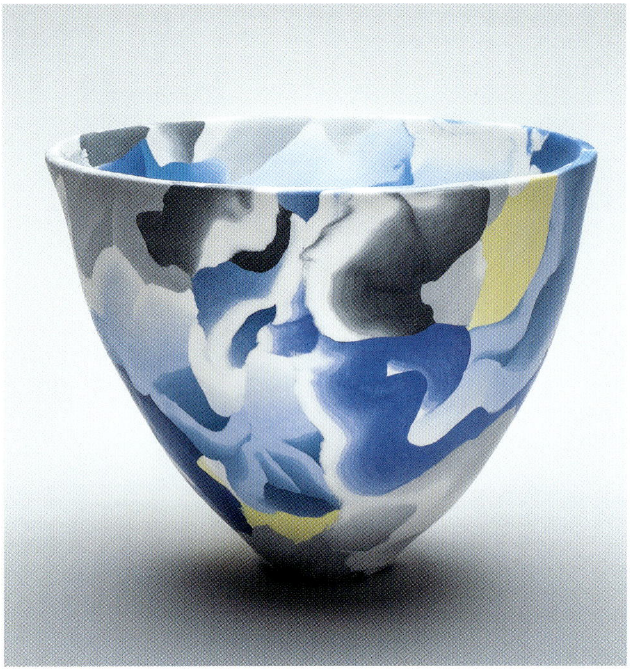

Yellow Cloud, 2020. Height: 6⅓ inches, diameter: 4¾ inches (16 x 12 cm)

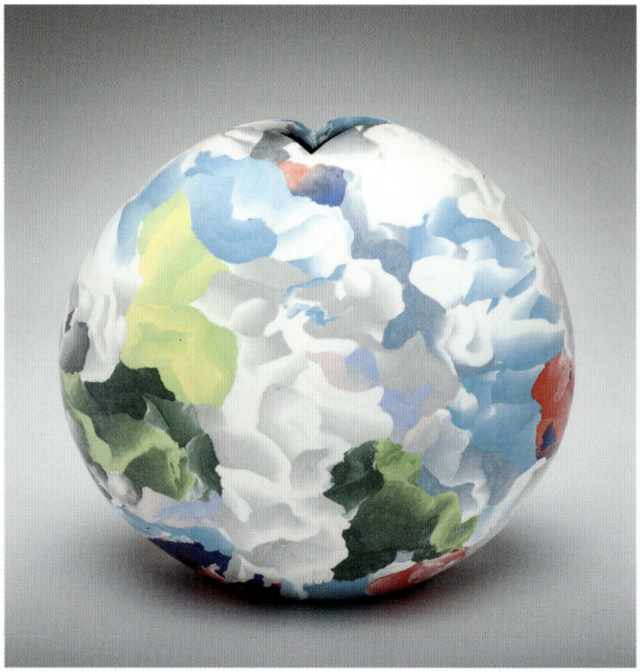

Summer Garden (silver kintsugi detail), 2021. Height: 6⅓ inches, diameter: 2½ inches (16 x 6.5 cm)

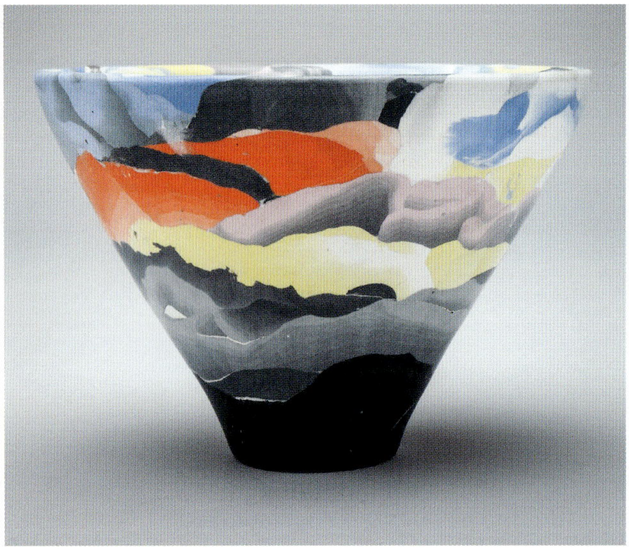

Small conical bowl (silver kintsugi detail), 2021. Height: 6⅓ inches, diameter: 4¾ inches (16 x 12 cm)

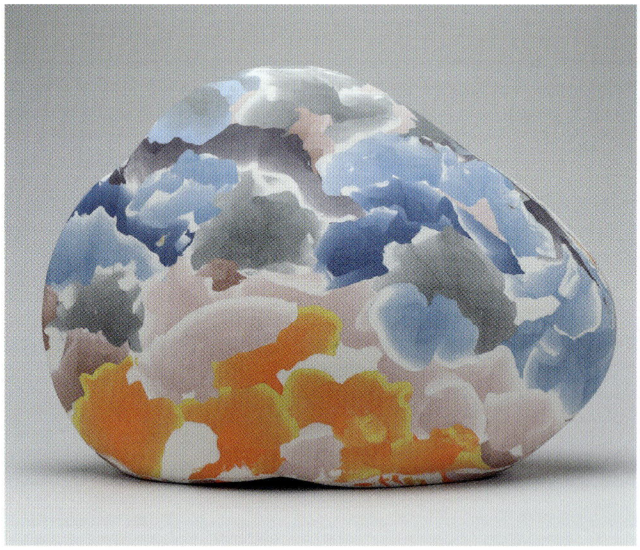

Breaking through the clouds, 2022. Height: 11 inches, width: 15⅓ inches, depth: 4¾ inches (28 x 39 x 12 cm)

AWARDS AND EXHIBITIONS

In her short career as a ceramist **Judy McKenzie** has exhibited widely in galleries and exhibitions in the UK, where she had a solo exhibition at Craft Co., Southwold, 2022, and was featured in "Color in Clay" at Watts Gallery, Surrey, 2021. She has also exhibited internationally in China and the US.

Judy McKenzie's steps for creating a composition.

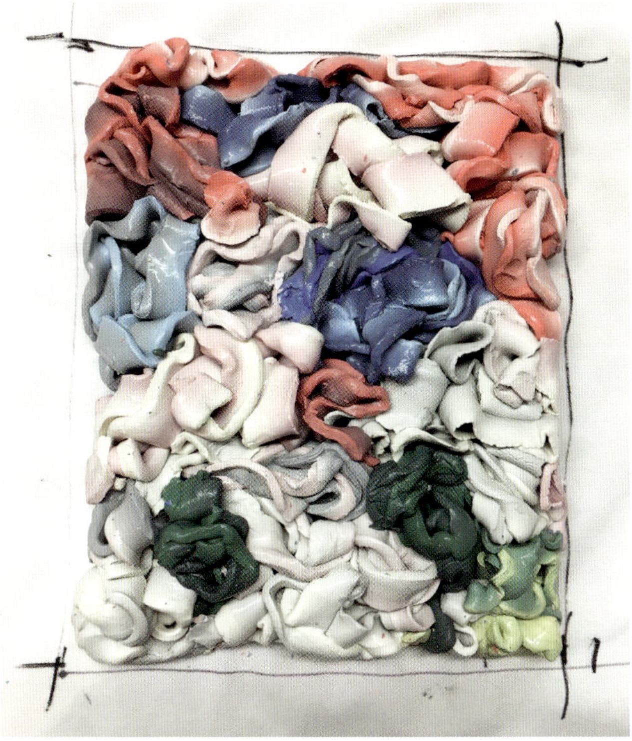

◀ Using a small harp, I slice off ribbons from a color-graded block of porcelain. These thin colored ribbons are then twisted and curved, and are placed where I want a particular color to appear in the finished panel.

Images on this page © Judy McKenzie

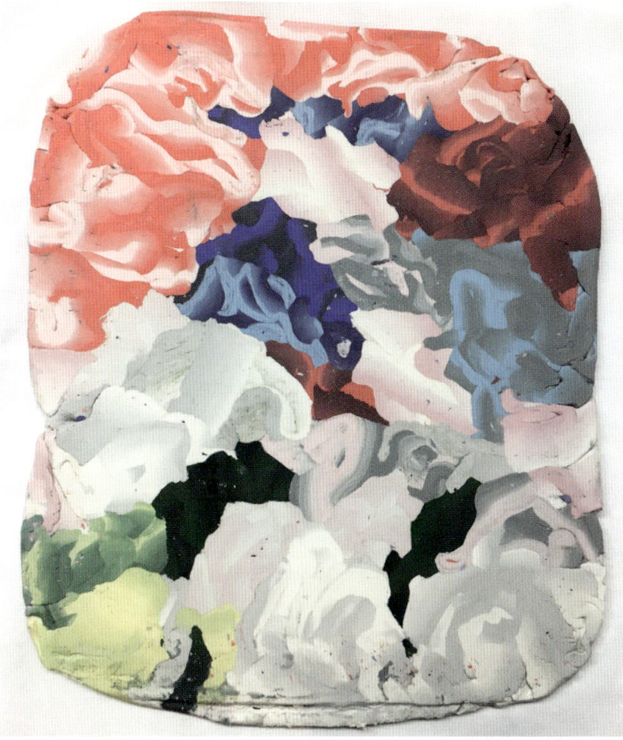

▲ Once all the ribbons have been laid down it is time to flatten them to create the panel. Using guides, I cover the panel with a cloth and roll to the required thickness. Air can be trapped at this stage, so I always start off much thicker than needed, so I can squash as much air out of the panel as possible. The finished thickness of the panel will be approximately ⅛ inch (3–5 mm). The reveal of the pattern under the cloth is always a nice surprise!

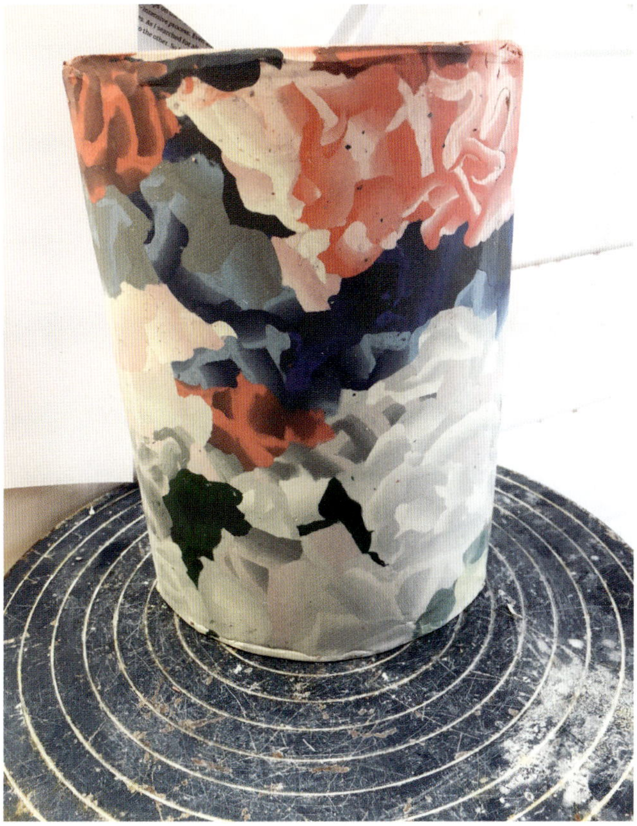

◀ The porcelain panel is now very floppy, and needs the support of a plaster mold to help shape it into the finished form. A two-piece mold is used. The sides are joined by scoring and slipping the edges, and the two halves of the mold are pressed together. When the porcelain is leather-hard, the form can be released from the mold.

Anne Mossman, Australia

After a full and rewarding career in management Anne Mossman opened a new chapter in her life and enrolled in a ceramics program at The Australian National University, Canberra (ANU), where she received her Diploma of Art (High Distinction) in 2007. It was at ANU when, in 2004, she spent several days studying under visiting artist Dorothy Feibleman, a well-known nerikomi specialist with an already international reputation (see pages 96–102). Seeing someone work with colored clay to create pattern blocks is fascinating to watch and by the end of the workshop Mossman was, by her own admission, hooked.

Early work focused on small-scale works with delicate geometric patterns followed by a period of experimentation with thin slices of nerikomi designs being slip cast into plaster molds. More recently she began using clean cylindrical forms as a "canvas" to showcase her more abstract designs. Mossman's subtle colors are often rimmed by a paper-thin band of black, accentuating the organic shapes created through her layering process while constructing her pattern loaves.

Color and the choices of various color palettes have become a primary focus for Mossman and she again is influenced by her surroundings.

ARTIST'S STATEMENT

I have found pattern-making by layering and manipulation to be a very accessible and spontaneous way of expression. I am intrigued by the shift of lines and the creation of patterns arising from quite subtle movements of the colored clay bodies . . . My influences come from geological formations which have undergone the same shifts and movements, but over eons. I'm originally from New Zealand, a country thrust up by relatively recent violent ruptures in the earth's crust, the scars of which are part of my visual lexicon.

My work is inspired by my immediate environment, my garden surrounds and the bush backdrop in the hinterland of the Gold Coast. I live on a large, densely treed ridge that lies between the Tallebudgera and Currumbin valleys on Yugambeh land. My ideas and palette are drawn from the disparate colors on some of the eucalyptus tree trunks in my surroundings. In the summertime the bark peels off in sheaths to reveal nude-like patches of new skin which is invariably smooth and lighter colored. I find the contrast in colors and tones stunning and ever-changing. In contrast to the eucalyptus is my developed sub-tropical garden which is a visual feast of striking, colorful bromeliads, bold architectural succulents, and grasses all enclosed and outlined by a meandering lawn.

Mossman has also introduced a more recent improvement to her nerikomi toolbox, the use of paper clay. She explains: "I have recently added paper to my clay to minimize the loss from cracking. The paper fiber helps mesh the myriad of joins in a vessel. I add toilet paper at a rate of three single sheets per 800gms of wet clay. It goes into the blender at the same time as the stain."

Finishing techniques include sanding at the bone-dry stage with fine steel wool, wet sanding with carborundum paper after the bisque firing, and wet sanding again after the high firing. The vessels are usually left unglazed. Mossman uses Keane's Lumina mid-fire porcelain fired to 1200°C (2192°F) – cone 6 – in oxidation.

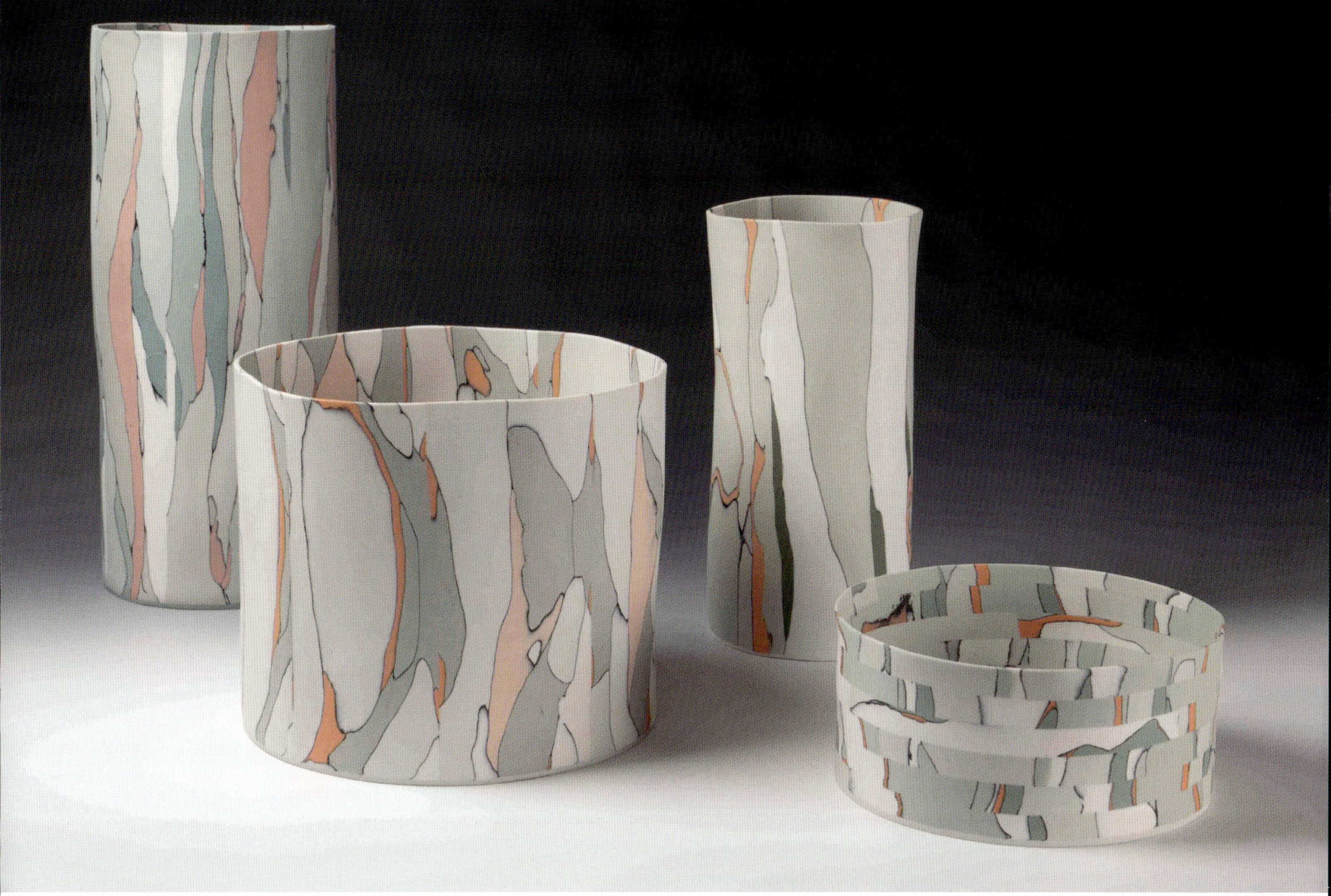

Eucalyptus Study 5, 2020. Height: 13¾ inches, width: 23⅔ inches, length: 28¾ inches (35 x 60 x 73 cm)

These images all © Anne Mossman

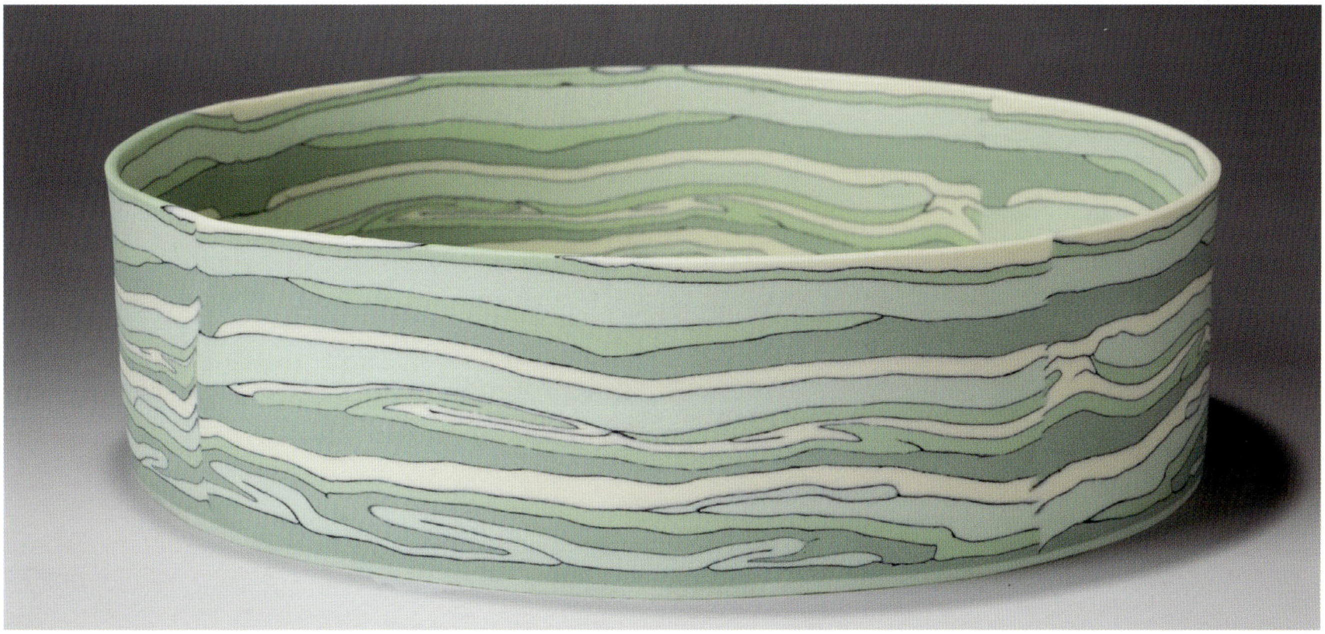

Fields of Green, 2019. Height: 2½ inches, diameter: 9 inches (6.5 x 23 cm)

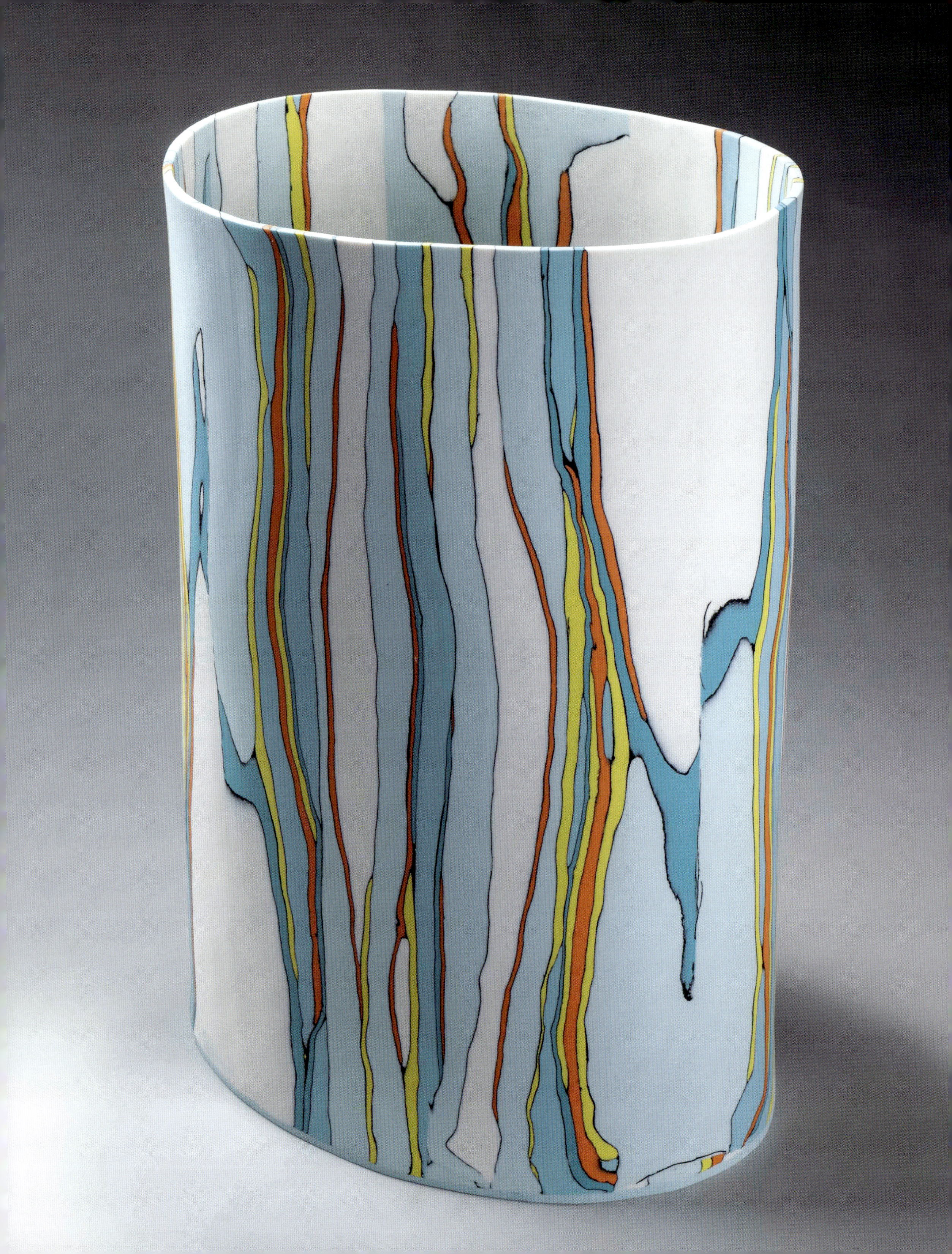

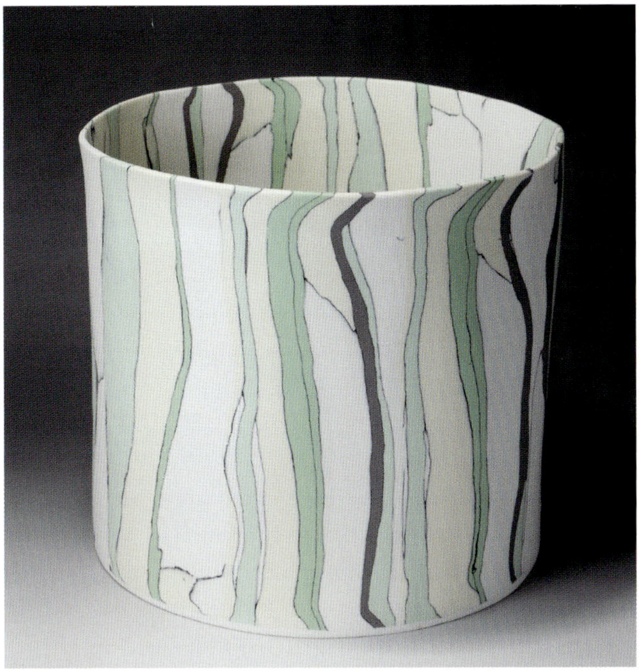

Eucalyptus Study 1, 2019. Height: 8⅓ inches, diameter: 8⅘ inches (21 x 22.5 cm)

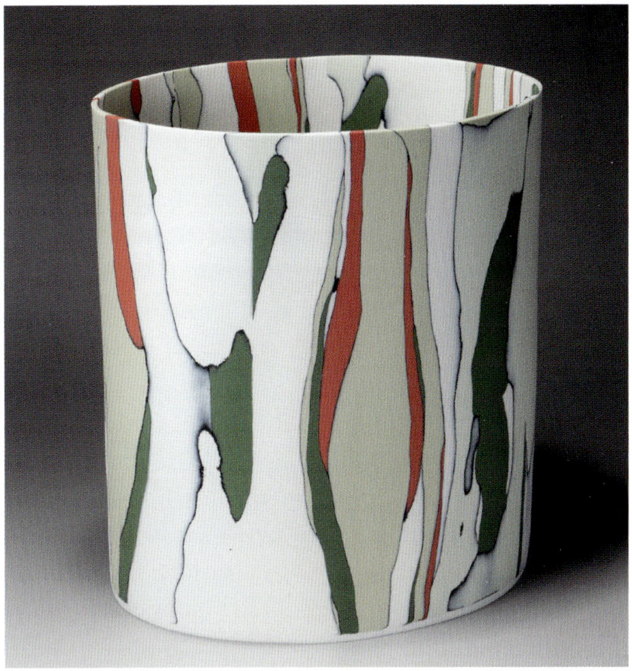

Eucalyptus Study 2, 2020. Height: 9½ inches, diameter: 7⅛ inches (24 x 18 cm)

AWARDS AND EXHIBITIONS

Anne Mossman has exhibited her work widely in Australia since 2008 and has received numerous awards. Among them are First Prize awards from the Gold Coast Potters Association in 2008, 2009, 2011, 2017. She is also the winner of Art Unlimited (2015) and the Port Hacking Ceramics Art Award (2016). Her work is included in the collections of the Stanthorpe Regional Gallery, Australia, and the Medalta Museum, Canada. Her work can be found in the Museum of Contemporary Art gallery shop, the Sturt Gallery and Studios and the Weswal Gallery, all in Australia. Her work has also been featured in the *Ceramic Arts Bulletin*, the *Museum of Contemporary Art Members' Magazine*, *Ceramics Monthly*, and *Canberra Times*. Mossman has also authored articles for the *Journal of Australian Ceramics* (November 2021 and November 2018).

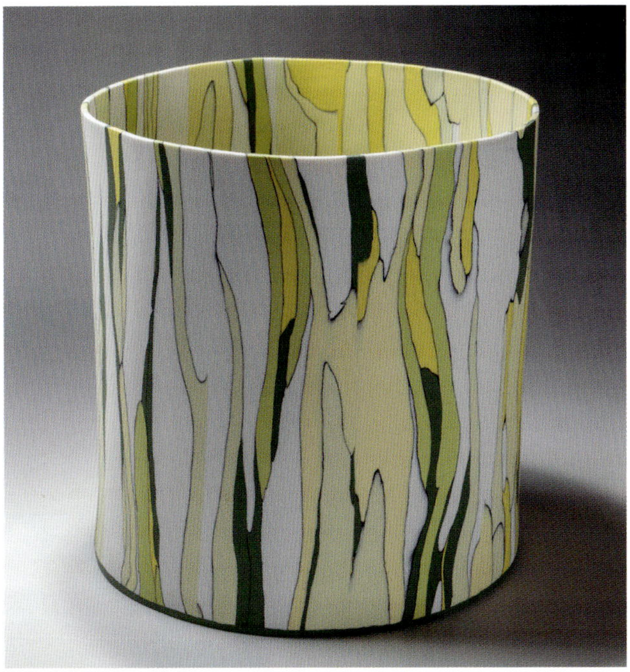

Eucalyptus Study 3, 2021. Height: 9 inches, diameter: 8⅔ inches (23 x 22 cm)

◀ *Seams of Bold*, 2021. Height: 11⅘ inches, width: 6 inches, length: 7½ inches (30 x 15 x 19 cm)

Aya Murata, Japan

The technique of nerikomi carries with it a long history of vessel-making, both functional and decorative. The idea of using it in a sculptural format is clearly a new 21st-century direction and the work of Aya Murata stands with that of Janny Baek, (see pages 70–3), showing two very personal directions.

Murata's creations notably showcase her expert technique but also a whimsical and even sometimes menacing sensibility which draws viewers in as they explore the fascinating details of her complex plant/animal-inspired creations.

ARTIST'S STATEMENT

All my works have a consistent theme, which is "Vitality." The vital energy takes many forms in the world we perceive as nature. Under severe survival conditions, some earned weapons such as claws and poisons, and some learned to beautifully mimic their surroundings. I try to express the beauty and harshness of nature through my works. Humans have their poisons as well, like anger or degrading one's value. Although it seems like a negative component, it helps us grow to become a better being. My goal is to create works that make one feel that they are alive and breathing.

Aya Murata began her studies in 1995 at the Kyoto Municipal Doda High School of Fine Art Handicraft where she focused on traditional Lacquer Ware. This was followed in 1998 by two years at the Kyoto College of Art where she studied ceramics. From 2001 to 2003 she taught ceramics to children and later taught ceramics techniques to people with disabilities. From 2003 to 2004 she studied at the Kyoto Prefectural Higher Technical Institute of Ceramic Work, creating sculptural hand-built work. After graduation, as is common with many artists, Murata began making tableware, employing the textures she had discovered from her sculptural work. Gradually her inlay patterns became more complex and the switch was made to using the nerikomi technique. Further experimentation brought her to a point where the "function" of the pieces was no longer important. On the advice of one of her teachers she finally committed herself to sculpture.

She writes: "Through this traditional technique of nerikomi, my work has become more creative and it has enabled me to add vitality. The patterns I create are inspired by cells, which are the foundation of vitality."

Murata's sculptures are made from porcelain and stains and are finished with a transparent glaze, fired to 1230°C (2246°F).

AWARDS AND EXHIBITIONS

Aya Murata began her solo exhibition career immediately after her training at the Kyoto Institute. The first show was at Gallery Maronie, Kyoto, which was followed by 15 solo shows in Japan and Taiwan. She has received many awards including the Paramita Ceramic Award, Paramita Museum, Mie, Japan, in 2022 and the Media prize and "new star selection exhibition" at The Museum of Kyoto, Japan, in 2018. Her work has also been shown in group exhibitions in Taiwan and Japan including two at the Ibaraki Ceramic Art Museum as well as one at the Kyoto Takashimaya. Murata has had residencies in South Korea, Finland, Denmark, the USA, Japan, and Taiwan. Her work is included in the collections of The Shigaraki Ceramic Cultural Park, Japan, and the New Taipei City Yingge Ceramics Museum, Taiwan, among others.

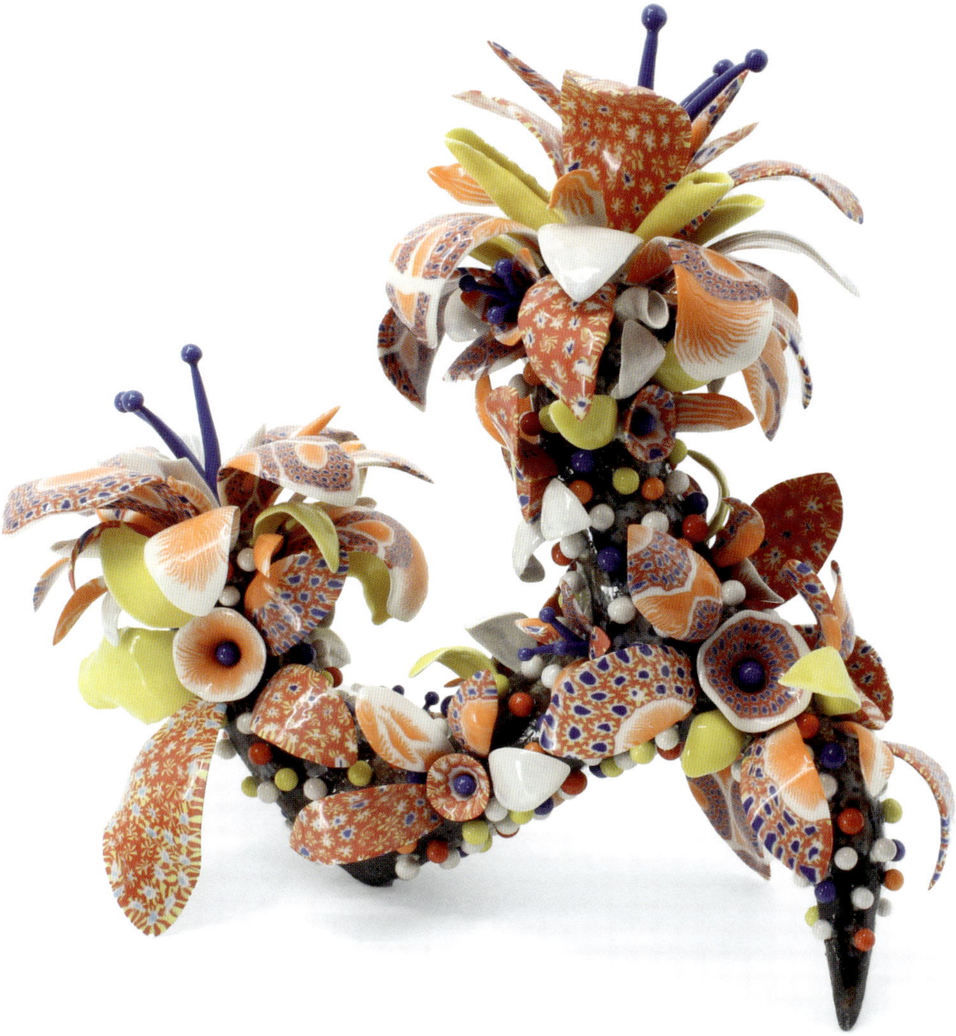

▲ *Cordyceps series 1,* 2022.
Height: 11⅘ inches, width: 13¾ inches, depth: 6¾ inches (30 x 35 x 17 cm)

These images all © Aya Murata

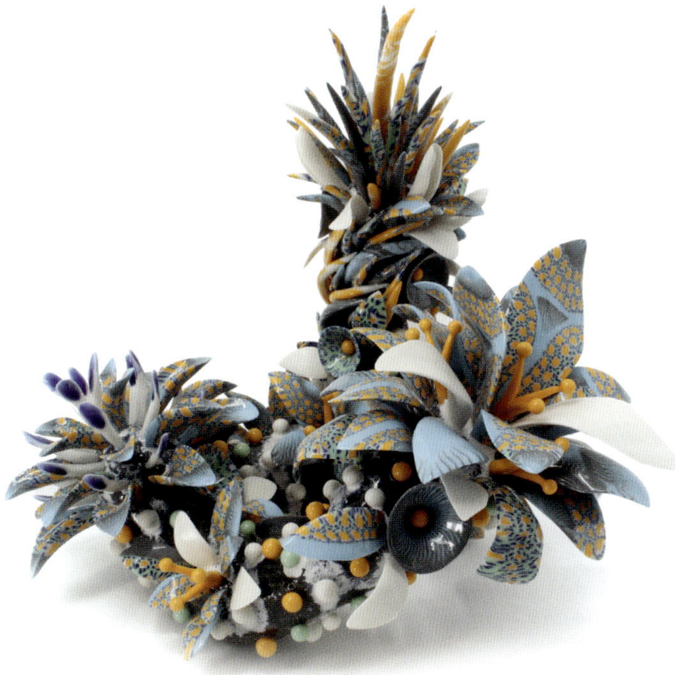

Cordyceps series 2, 2022.
Height: 9½ inches, width: 9⅞ inches, depth: 7⅘ inches (24 x 25 x 20 cm)

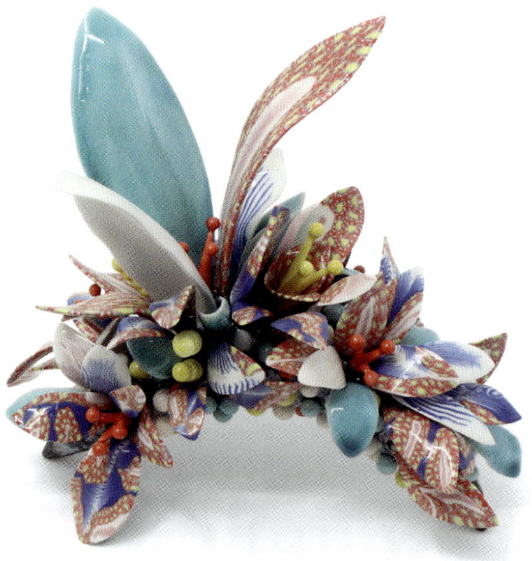

Caterpillar is dreaming series 3, 2022. Height: 9½ inches, width: 9 inches, depth: 4⅓ inches (24 x 23 x 11 cm)

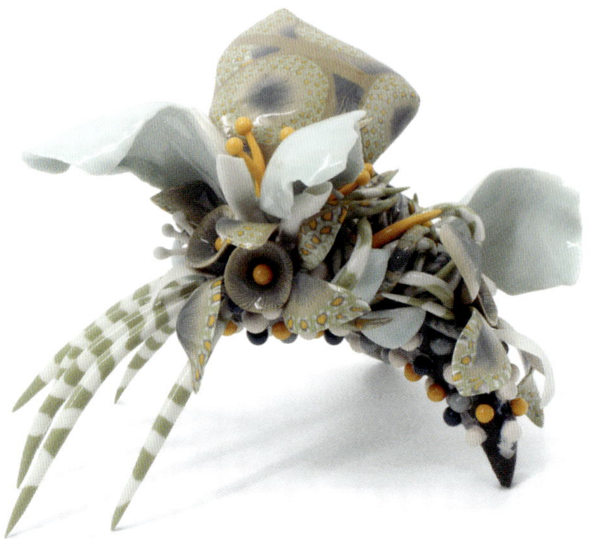

Caterpillar is dreaming series 4, 2022. Height: 6⅓ inches, width: 8⅔ inches, depth: 5½ inches (16 x 22 x 14 cm)

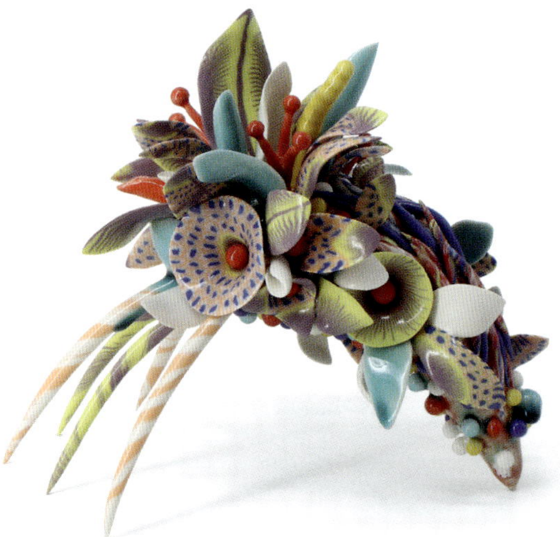

Caterpillar is dreaming series 5, 2022. Height: 7⅘ inches, width: 9 inches, depth: 6⅓ inches (20 x 23 x 16 cm)

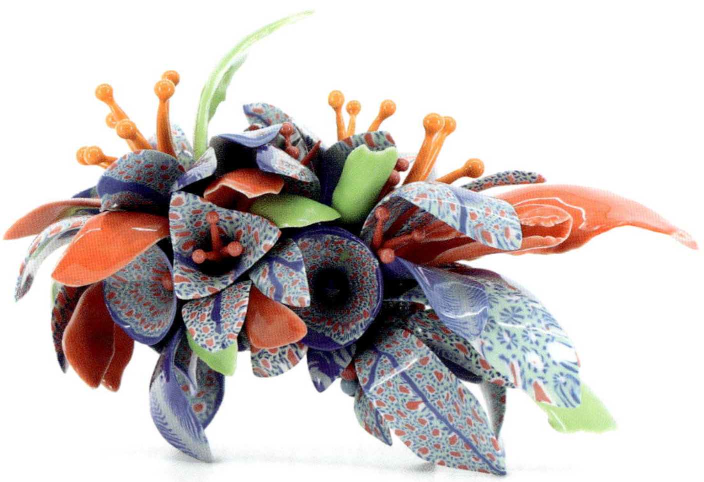

Caterpillar is dreaming series 6, 2022. Height: 6⅓ inches, width: 9 inches, depth: 5⅛ inches (16 x 23 x 13 cm)

Jongjin Park, South Korea

The fascinating and beautiful sculptures of Jongjin Park lie outside of the traditional definitions of nerikomi and neriage. While still using colored porcelain he has developed a completely new method of working with clay. The distinctive multilayered forms suggest something quite fragile, although the opposite is the case. The unexpected material he uses is common paper towels. Differently colored porcelain slips are brushed onto the paper sheets in a repetitive manner, creating stacks of sheets, sometimes in the many hundreds. While being fired to 1280°C (2336°F) the paper burns out leaving only the paper-thin layers of slip which often bond together while remaining delicate and fragile to the eye.

Park began his study in ceramics in 2000 at Kookmin University, receiving his BFA and MFA degrees in 2007 and 2009. This was followed by work as a researcher at the National Museum of Korea where he gained a wide understanding and experience of Korean traditions. He then obtained a second Master's at Cardiff Metropolitan University in Wales. It was there that Park was challenged to break away from his traditional training and open himself up to unexpected influences. Upon seeing an exhibit of ancient jewelry in the National Museum of Wales he began exploring the ideas of materiality and tension.

ARTIST'S STATEMENT

Specifically, I was drawn to the glossy surface and cold mood created by metal in perfect balance with the warm preciousness of gold . . . In the twisted metal shapes – created by traditional techniques of hammering, bending and fusing – I was able to realize an inherent property of the material. From then on, I started to be more experimental with materiality.

While experimenting with various physical properties during the course of study, a stratigraphic image of the coast of Cardiff emerged from a pile of paper that piled up for recycling. I came to think again about the various experiences and the scale of time as an international student and the power it can have. I thought that if I could reproduce this figure in ceramics, it would be a form that could express my present self, and I came up with the "Artistic Stratum" series through various experiments.

This seed of an idea was developed and modified over four years through an extensive series of experiments with his new technique. In addition to the steps outlined above, Park discovered that he could use electric tools to cut and grind the fired shapes to modify their overall forms. Although the systematic stacking of slip-coated paper into geometric piles seems calculated, Park admits to other factors which can influence his methods: "The process of my work is about layering time as formative expression. Working becomes a form of meditation. It's just me and my thoughts. Sometimes, dramatic emotional shifts are manifested in strong color choices or intentionally fabricated gaps between layers."

After returning from Wales Park began teaching at Seoul Women's University where he was appointed assistant professor of crafts in 2021. He has also recently completed his PhD in the Department of Ceramics, Graduate School of Techno Design, Kookmin University in Seoul.

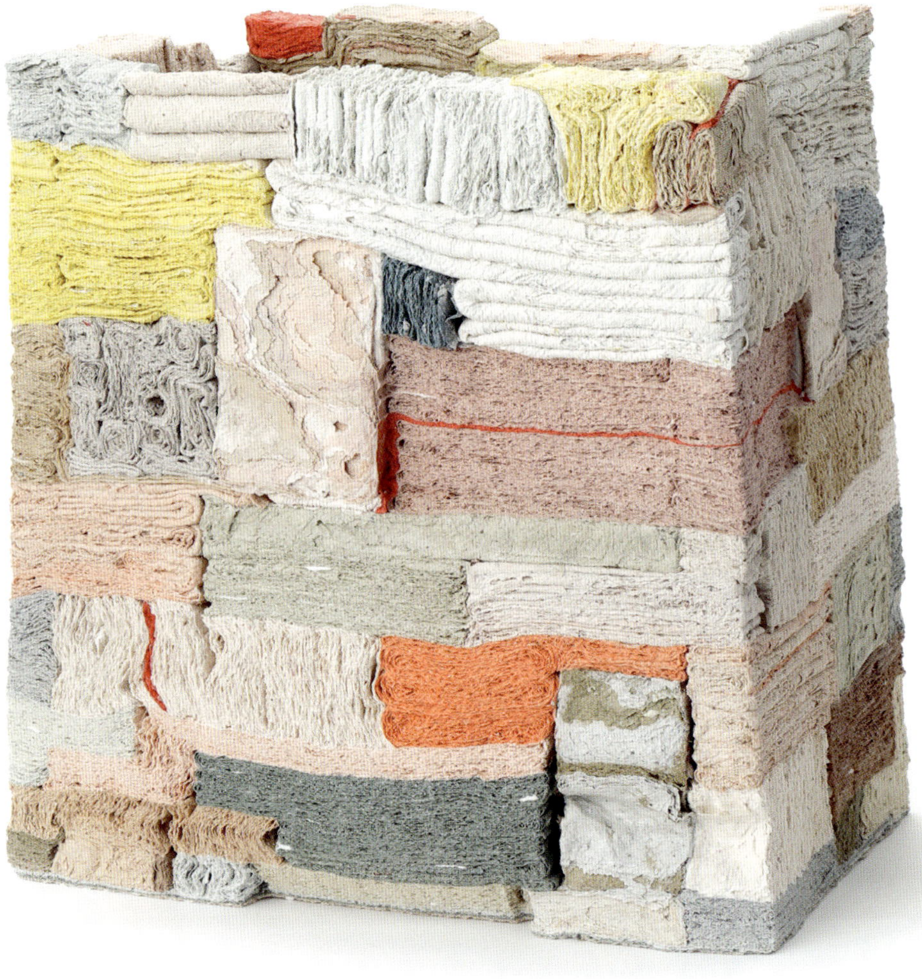

Artistic Stratum – Patch Vessel #2, 2021. Height: 13⅓ inches, width: 11⅖ inches, depth: 9 inches (34 x 29 x 23 cm)
Image: © Jongjin Park; photo by Gumhye Ye

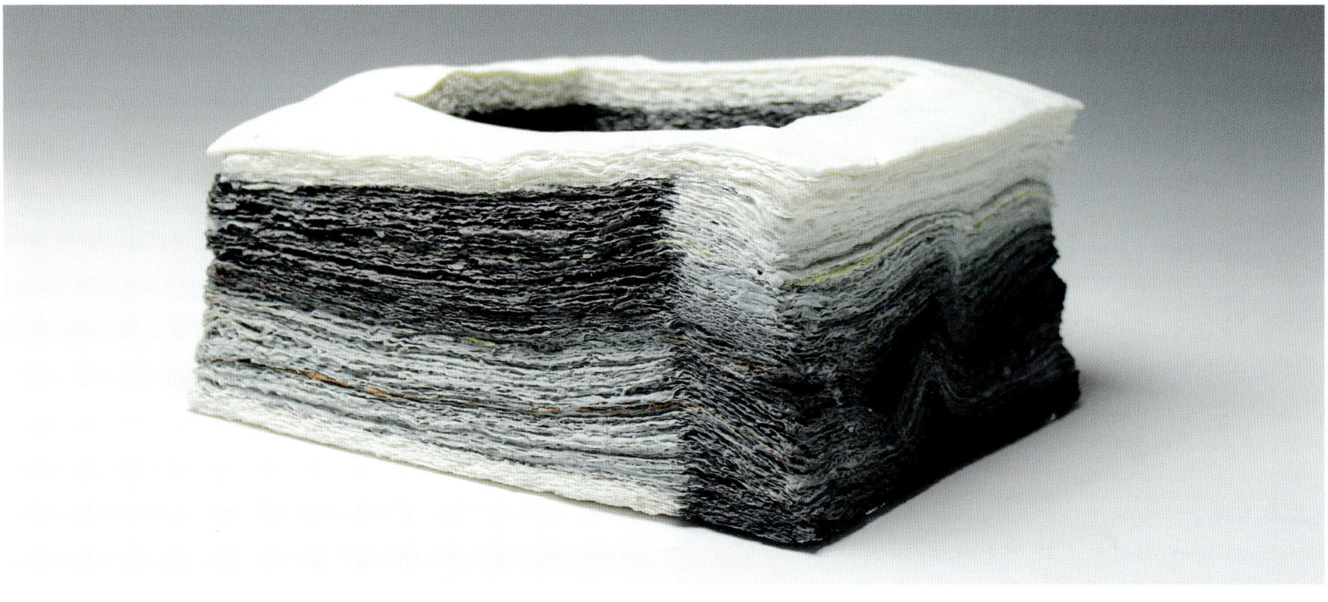

Artistic Stratum – Horizontal Vessel #1, 2014. Height: 5½ inches, width: 7⅛ inches, depth: 7⅛ inches (14 x 18 x 18 cm)
Image: © Jongjin Park

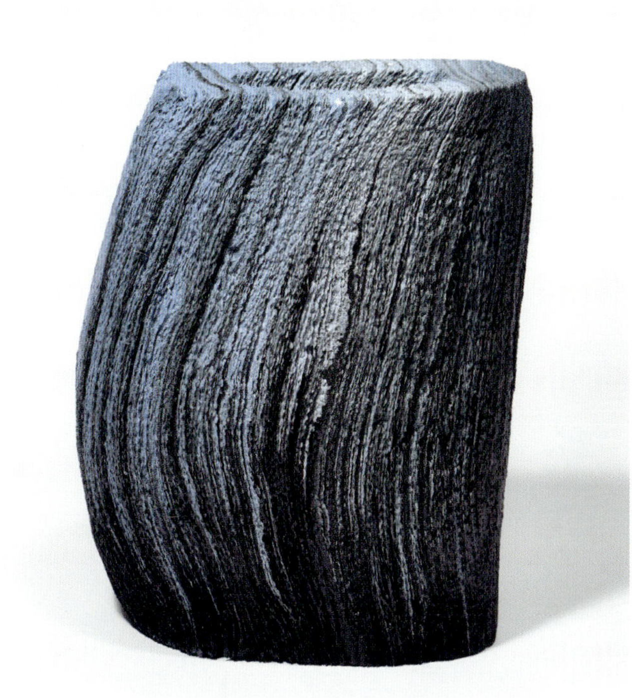

Artistic Stratum – Vertical Vessel #1, 2018. Height: 8⅔ inches, width: 7⅛ inches, depth: 4¾ inches (22 x 18 x 12 cm) *Image: © Jongjin Park*

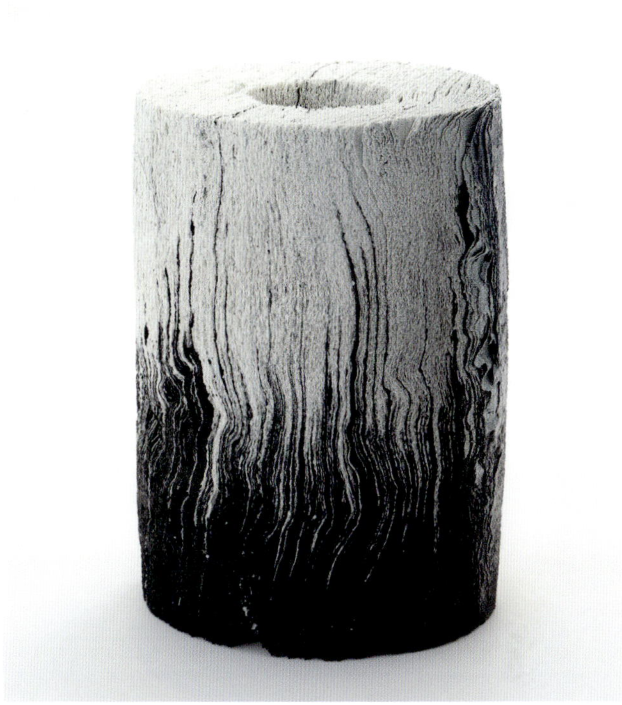

Artistic Stratum – Vertical Vessel #2, 2021. Height: 12½ inches, width: 8⅓ inches, depth: 7⅘ inches (32 x 21 x 20 cm) *Image: © Jongjin Park; photo by Gumhye Ye*

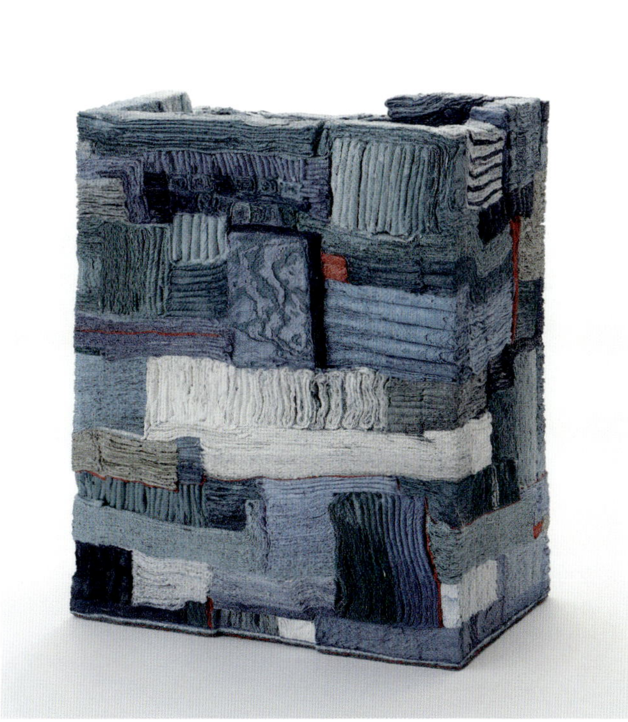

Artistic Stratum – Patch Vessel #1, 2021. Height: 15¾ inches, width: 11⅘ inches, depth: 8⅓ inches (40 x 30 x 21 cm) *Image: © Jongjin Park; photo by Gumhye Ye*

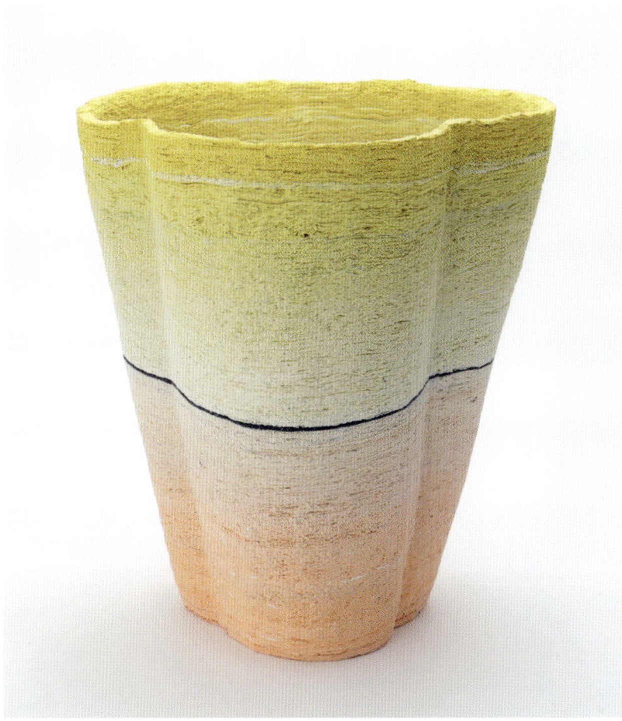

Artistic Stratum – Vertical Vessel #3, 2022. Height: 14 inches, width: 13 inches, depth: 11⅘ inches (36 x 33 x 30 cm) *Image: © Jongjin Park; photo by Gumhye Ye*

Jongjin Park's making process for *Artistic Stratum – Vertical Vessel.*

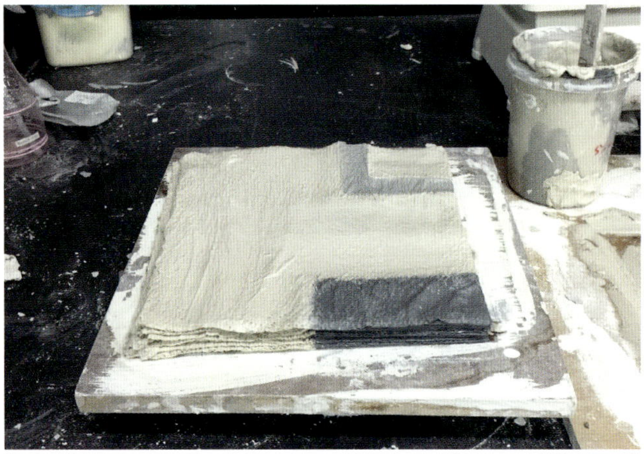

Apply slip over the whole surface for the bottom part.

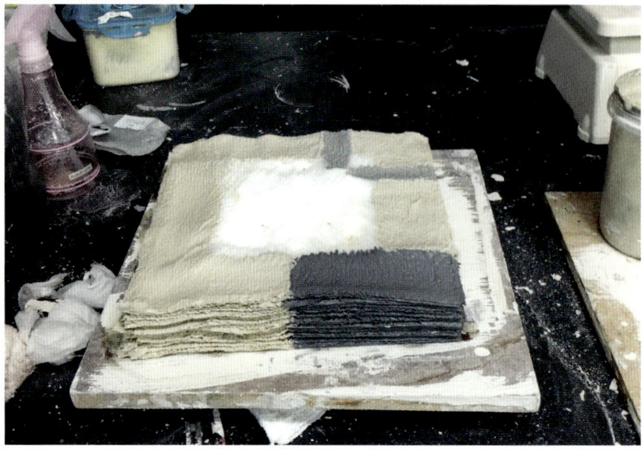

After that, apply only to the edge of the part that becomes the wall of the vessel shape (the white square in the center is the unpainted paper towel).

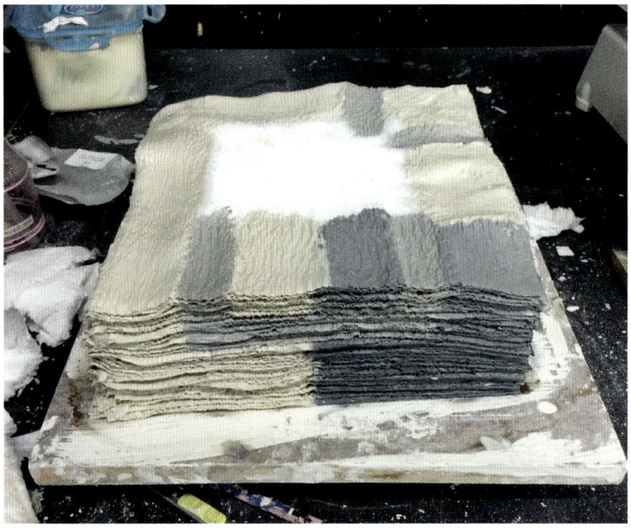

Paint color slip on each layer according to the design.

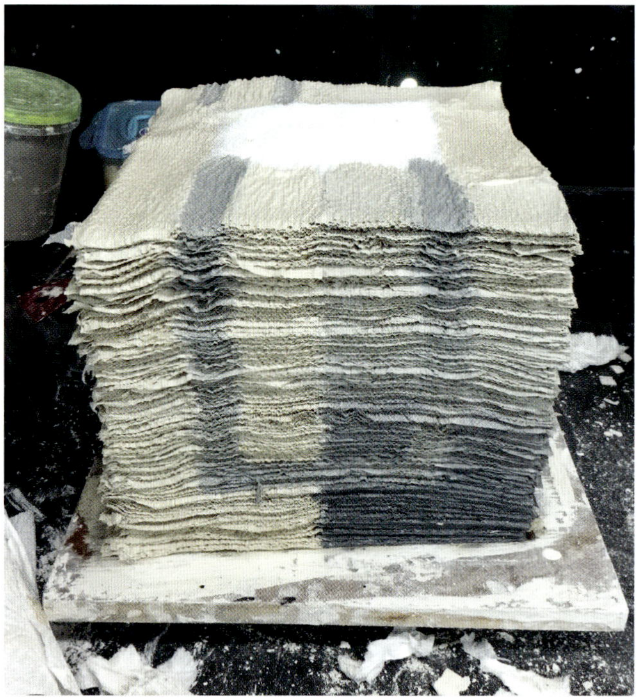

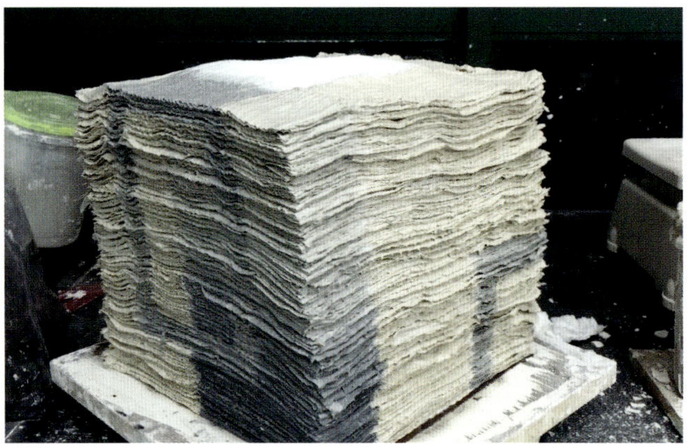

Keep building up layers until your complete form has taken shape. Consider while building the form that there will be 30–40 percent shrinkage after firing. After drying, fire to a high temperature (1250C°) to make hard porcelain, and then use a power tool to grind and carve the inside and outside. Then spray transparent glaze on the surface and fire a second time at the same temperature as the first time – or slightly lower – so that no further shape change occurs.

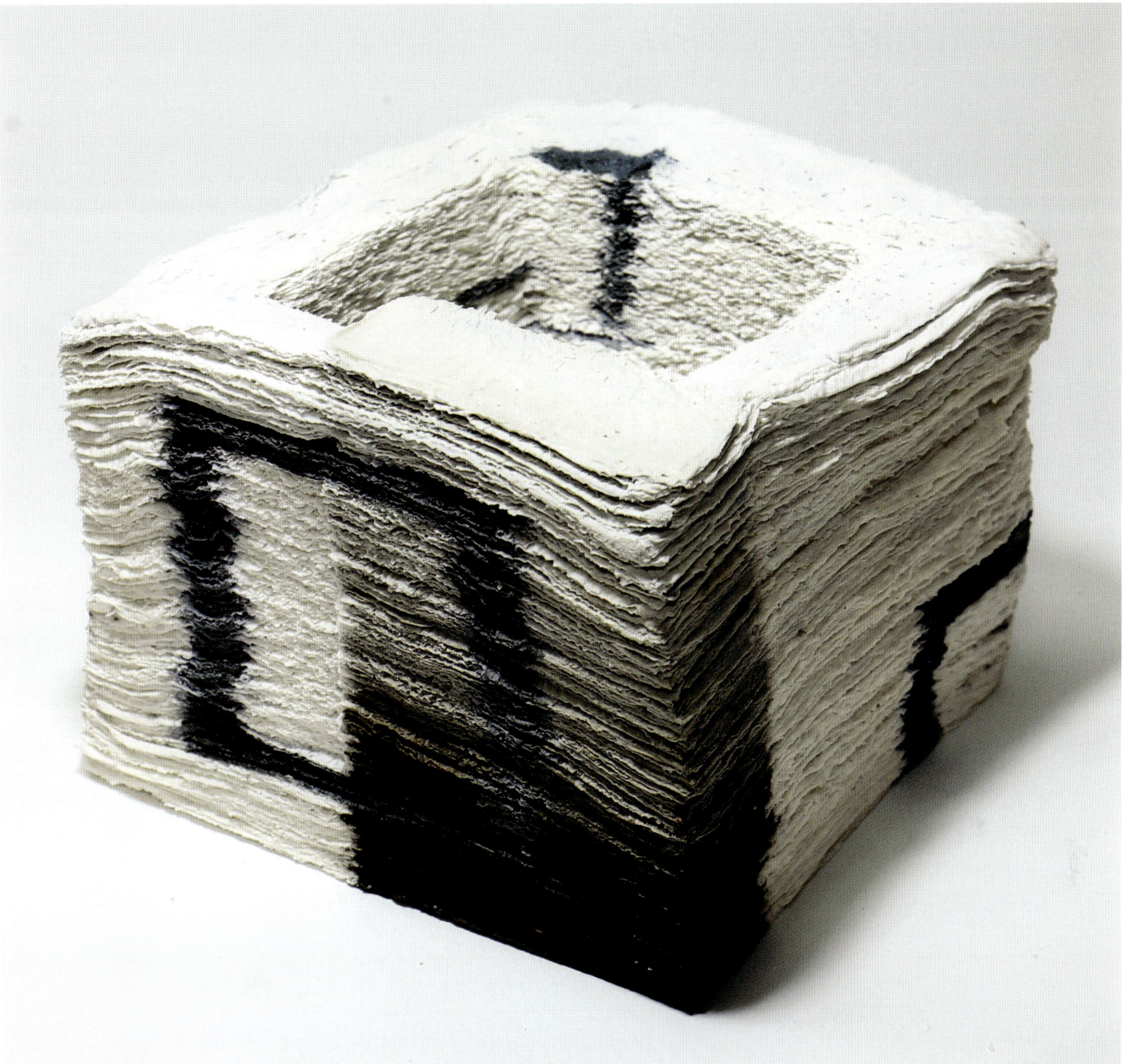

AWARDS AND EXHIBITIONS

Jongjin Park has had several solo exhibitions of his work in both London, England, and Seoul, South Korea. His group exhibitions include the Salon Art + Design Art Fair, New York; Ceramic Now, Paris; Collect, London; as well as numerous others in Spain, Italy, France, Lebanon, Greece, Japan, China, Taiwan, Singapore, Thailand, the UK, Hungary, Germany, Australia, the Netherlands, Belgium, and the USA. He is also the recipient of numerous international and national awards and prizes including the Grand Prize – The 1st Gwangju White Porcelain Competition, 2008; the Gold prize – The 7th Cheongju International Craft Competition, 2011; and Winner – Lexus Creative Masters Award, 2018. His work is included in the collections of the GYEONGGI Ceramic Museum and the Icheon World Ceramic EXPO Center, both in South Korea, among others. In addition, Park's sculptures have been written about in a long list of publications.

David Pottinger, Australia

Well known among Australia's ceramics royalty is David Pottinger, born in Queensland and now living and working in Melbourne. After initial training at Kelvin Grove College, David Pottinger attained a Master's Degree of Visual Arts at the Melbourne Institute of Technology. Early in his career he established a recognizable style which was, and still is, unique in the world of nerikomi ceramics. Among the wide spectrum of colors available to stain white porcelain clay Pottinger has chosen subdued, subtle, and quite gorgeous palettes with which to create his forms.

The highly detailed patterns often cover the entire piece, although "cover" isn't an accurate term here since the patterns and colors permeate the thickness of the walls of his vessels – the very definition of nerikomi. The initial stacking and re-stacking of his loaves appears to be slightly random at times, given the specific palette with which he has chosen to work. Ultimately this randomness is tamed but still shows through, reminding us of nature and her infinite expressions. Slicing through these stacked layers, at a variety of angles, is followed by re-stacking of those slices in sometimes complex configurations which yield the compelling yet quiet designs seen in his work. The use of small-scale abstract mirror images is one of his signature techniques.

Colin Batrouney, writing of Pottinger's work states:

> *This patterning is in no way regular, but rather governed by the technique of nerikomi itself. This technique will often see the random tug, pull and drag of patterns around vessels as the clay shifts in response to the layering and joinery process. This layering and patterning allows a deep contemplation of the finished vessels and suggests that the forms are constantly shifting through a process of plaiting, twining and weaving in, out and through the pots.*

Time spent in nature has heavily influenced Pottinger who has mentioned studying the growth rings on trees, tidal lines on beaches, sedimentary rock formations, as well as how time makes an object. His color choices reflect this interest and, all being derived from the earth's oxides, exhibit the muted, subtle yet complex shades of nature. Other influences on his work have been Ikat weaving and basketry and glass techniques.

ARTIST'S STATEMENT

My work is an exploration of form through the interplay of color, tension, rhythm, and pattern in porcelain, utilizing the ancient technique of nerikomi . . . My work is inspired by the palette of nature, the colors of the sky and trees, the water and the earth, the rhythms and shifts of the winds as the seasons change from winter to spring. This work is also suggestive of the paradox of nature: constantly shifting, moving, growing, and dying and yet remaining complete and whole.

The complexity of the vessels invites contemplation akin to the reading of a Zen tea bowl before drinking from it. The work strives for a deeper connection from the hand of the maker to the completed forms. To quote Sōetsu Yanagi from *The Unknown Craftsman*, "The maker of a pattern draws the essence of the thing seen with his own heartbeat, life to life."

Pottinger's vessels are made from Cool Ice Porcelain from Melbourne's Clayworks with the addition of 10 percent paper fibers. It is colored with a variety of metal oxides including iron, cobalt, chromium and magnesium. He joins slices from his highly patterned and multilayered pattern blocks to form a large sheet which is compressed and formed into a cylinder. This is then paddled into shape at the leather-hard stage. Pieces are wrapped with a lightweight paper and dried slowly for several weeks, after which they are sanded and bisque fired. After a light wet-sanding, a thin coat of transparent glaze is applied to the interior only, followed by a glaze firing (to 1220°C [2228°F]) with final sanding and polishing post firing.

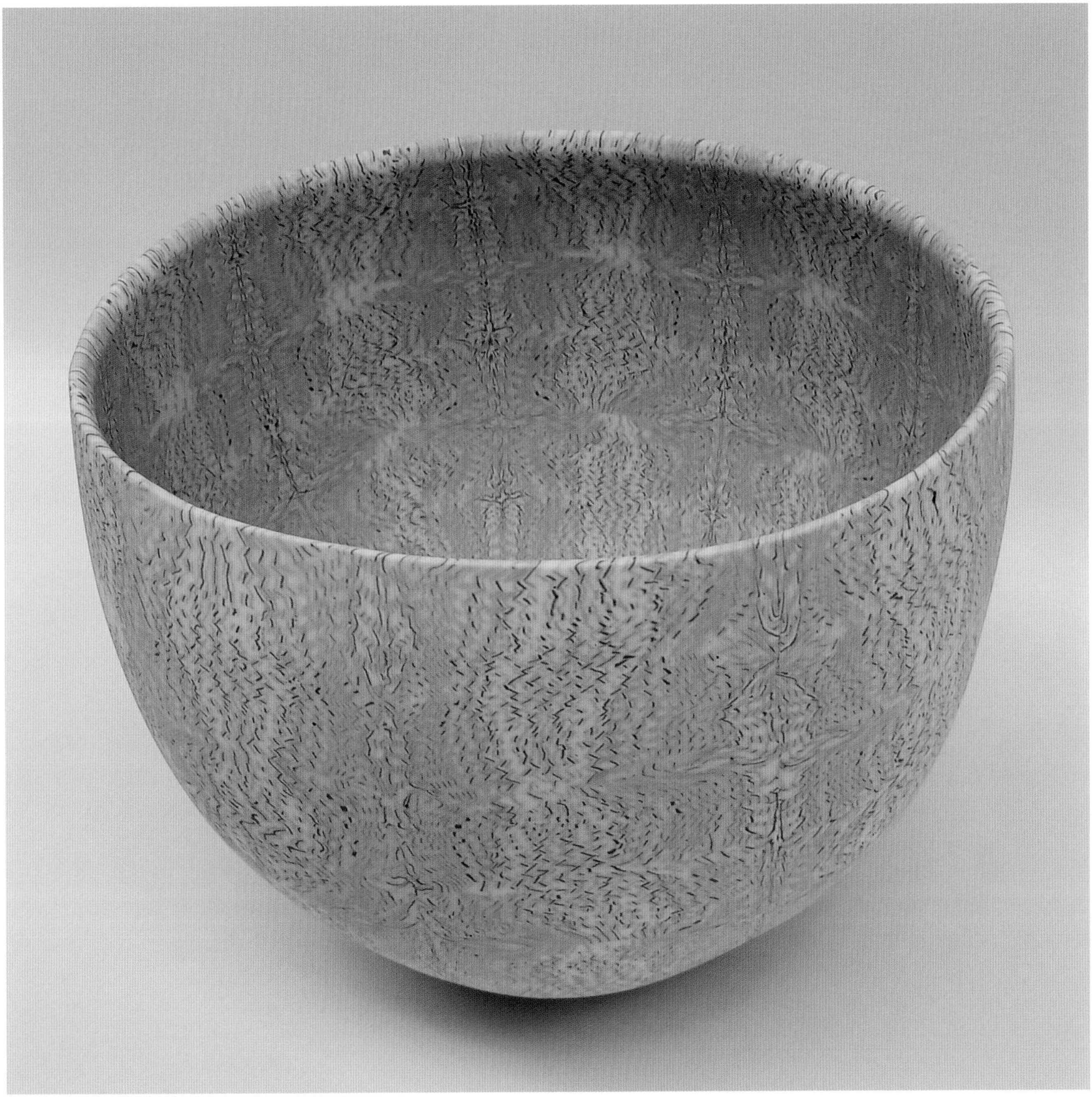

***Untitled Bowl*, 2019. Height: 6⅞ inches, width: 8⅓ inches (17.5 x 21 cm)** *Image: © David Pottinger*

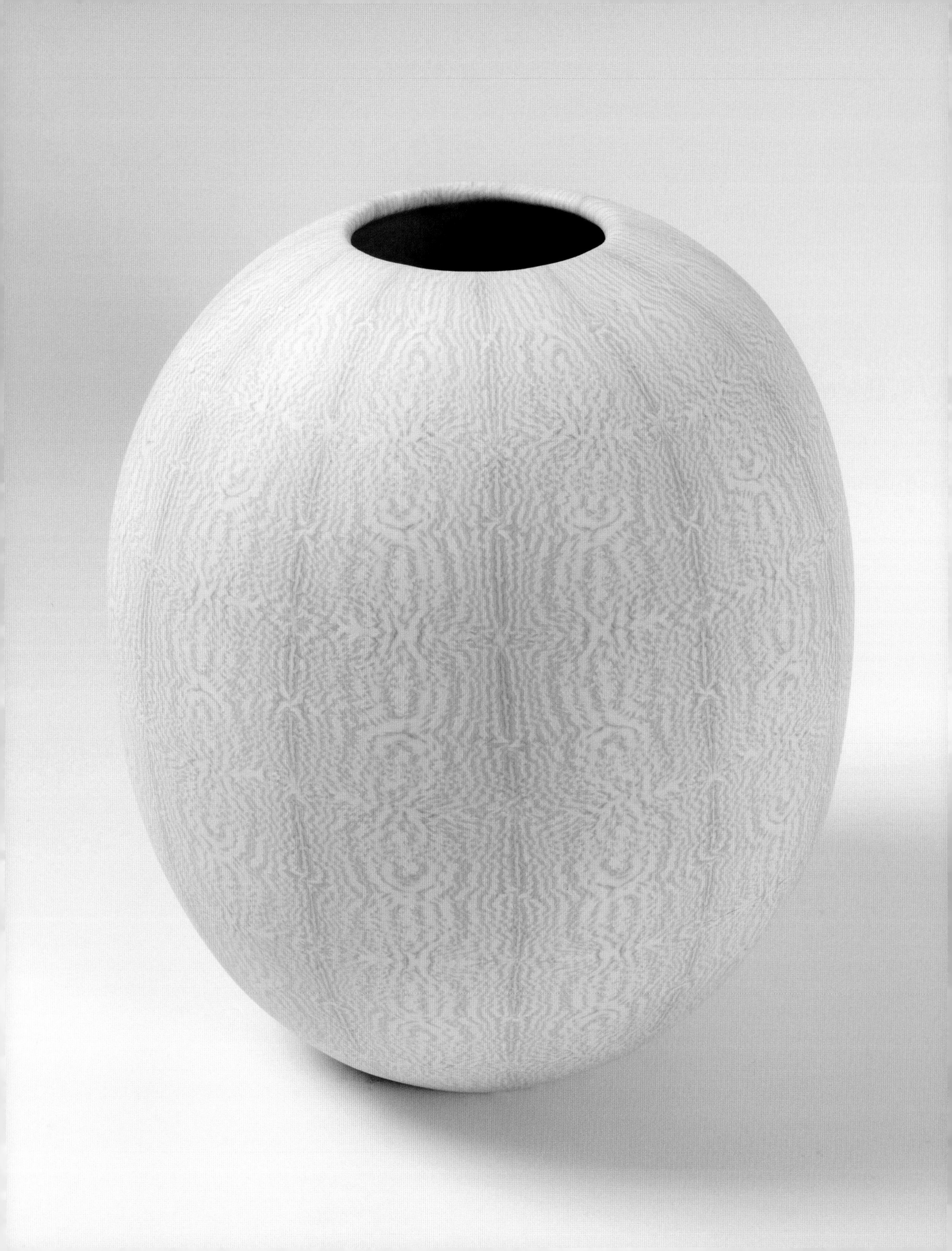

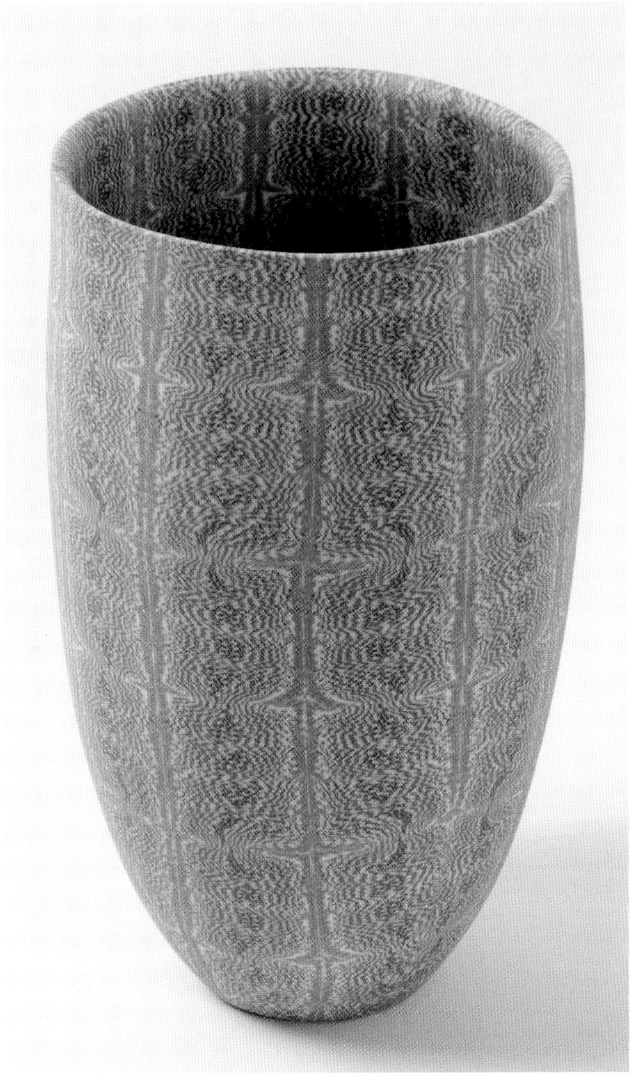

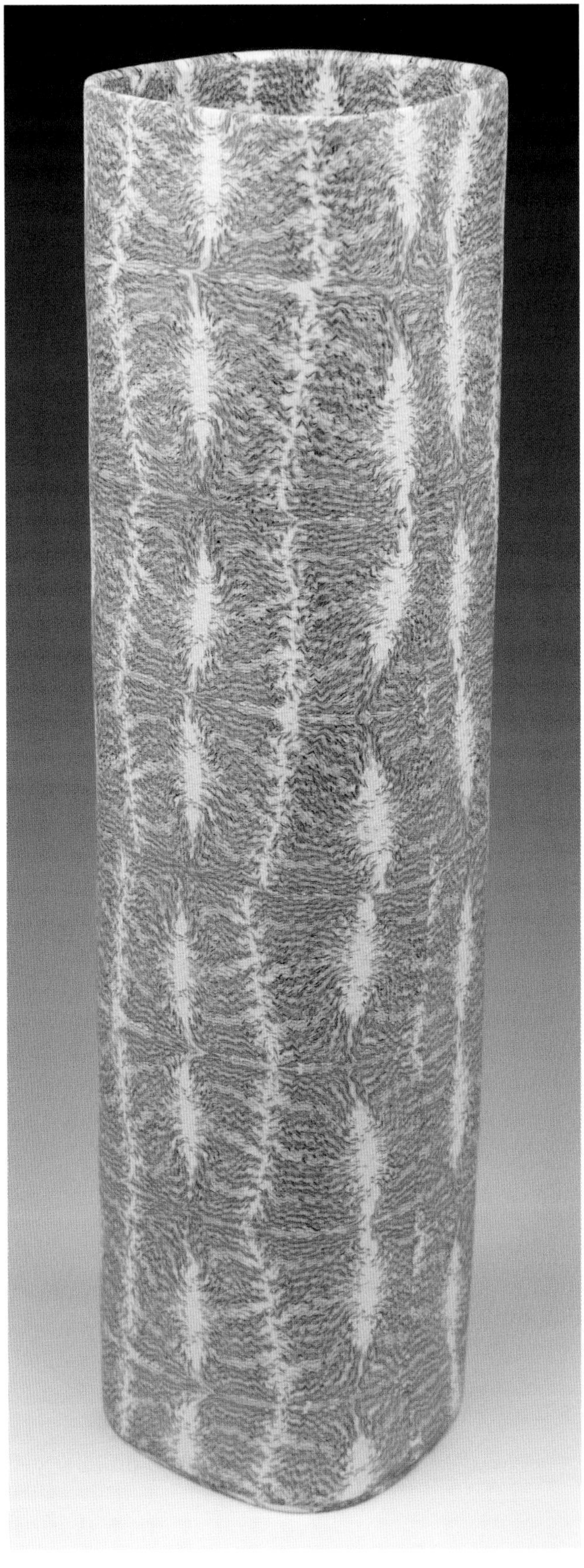

▲ *Vessel,* 2021. Height: 8⅘ inches, width: 5¾ inches (22.5 x 14.5 cm) *All images on pages 148–50 © David Pottinger; photo by Tim Gresham*

▶ *Totem,* 2019. Height: 13¾ inches, width: 4 inches (35 x 10 cm)

◀ *Blossom Jar 2*, 2022. Height: 8⅔ inches, width: 9½ inches (22 x 24 cm)

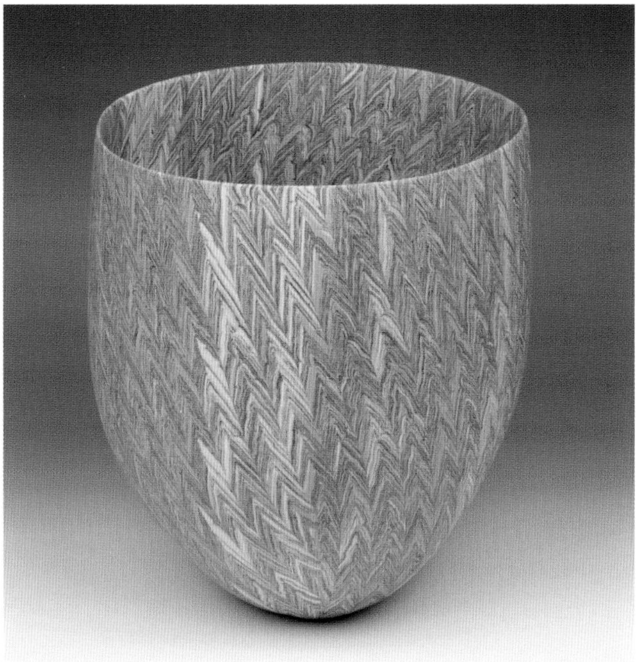

Bowl #2, 2016. Height: 11⅘ inches, width: 9 inches (30 x 23 cm)

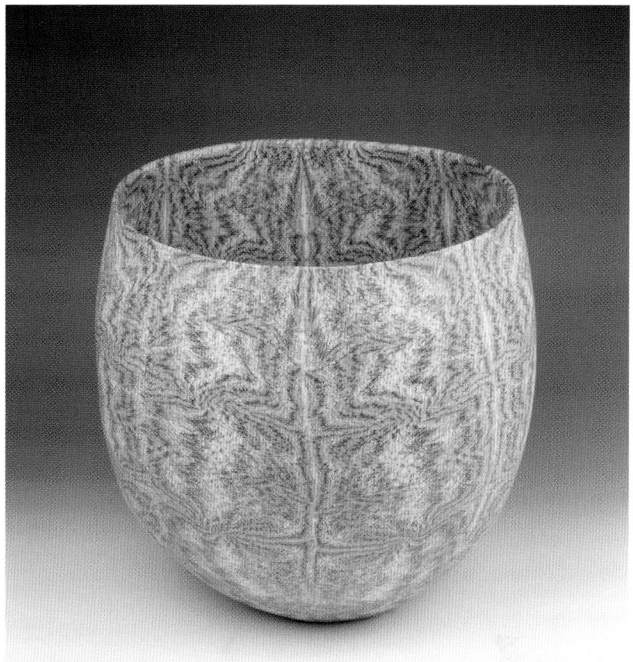

Bowl #1, 2016. Height: 15 inches, width: 13¾ inches (38 x 35 cm)

AWARDS AND EXHIBITIONS

David Pottinger's work has been shown in over 15 solo exhibitions including recent shows at Craft Victoria and Gallery Smith, both in Melbourne, and two at the Robin Gibson Gallery in Sydney. Among the many group exhibitions in which he was invited to participate are SOFA Chicago, 2007; the "Exhibition of Australian Ceramics," Tribute to Janet Mansfield OAM 1934–2013, Mossgreen Gallery, Melbourne; "Collect Exhibition," the Victoria and Albert Museum, London, 2008; "Porcelain," Scottish Gallery, Edinburgh, Scotland; "Voices," Skepsi Gallery, Malvern; and "Collect 2011," Scottish Gallery at Sacchi & Sacchi Gallery, London. Pottinger also received an Honorable Mention at the 2021 International Ceramics Competition Mino, Japan. Pottinger's work has been reviewed in many publications and is included in the permanent collections of the National Gallery of Victoria, Melbourne; the National Gallery of Australia, Canberra; and the Shepparton Art Museum, Victoria.

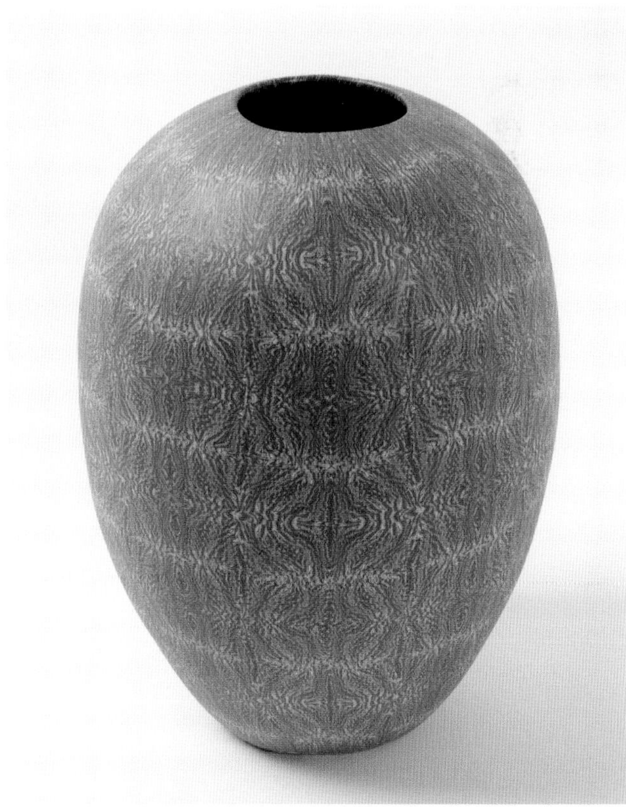

Blossom Jar 1, 2022. Height: 9⅔ inches, width: 6¾ inches (24.5 x 17 cm)

Kanako Sahashi, Kenya

Another quite unique artist in this group is self-taught potter Kanako Sahashi. She has overcome many obstacles that most have not encountered and her work is all the stronger for it. Living in a country where clay, glazes, and even tools are not readily available has actually connected her with the worldwide community of early potters whose work fills our art historical museums and provides us all with the traditions and aesthetics that comprise the foundation of our common ceramics heritage.

Sahashi grew up in Japan in a home where she was surrounded by her parents' collection of antiques and pottery and developed an interest in ceramics. She learned about the nerikomi technique later and was fascinated to try it herself but only found the opportunity to do this after moving to Kenya, where she was able to study for a few months with Kenyan potter Waithira Chege. There she learned the basics of ceramics and later, after setting up her own studio in Nairobi, began to teach herself the nerikomi technique, which she had longed to do.

After working with traditional clay bodies Sahashi discovered clay from the soil at the foot of Mount Kenya. Used alone for its natural color as well as with Japanese stains, which when added provide wonderfully muted tones, she has created a body of work that is beautiful and unique. The naturally groggy texture of this Kenyan clay must also be noticed. The vast majority of the work currently being done in the nerikomi/neriage world is created with smooth fine-grained clay bodies. Sahashi has added this local clay's appealing natural texture to her work which also sets it apart.

ARTIST'S STATEMENT

Working with the nerikomi technique is a meditative act. I am fascinated by unique patterns that come from co-creating my calculations and clay's fluctuation and by the difficulty of shaping the slabs into vessels that fit daily life.

I get inspiration from the scenery I encounter in the city. I am particularly attracted to the overlap of nature and manmade artifacts, such as ivy-covered exterior tiles and iron doors, or the harmony of plants and people's geometric clothes on the street. I decide on colors, patterns, and shapes based on my imagination of how they will fit in the rooms of my close friends. I also consider what patterns can be expressed only with the nerikomi technique.

It is a miracle and a great pleasure for me to make clay from the original soil at the foot of Mount Kenya, create patterns, and finish the vessels with my own hands. I am on a journey to find patterns and shapes that take advantage of the colors and textures of the local soil.

My contact with the clay is fleeting in the context of the universe's workings. Still, the possibility that my vessels will exist somewhere in the world even after I am gone is a great driving force for me to make vessels.

In traditional nerikomi style Sahashi creates pattern blocks from which she cuts slices which are then added to one another, forming a larger sheet. This is then wrapped around plaster or bisque-fired molds to create the initial vessel shapes. She then uses a hand wheel to gradually modify the form into its final shape. Once completely dry she cleans the surface with steel wool. After a bisque firing, it is cleaned again with sandpaper, glazed – for vases often inside only – and high fired in an electric kiln to 1240°C (2264°F) – cone 7.

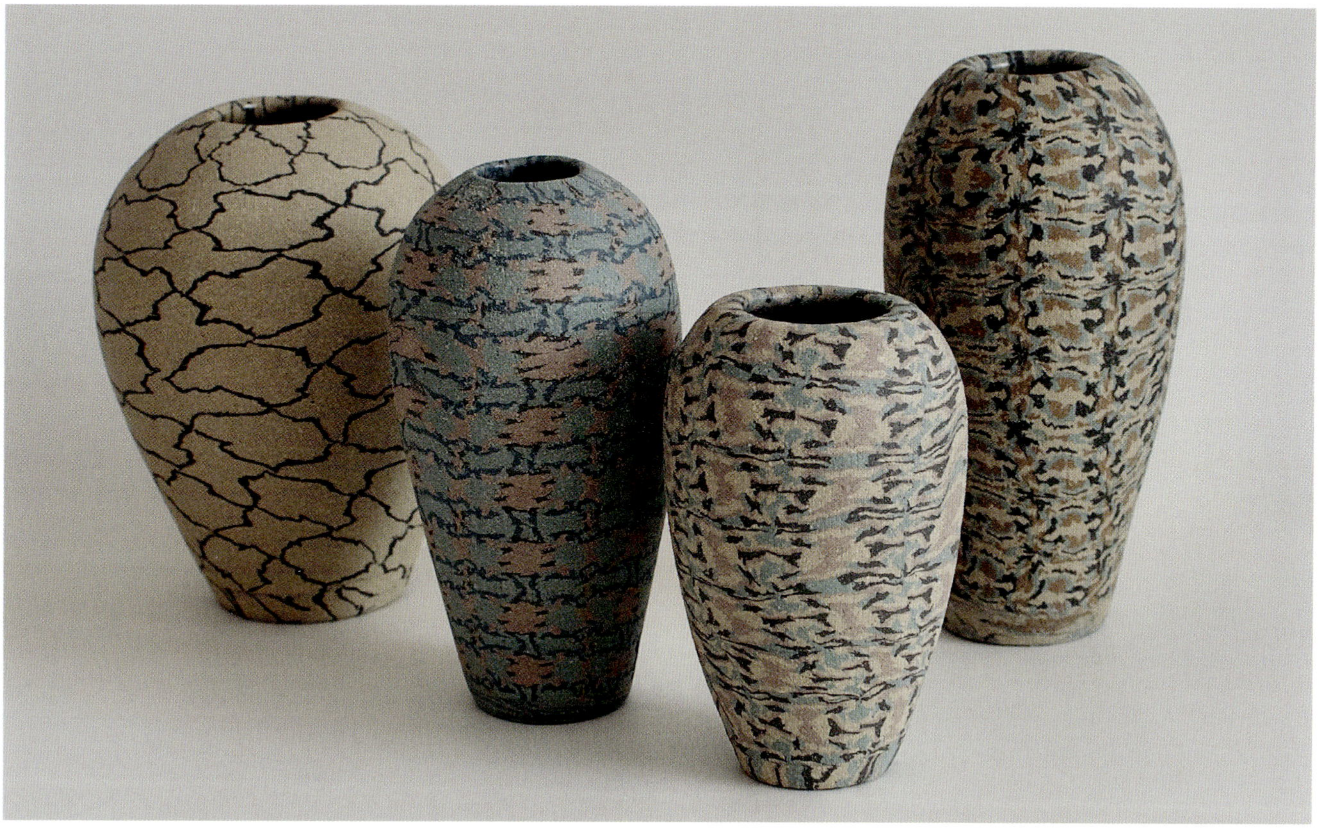

untitled #4, 2022. From left to right: height: 5½ inches, diameter: 4⅛ inches (14 x 10.5 cm); height: 5½ inches, diameter: 2¾ inches (14 x 7 cm); height: 4½ inches, diameter: 2¾ inches (11.5 x 7 cm); height: 6 inches, diameter: 3⅛ inches (15.5 x 8 cm) *All images: © Kanako Sahashi*

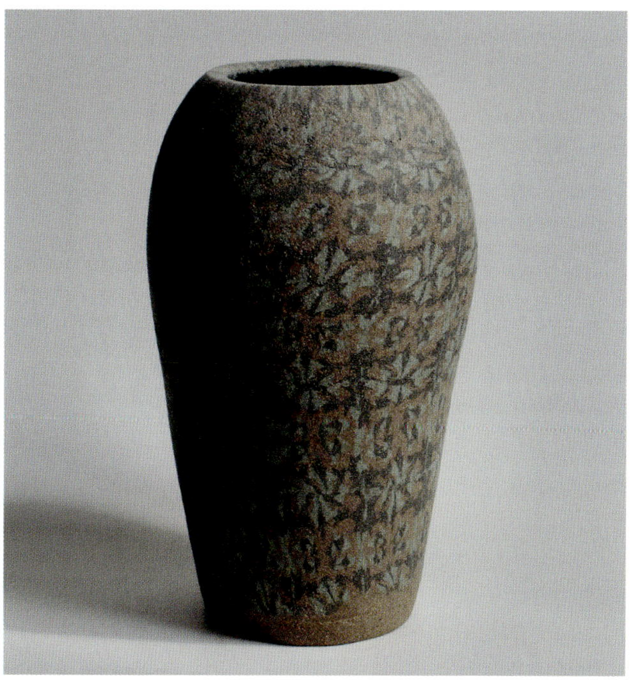

untitled #1, 2022. Height: 7½ inches, diameter: 4 inches (19 x 10 cm)

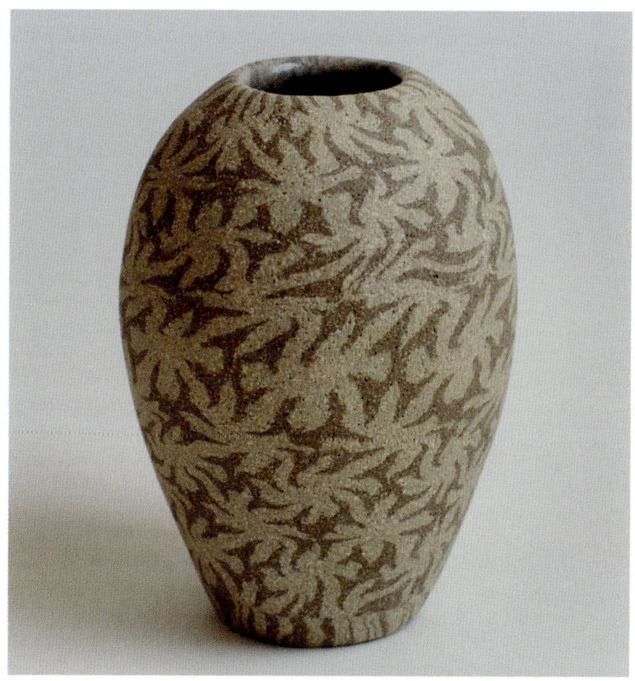

untitled #3, 2022. Height: 4½ inches, diameter: 3 inches (11.5 x 7.5 cm)

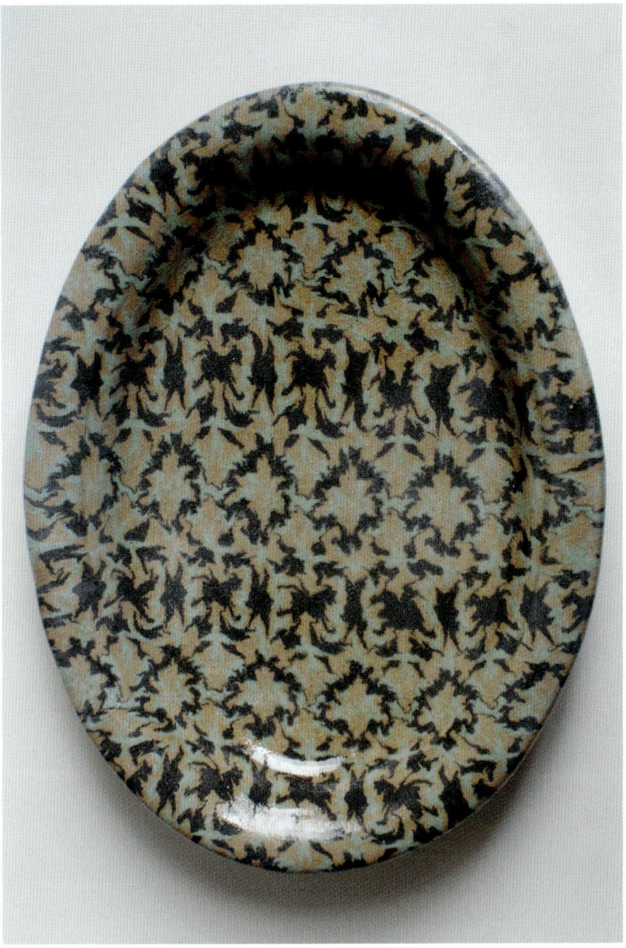

untitled #5, 2022. Height: 1⅛ inches, width: 5¾ inches, length: 10⅗ inches (3 x 14.5 x 19.5 cm)

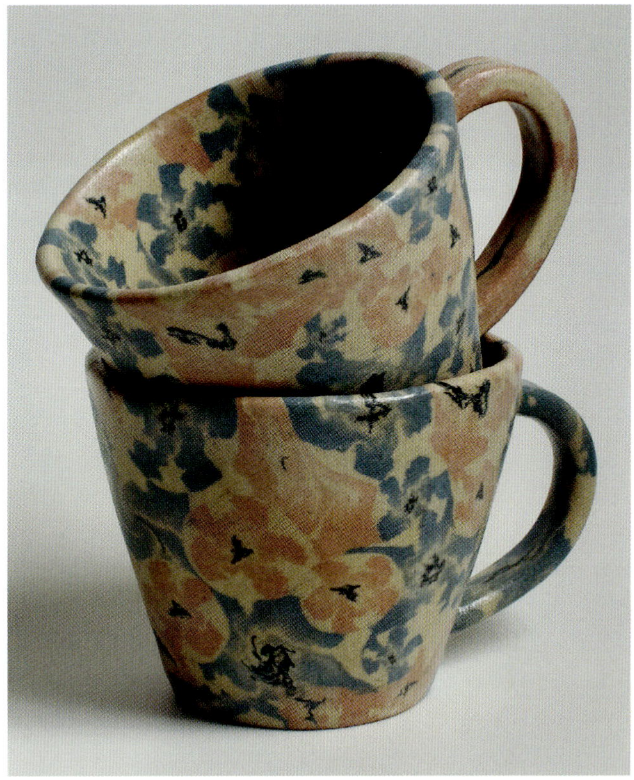

untitled #6, 2022. Height: 3 inches, diameter: 3½ inches (7.5 x 9 cm)

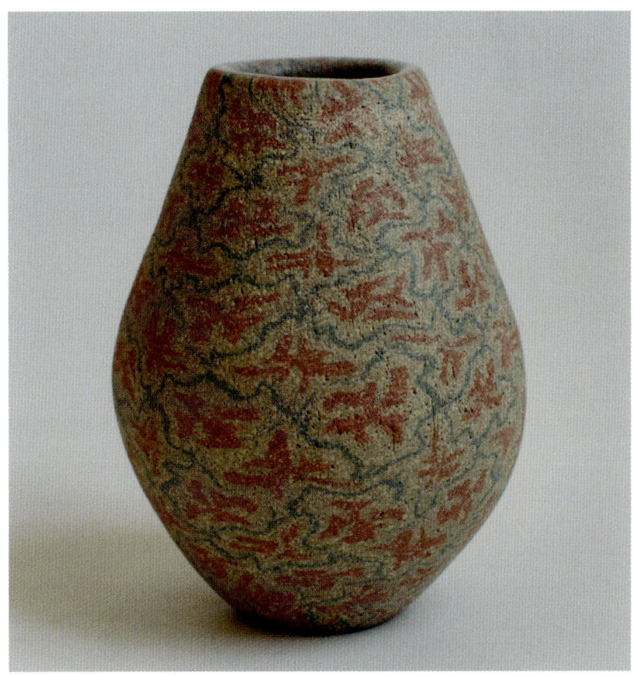

untitled #2, 2022. Height: 4 inches, diameter: 2¾ inches (10 x 7 cm)

AWARDS AND EXHIBITIONS

Kanako Sahashi only started making work in 2019 and held a private solo show for acquaintances in Japan in 2021. Her work is carried locally in Kenya and she looks forward to exhibiting more widely in the future.

Lorraine Shemesh, USA

The boldly patterned neriage and nerikomi vessels by Lorraine Shemesh, when first encountered, stand out and announce their presence in a way that is unique and intriguing. Their back story is unique as well. Shemesh is better known as a painter, getting her start at Boston University (BFA in Painting, Magna Cum Laude 1971) followed by the Tyler School of Art in Rome (1971, 1972) and an MFA from the Tyler School of Art in Elkins Park, Pennsylvania (1973). Numerous grants and awards soon followed and in addition to her painting Shemesh also became a professor at both the Rhode Island School of Design (RISD) and Amherst College.

While teaching at RISD she encountered the Japanese technique of nerikomi in the ceramics department. Years later Shemesh studied ceramics in a communal studio at Parsons School of Design in New York. Later she had two residencies in Maine and she served as artist-in-residence at Dartmouth College in New Hampshire where she focused on ceramics. Since geometric patterns played a prominent role in her paintings, both in the costumes worn by the dancers and swimmers, it is no surprise that she has also been inspired to explore the same motifs in colored clay.

ARTIST'S STATEMENT

Early on, I was trained as a painter and drawn to water as a subject because of its myriad of patterns in constant flux. I conceived figure compositions from the inside out, a result of deeply coming to terms with the architecture of the human body. As the paintings developed, they became increasingly involved in a dialogue between figuration and abstraction induced by my first experiences with ceramics.

Pattern, whether it is planned or improvised, is memorable because it has the capacity to transform things. The process of developing various repeated patterns in order to create movement, rhythm, and harmony, presents the possibility of a visual metamorphosis in the work at hand. Initially, I was attracted to patterns that occur in nature, striated light patterns, and both geometric and improvisational shapes found in African Kuba cloth.

When I began to work with clay, I was attracted to the fact that the process employs all the elements: earth, air, fire, and water. The tactile experience of holding unfamiliar materials in my hands shifted the way I was thinking and painting at the time. It became clear to me that I did not want to create pattern only on the surface of each piece as an afterthought. I chose, instead, to build vessel forms employing light and dark clay bodies which are coiled and then woven or twisted together to create repeated patterns. In this manner, the two colors are completely integrated into the interior and exterior of the vessel. The pieces are, in part, inspired by layered rock formations in nature, whose various colored stripes are the earth's memory and witness to the changes in the environment that have occurred over time.

Then there is the emotional component to the weaving together of the light and dark of the human experience. Clay has a memory, is easily broken, sensitive to touch and temperature, capable of fragility and strength, and greatly affected by the atmosphere in which it finds itself. Much the same thing can be said for the body and the ephemeral nature of life itself.

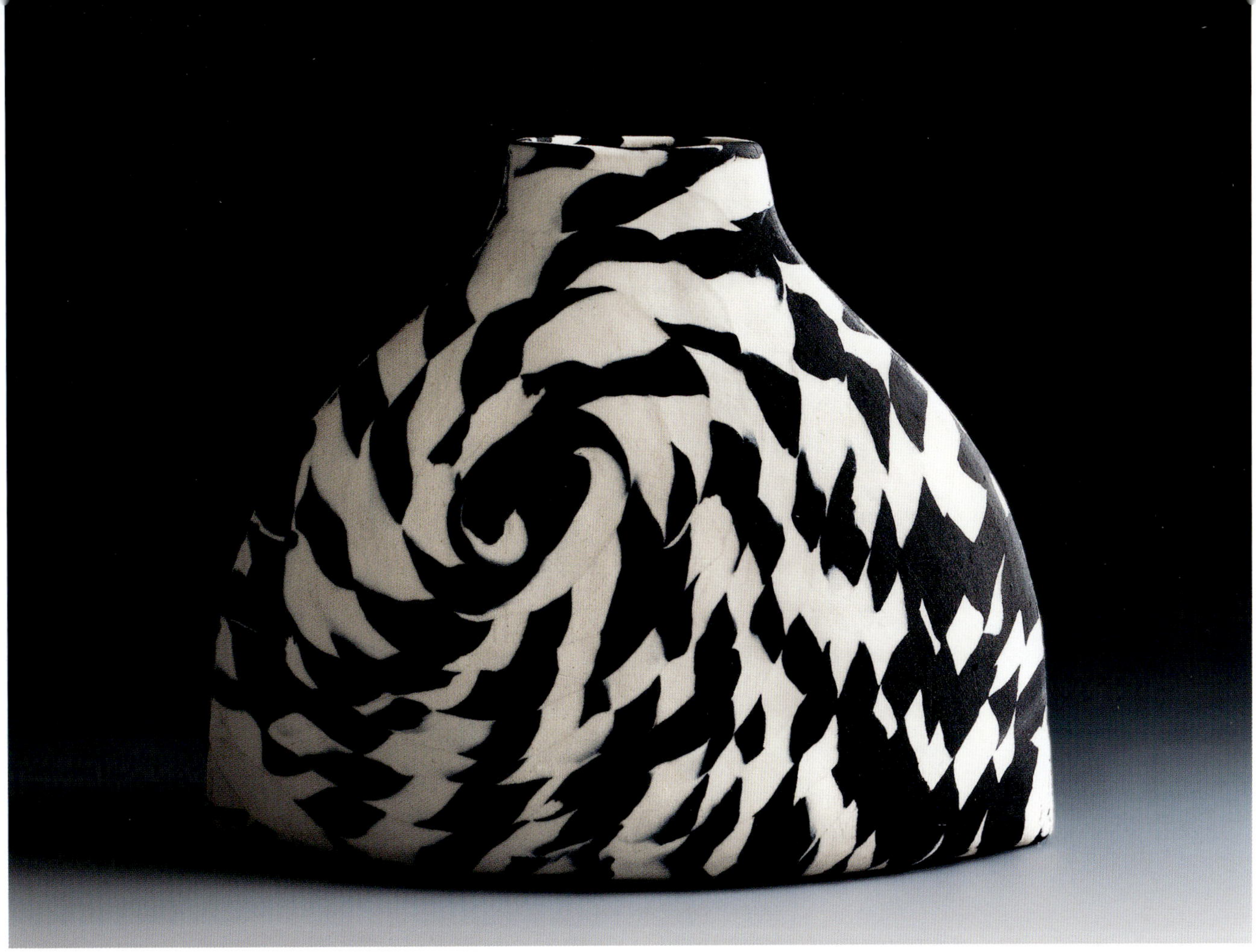

Nerikomi bottle, 2021. Height: 8½ inches, width: 10¼ inches, depth: 3¼ inches (21.6 x 26 x 8.2 cm)

Black-and-white wide twist nerikomi bottle, porcelain. *All images: © Lorraine Shemesh*

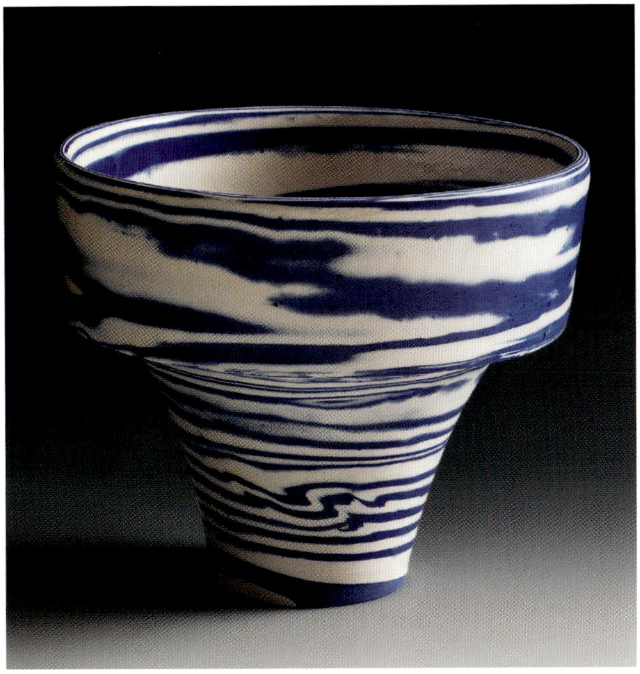

Neriage vessel, 2021. Height: 7½ inches, width: 8¼ inches, depth: 8½ inches (19 x 21 x 21.6 cm)

Blue and white swirl ledge neriage vessel, porcelain.

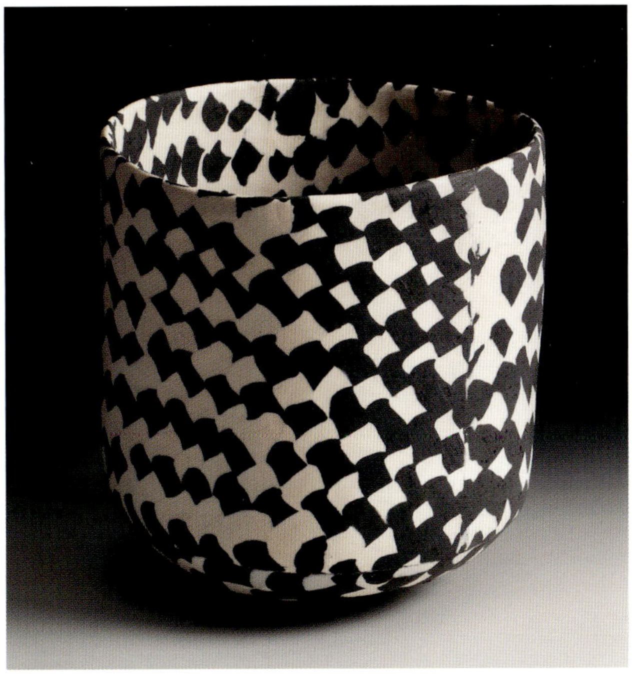

Nerikomi vessel, 2021. Height: 10¾ inches, width: 9⅝ inches, depth: 9 inches (27.3 x 24.4 x 22.8 cm)

Black-and-white checked nerikomi vessel, porcelain.

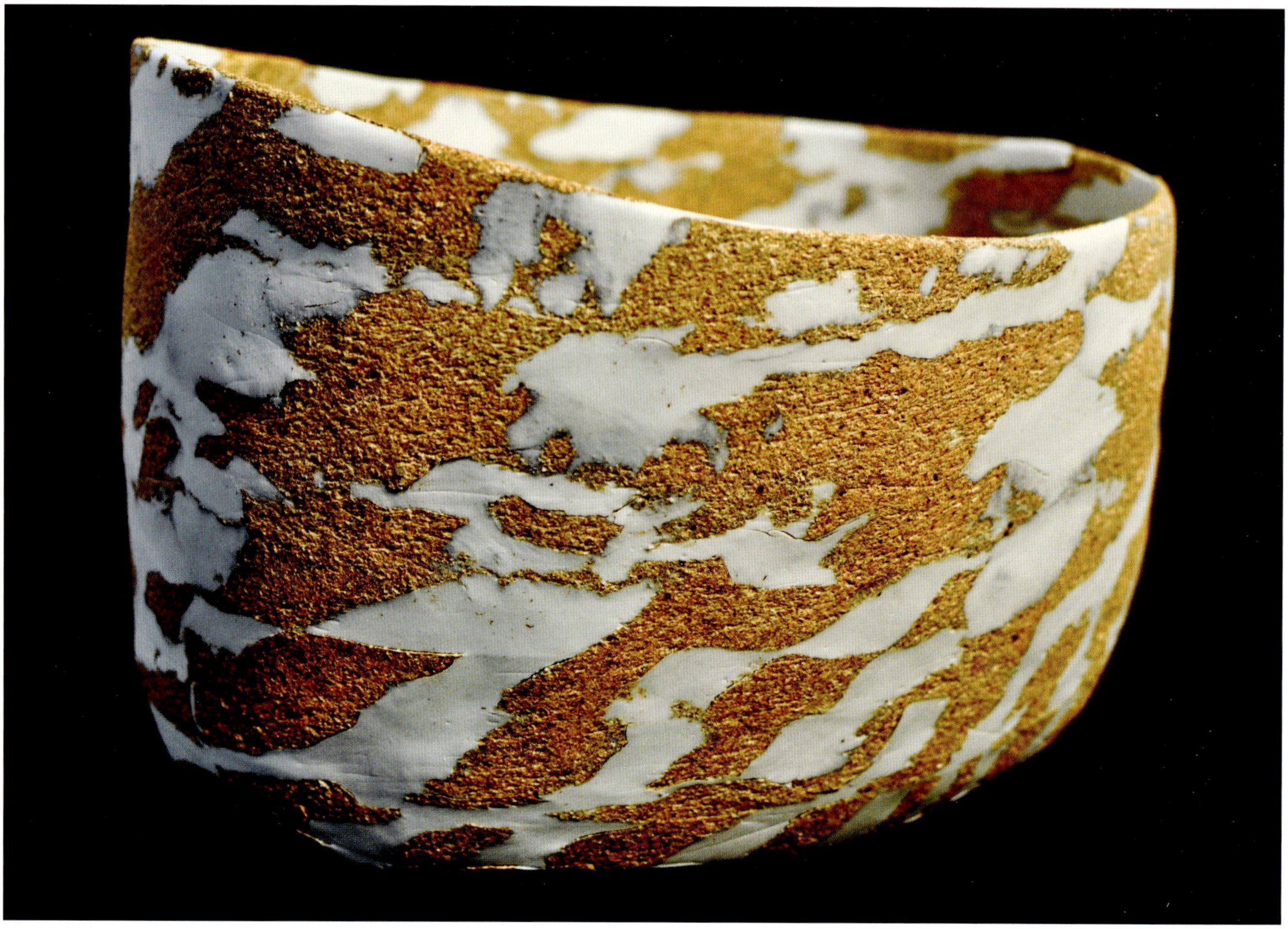

Vessel, 2011. Height: 8 inches, width: 10 inches, depth: 8¾ inches (20.3 x 25.4 x 22.2 cm)
White and toast oval twisted vessel, porcelain and stoneware.

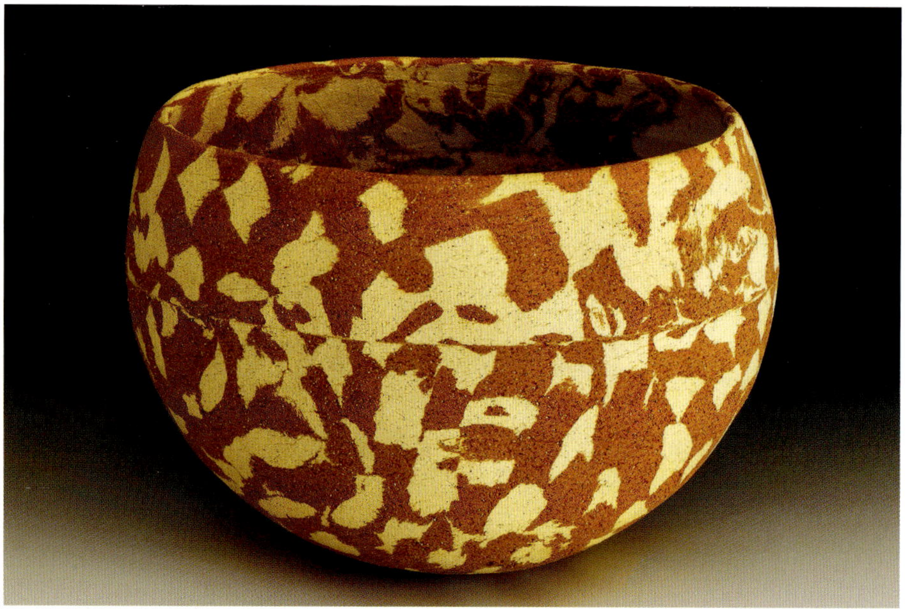

Vessel, 2017. Height: 7¾ inches, width: 10 inches, depth: 10½ inches (20 x 25.4 x 26.7 cm)
Cream and rust-checked stoneware.

Shemesh has made use of both nerikomi and neriage techniques in her ceramics, the latter taking inspiration from the natural world as she has mentioned. Both smooth-bodied porcelain, in black, white, and blue, and textured grogged stoneware clay bodies have been explored. One striking piece shown here contrasts the smooth, nearly translucent white porcelain against a highly textured golden-brown clay body, adding yet another dimension to an already dynamic vessel. Holding it in one's hands would be a feast for the senses.

Shemesh's ceramics are fired to 1200°C (2192°F) – cone 6 – and 1320°C (2408°F) – cone 10.

AWARDS AND EXHIBITIONS

Lorraine Shemesh's work has been exhibited in solo and group shows both in the United States and internationally and her work has been reviewed or reproduced in *The New Yorker*, *Art in America*, *The New York Times*, and *Harper's Magazine*, among many others. The recipient of numerous scholarships, fellowships, awards, and prizes, Shemesh has also been invited to be a visiting artist at colleges, universities, and institutes in the USA. Her work is in the permanent collections of the Hood Museum, the National Academy of Design, New York, the DeCordova Museum, Massachusetts, the Museum of The City of New York, the Museum of The Rhode Island School of Design (RISD), among others in New Hampshire, Massachusetts, New York, New Mexico, Connecticut, Ohio, Idaho, Illinois, North Carolina, California, and Basel, Switzerland.

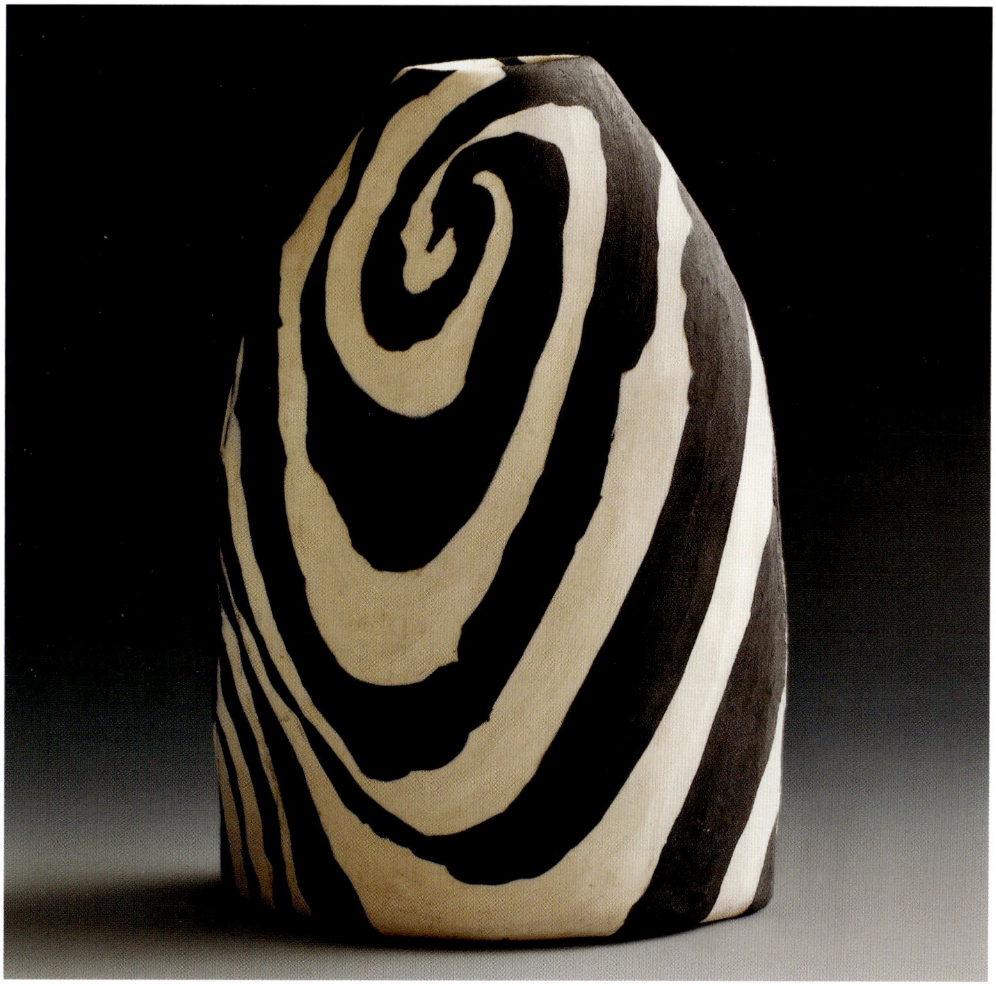

Nerikomi bottle, 2021. Height: 10⅝ inches, width: 7⅝ inches, depth: 2¾ inches (27 x 19.4 x 7 cm)

Black-and-white tall swirl nerikomi bottle, porcelain.

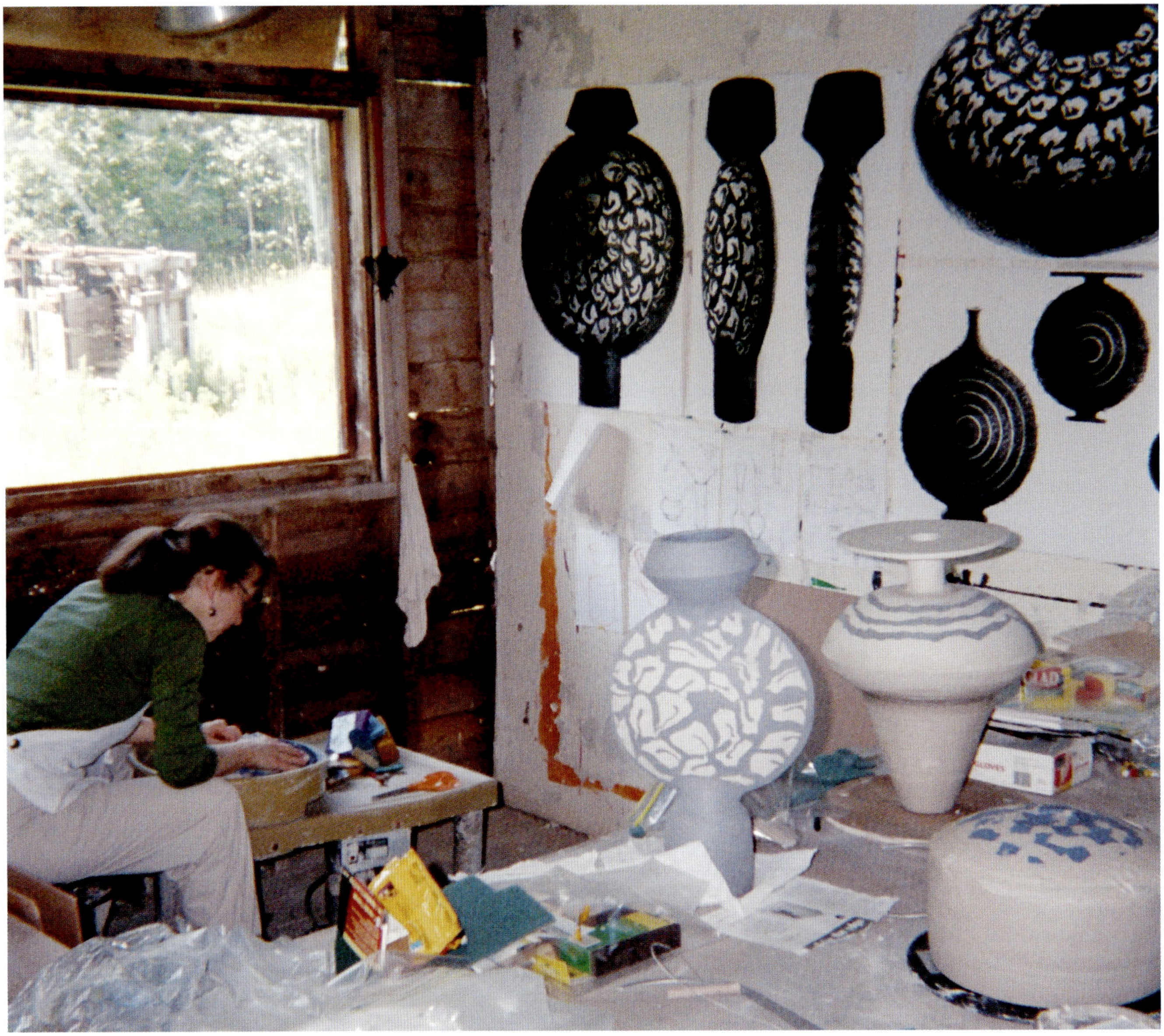

▲ Watershed Studio Shot, 2005.

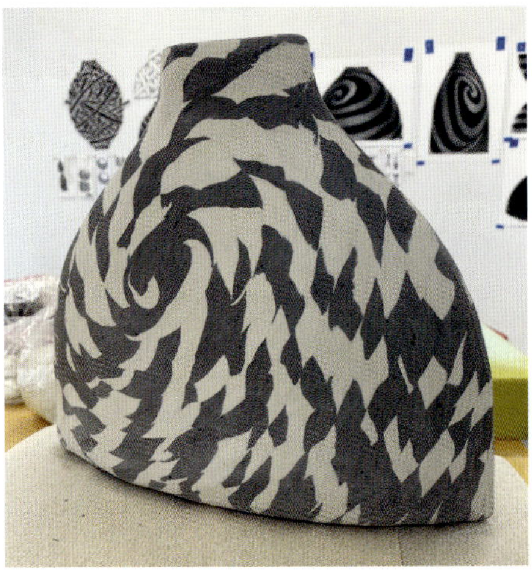

◀ Dartmouth Artist-In-Residence Studio, 2021.

Tsuneharu Tanaka, Japan

The pure form of neriage, wheel throwing with colored clay, is beautifully demonstrated in the works of Tsuneharu Tanaka. Focusing on the seemingly infinite variety of vase shapes, he has also incorporated an appealing surface-texturing technique into his production style. A typical piece starts out as a vertical stack of disks made from three types of clays, black/gray with white, or red/browns with white, which are then centered and compressed on the wheel. The resulting cylindrical form is then opened up with a "*sasibou*" wooden tool. Traditional throwing techniques are used to create the final form. When the clay is leather-hard it is scraped with a metal rib to bring out the swirling multi-clay patterning distinctive of neriage and then textured with a metal tool, the "*tobikanna*," as the wheel is turning. In English this type of texture is called "chattering." This texturing adds yet another dimension to Tanaka's work and has become his signature style.

After graduating from Fukuoka University, Tanaka went on to study at the Arts Ceramics University, Potter's Wheel Department. In 2004 he opened "Togen Ushikubi," a pottery workshop where he taught students ranging from children to the elderly. In addition, from 2005 to 2016 he created and taught at "Four Seasons Workshop" at the Onojo City Lifelong Learning Center. In 2006 he also created "Iroha Ceramics," a pottery class held at health facilities for the elderly with the goal being physical rehabilitation and improvement of cognitive function through enjoyable manual experiences and practice. In 2009 he renamed his workshop "Hasemusi Kiln," from the old name of the area where he now lives.

In 2020 Tanaka suspended classroom operations and instead started his own pottery production activities. In this relatively short time he has carved out a niche for himself in the highly competitive Japanese ceramics world.

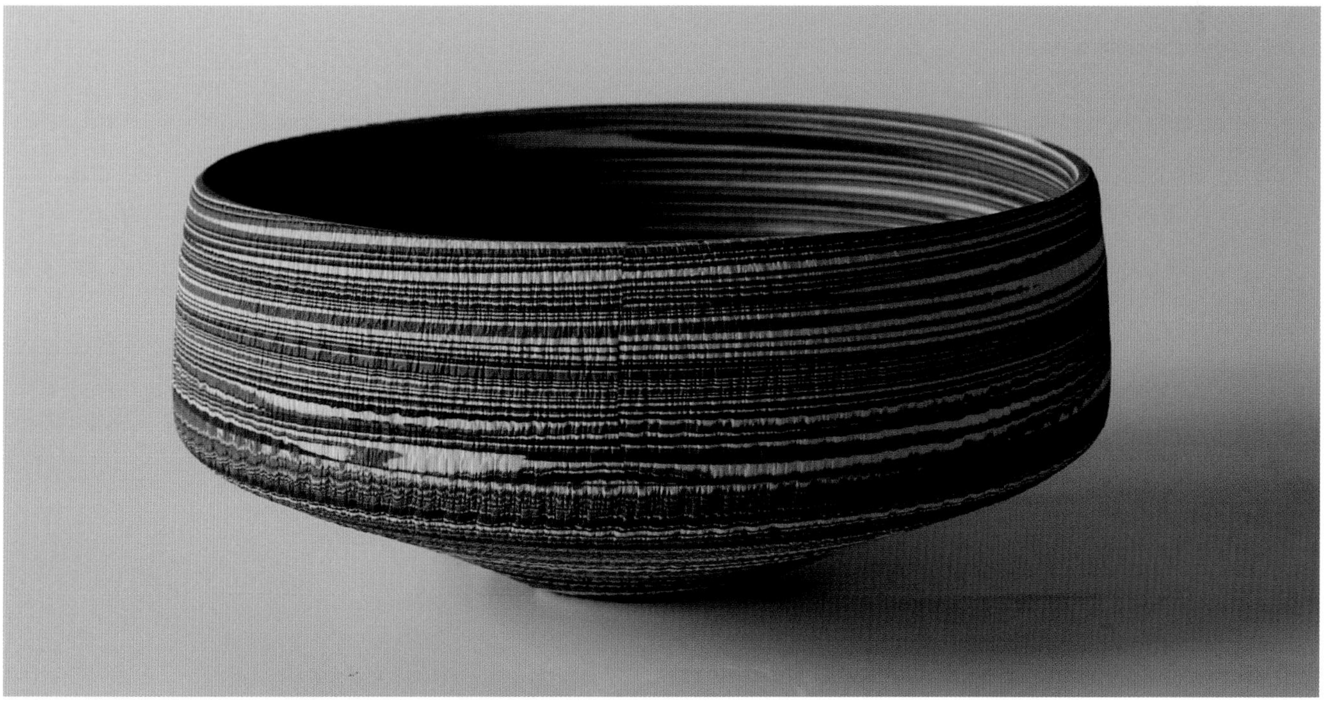

Untitled piece 2022. Height: 4 inches, diameter: 8⅓ inches (10 x 21 cm) *These images all © Tsuneharu Tanaka*

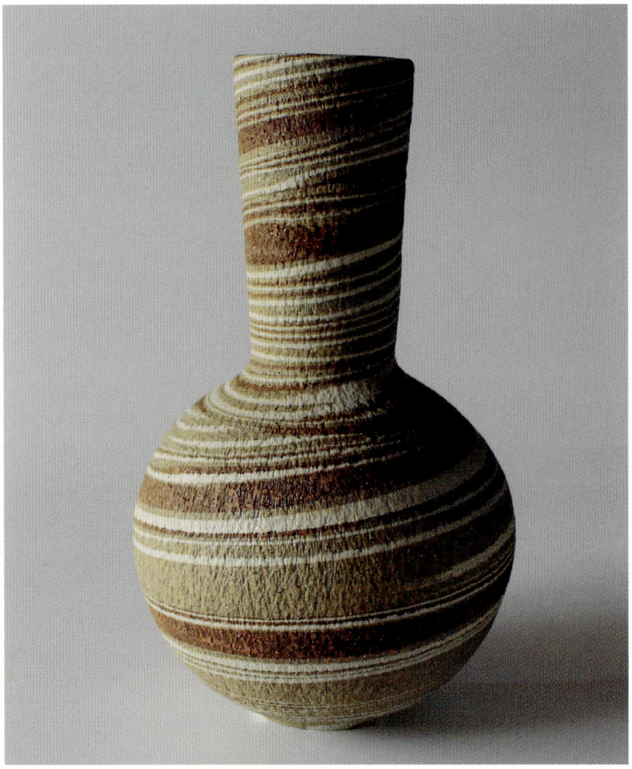

Untitled piece 2022. Height: 9⅞ inches, diameter: 6 inches (25 x 15 cm)

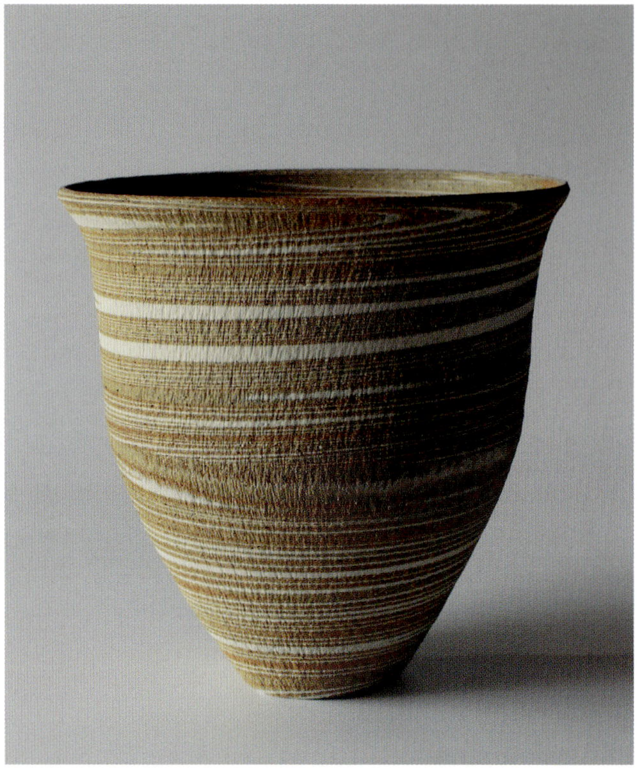

Untitled piece 2022. Height: 7½ inches, diameter: 7⅛ inches (19 x 18 cm)

ARTIST'S STATEMENT

I am deeply grateful to have encountered pottery, to be fascinated by it, and to have it as part of my life. Through my work, I feel great joy in being in contact with people from all over the world, and I will continue to use this as a source of inspiration for my creative activities.

Tsuneharu Tanaka's ceramics are fired in oxidation in a gas kiln to 1245°C (2273°F) and are sold directly in his workshop gallery.

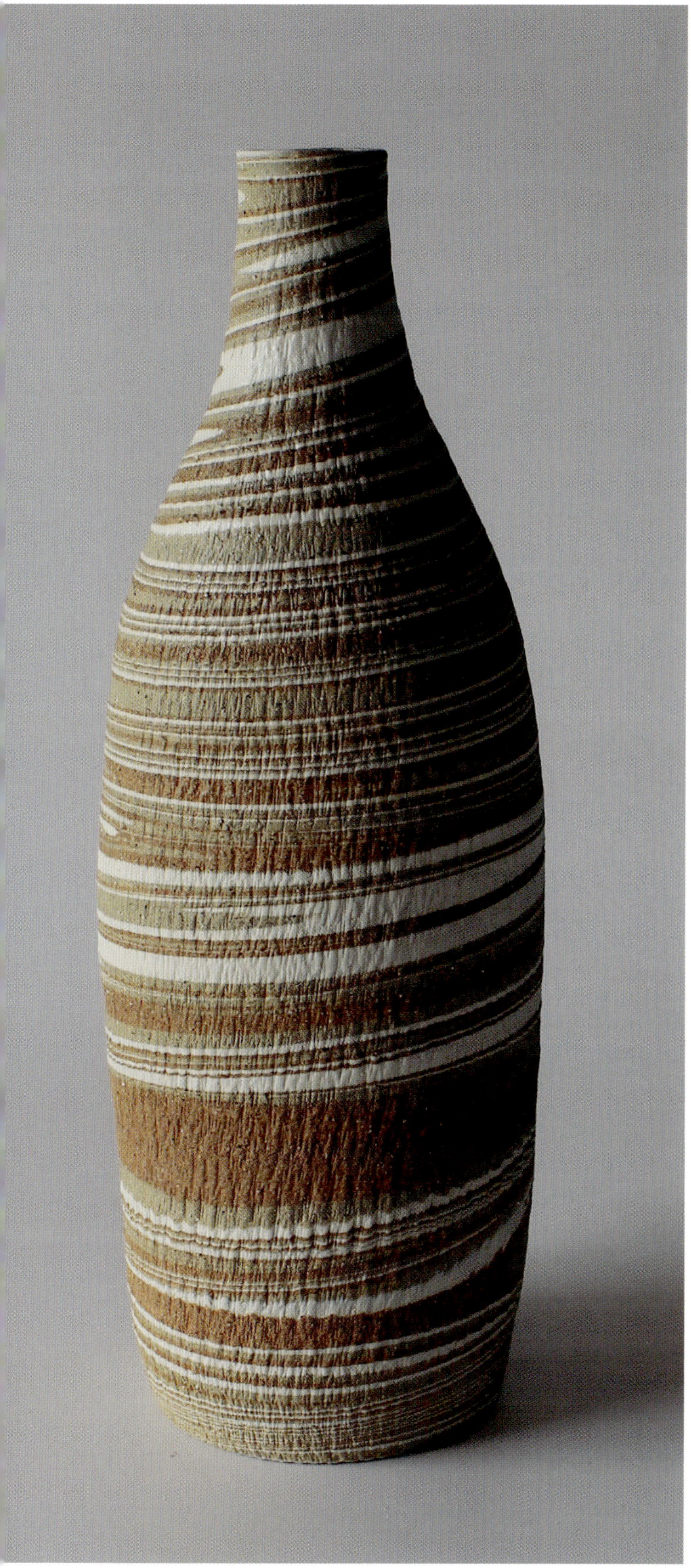

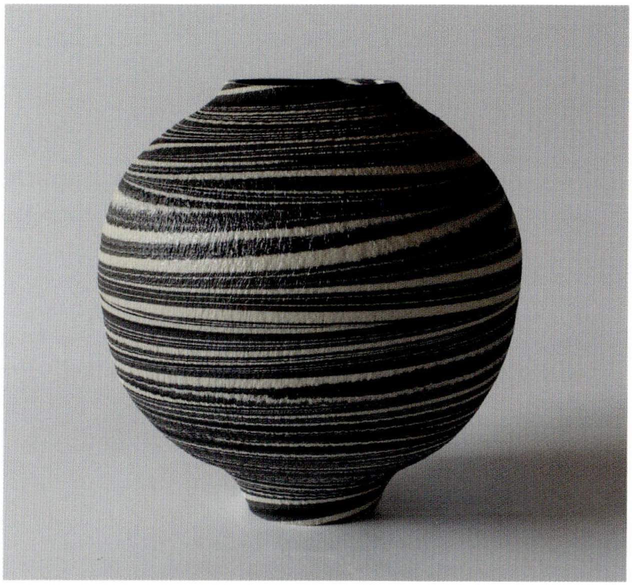

Untitled piece, 2022. Height: 6¾ inches, diameter: 6⅓ inches (17 x 16 cm)

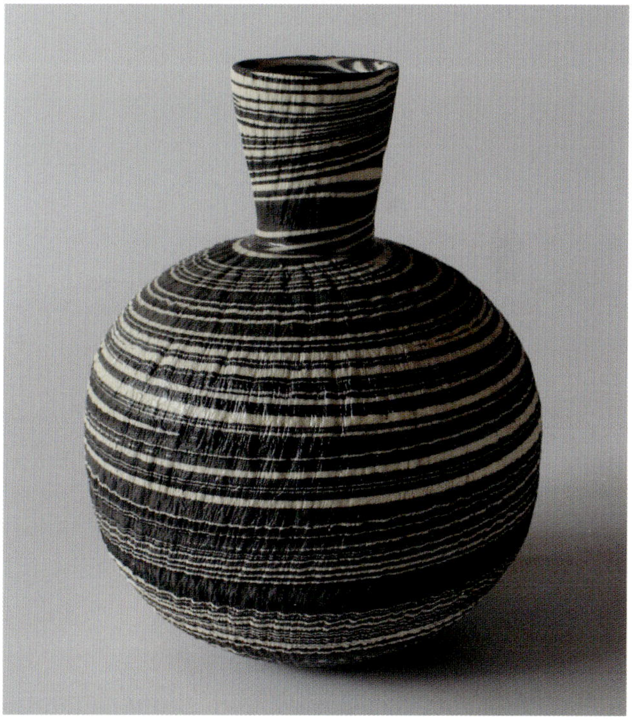

Untitled piece, 2022. Height: 6¾ inches, diameter: 5⅛ inches (17 x 13 cm)

◄ Untitled piece, 2022. Height: 11⅖ inches, diameter: 4⅓ inches (29 x 11 cm)

Studio shots of Tsuneharu Tanaka's process.

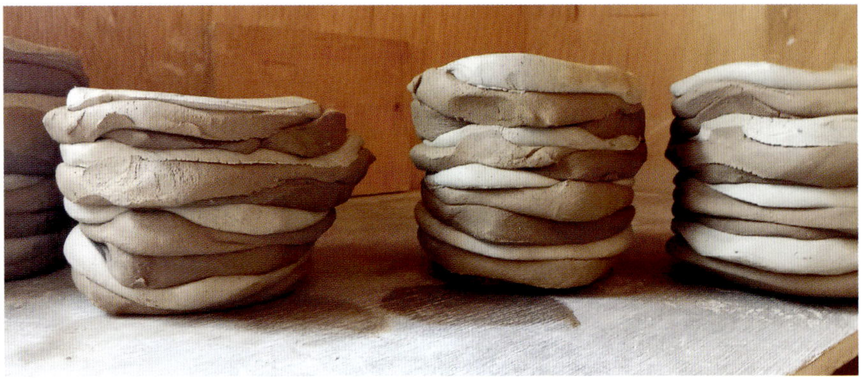

▲ Stacks of various clays in preparation for throwing.

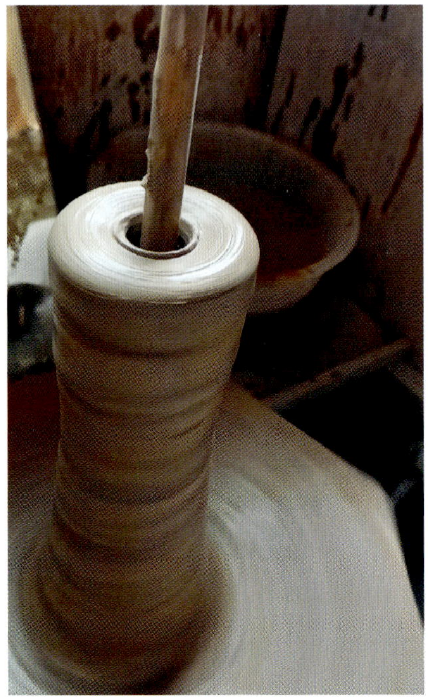

▶ After the stacks are centered on the wheel the solid mound is opened with a "*sasibou*" tool.

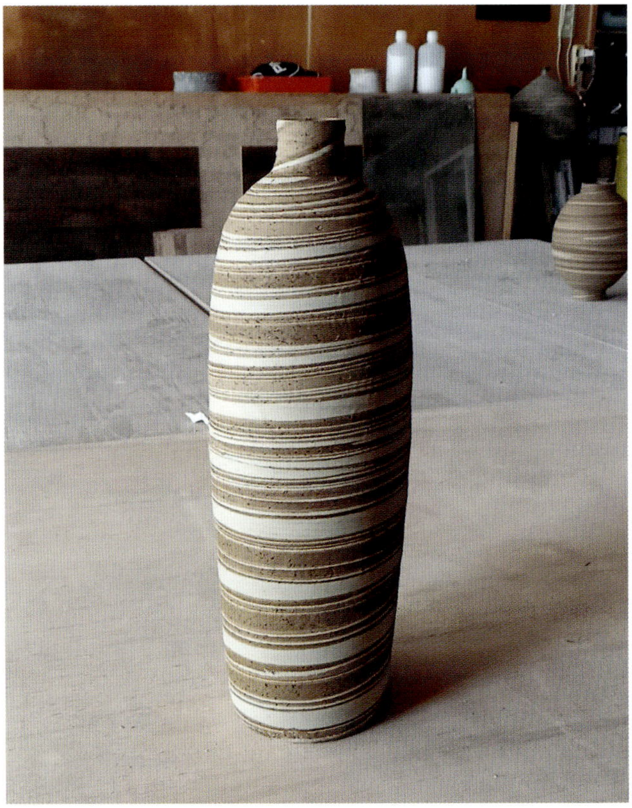

The leather-hard vase has been scraped clean to reveal the distinctive neriage patterning.

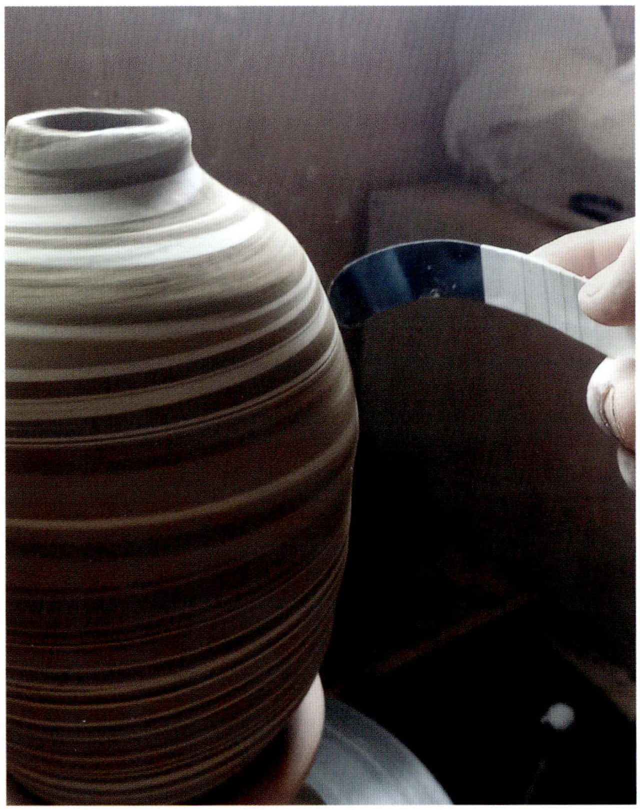

The leather-hard vase is supported on a chuck on the wheel and with a "*tobikanna*" tool, the texture is created over the entire surface as the wheel turns.

Larissa Warren, Australia

If an artist's environment can influence her creative output then the works of Larissa Warren comprise a beautiful example. Several years ago she moved her family and ceramics practice to the semi-rural town of Tamborine Mountain in the Scenic Rim, Queensland, Australia. This is an area with a rich history of potters, especially women potters, who worked for several decades starting in the 1940s. All used a variety of local clays which were varied and abundant due to the geological history of the area. Warren began an extensive research project of the area's history ("Wild Women, Wild Clay") including the rediscovery of many of these clay deposits and subsequent technical experiments on these quite varied clays.

This research project led to a group of vessels comprising the exhibit "Sedimentary Stories" (2021).

ARTIST'S STATEMENT

I must admit I have become quite spoilt with access to each of the original clay sources . . . with each clay source being an almost pure deposit with few organics to be removed . . . Apart from one slightly lower-firing terracotta, each clay has successfully fired to cone 10 with colors ranging from white, pink, and blue in the south side to rich orange and red iron-based clays along the north and east escarpments.

I like the hidden and forgotten stories that a place holds in its geology and its people. Often buried by time these stories are lost to shifts in generations and the changing use of land, but places contain layers of meaning always bound to history and that is what my work explores. This series ["Sedimentary Stories"] celebrates the unexpected. Digging into the lineage of female ceramic makers from Tamborine Mountain, where I also live and work, led to the rediscovery of forgotten local wild clays. Contrasting these with commercially produced clays reveals their unpredictable nature and, in many ways, also speaks to the nature of the women who used them – tough, resilient, and unique.

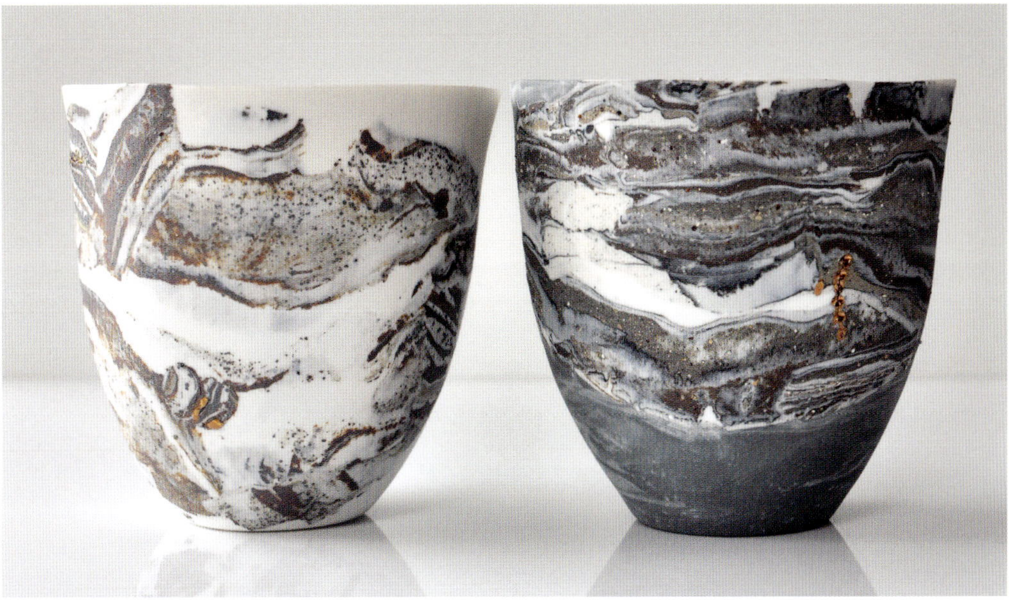

Vases, 2020.
Each vase: height: 4⅓ inches, width: 4⅓ inches (11 x 11 cm)
Black-and-white vases from "Layers of Place" exhibition, 2020. Australian porcelain, ceramic stains, locally sourced stoneware clay, fired in reduction to cone 10. Additional copper luster. Slip cast with nerikomi inlaid through the walls (seen inside and out). *These images all © Larissa Warren*

Warren is clearly connected to her environment and its history: "As an artist, I strive to ensure I walk softly on this earth and that my practice works in a manner that respects and recognizes the important spiritual connections the Wangerriburra people, past and present, have to these lands."

Warren tells an amusing story of how, as an eight year old, she remembers eating ice cream with a spoon out of a "heavily grogged stoneware bowl at my grandmother's house." She found the experience almost distressing, comparing it to "nails on a chalkboard." Little did she know that as an adult she would be drawn to just such textures in her distinctive ceramics. In addition to the variety of local clays, with their colors and textures, she also uses porcelain for its contrasting smooth and translucent qualities. Staining the porcelain with ceramic stains is also integral to Warren's practice, the colors often influenced by her surroundings and even the weather.

Warren began her education at Queensland College of Art with a BFA in 1998 followed by a Bachelor of Education Secondary Art & History at Griffith University, 2000. She is a registered Art and Media teacher with more than 20 years of experience in developing and writing art programs. Her broad experience with ceramic techniques such as hand-building, slip casting, and wheel throwing have provided variety and depth to her teaching workshops.

From an exhibit in 2019, "Scripted by Nature," Warren writes: "It is always my intention to challenge some of the established principles associated with the use of clay, and particularly, porcelain. Starting with a fascination of patterns and contrasting texture, I embed hectic nerikomi pattern slices of color, stoneware, and raw minerals into fine porcelain forms."

This is done by embedding such slices into slip cast forms as well as adding them to wheel-thrown pots, both porcelain and stoneware. She continues: "I layer colored porcelains, wild clays and minerals that I've collected, into intricate patterns which resemble familiar landscape features. Technically, fusing stoneware into porcelain is very rare in the ceramic world due to their vast differences in properties, firing, and shrinkage. It has taken me years to perfect this personal process and I take pride at the end result."

In ceramics there is quite often a scientific side of the personality of the artist alongside the aesthetic. Larissa Warren exhibits both capacities in her work. Partly because she uses such a varied range of clays, many being locally dug, she engages in extensive testing to discover technical information which will be needed as she adjusts drying and firing cycles to limit potential losses. Careful notes are taken and the knowledge obtained must be constantly held in the background as she takes the role of the artist and creates work with these diverse materials.

AWARDS AND EXHIBITIONS

Larissa Warren has had several solo exhibitions of her work in Australia, including at Gallerysmith, Melbourne; Art Nuvo Gallery, Buderim Qld. and Suki & Hugh Gallery, Bungendore NSW, and has been included in numerous group shows in Australia and the USA. She is also the recipient of several awards and grants including the 2018 Siliceous exhibition Overall Winner Ceramic Arts Qld. 2018 and the 2019 Clunes Ceramic Award, People's Choice winner and special award acquisition 2019. She has been featured on national radio, on "The Potters Cast" (podcast), and in magazine articles and workshops in Australia.

▶ **Terra sphere, 2019. Diameter: 5¾ inches (14.5 cm)**

Terra sphere from "Scripted by Nature" exhibition, 2019. Porcelain, ceramic stain, stoneware, mother of pearl. Oxidized firing to cone 10.

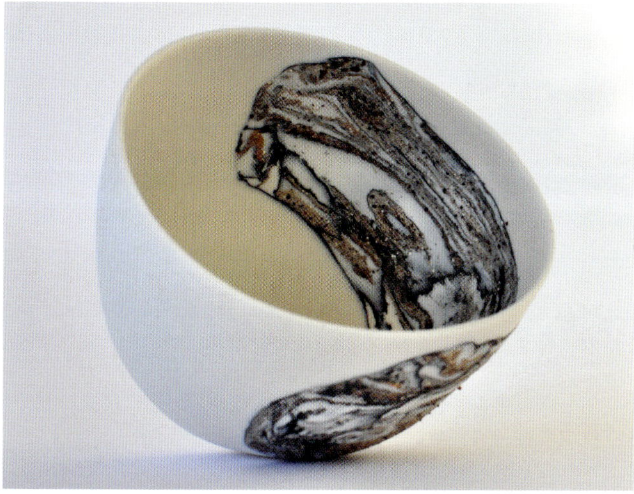

Untitled cup, 2017. Height: 4¼ inches, width: 3⅝ inches (11 x 9.5 cm)

Porcelain slip, nerikomi inlay – a mixture of commercial Australian stoneware, stains, minerals such as trachyte, ilmenite, and quartz.

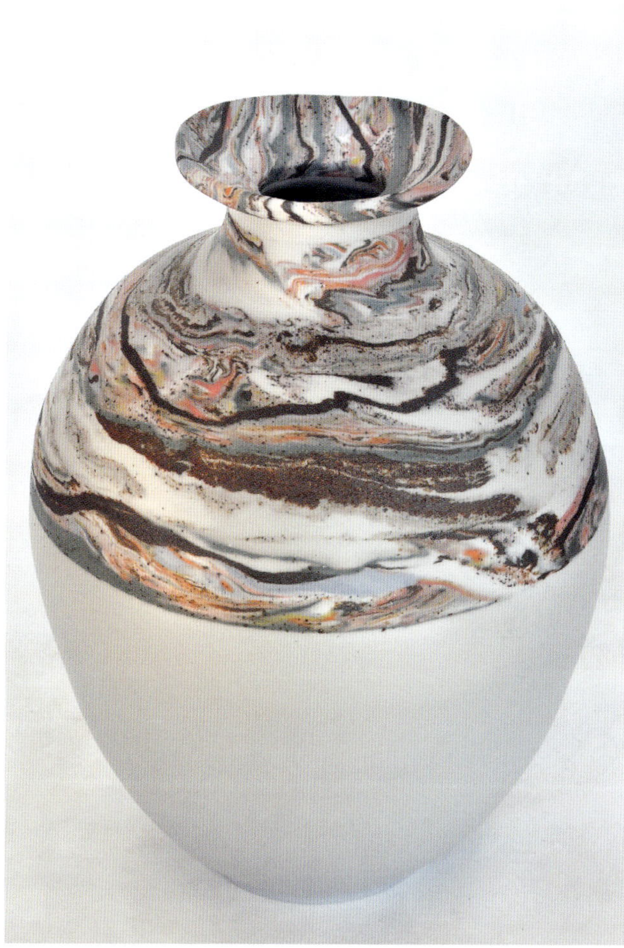

▶ **Vase, 2022. Height: 13⅓ inches, width: 9⅞ inches (34 x 25 cm)**

In the Fold, large wheel-thrown porcelain vase from "Lucid Dream" exhibition, 2022. Porcelain, local Tamborine Mountain clays, ceramic stains, fired in reduction to 1282°C (2340°F).

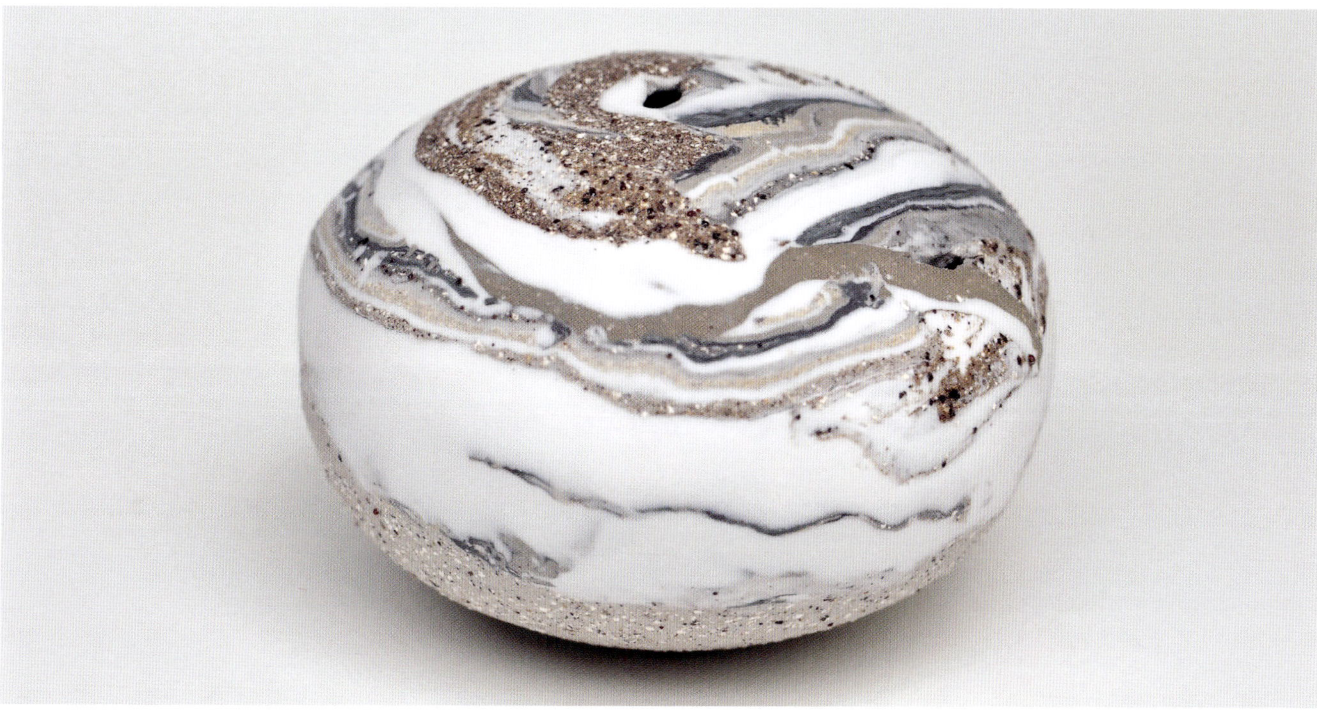

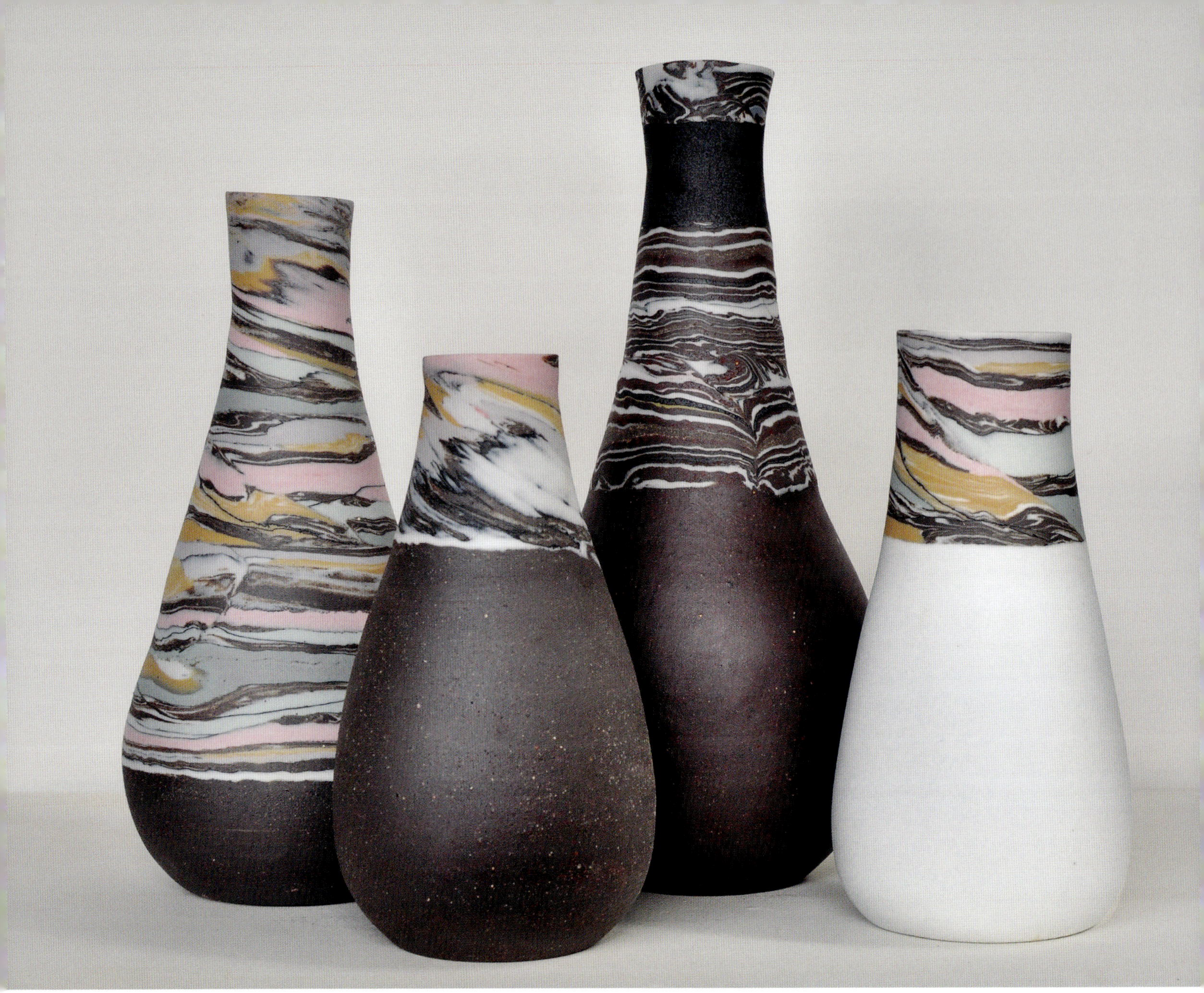

Bottles, 2021. Various measurements, approx.: height: 10⅗ inches, width: 11⅘ inches (27 x 30 cm)

Bottles from "Sedimentary Stories" exhibition, 2021. Locally sourced stoneware clay, Australian porcelain, ceramic stains, fired in reduction to cone 11 (1293°C [2359°F]); wheel-thrown.

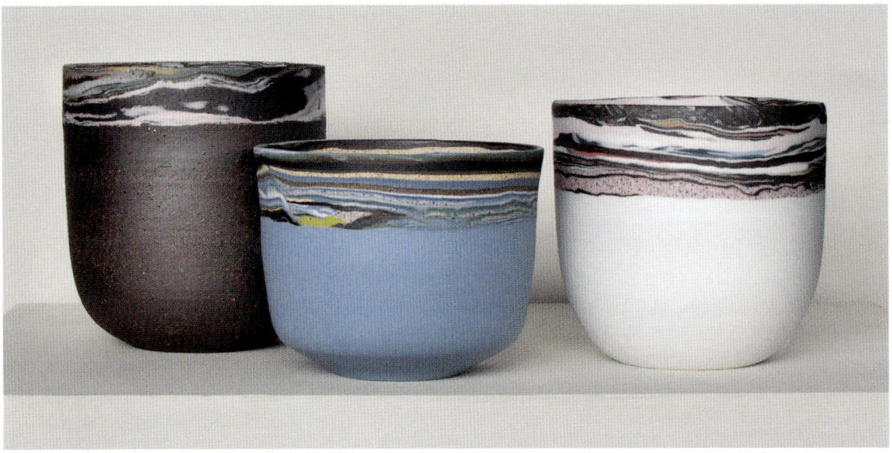

◀ **Collection of cups, 2021. Various measurements, approx.: height: 6 inches, width: 15¾ inches (15 x 40 cm)**

Collection from "Sedimentary Stories" exhibition 2021. Wheel-thrown locally sourced stoneware clay, porcelain, ceramic stains, copper luster, fired in reduction.

Dorothee Wenz, Germany

A life in ceramics seems to have been fated for Dorothee Wenz since she was a child. Her family vacationed on the island of Bornholm in Denmark where there were a number of ceramicists. The yearly routine was to visit each studio and young Dorothee was especially fascinated with the work of Hans Munck Andersen (profiled on pages 66–9) who worked with colored porcelain. However, it wasn't until later that she rediscovered her passion.

After graduating from high school she chanced upon a ceramics exhibition of British artist Elizabeth Fritsch and was entranced. "I was mesmerized by the shapes, the colors and the tender surfaces," she revealed. From 1990 to 1995 she studied at Johannes Gutenberg University in Mainz, first in art education and later switching to ceramics studies at the Art Academy. Her passion from the start was with hand-building vessels. Once again Wenz encountered works made from colored clay and soon recognized that she had found her medium. After receiving her diploma she began what was to become her life's work.

ARTIST'S STATEMENT

What has been exposed to weather and other influences over long periods of time, shaped and pressed in layers of rock over the ages, has its own beauty. To creatively follow the work of nature and the elements and what wind, sand and water take over, to translate this beauty into ceramics and in doing so to explore the limits of feasibility, that fascinated me from the very beginning. Coloring, building layers, grinding, polishing. Each phase has its own appeal, takes its time and devotion to material and form. It is a dynamic process whose progress I observe with curiosity and excitement, continuously surprised by what I discover in the making.

Inspired by this natural beauty Wenz has developed a signature style of casually layering colored clays, often using earth tones but exploring more vibrant colors as well, into nerikomi loaves. Slices are cut and joined using traditional hand-building techniques to create her asymmetrical forms which, again are inspired by nature. Of her shapes she discloses:

> *The minimal forms, usually asymmetrical and curved, have taut contours. Though individually balanced in volume and form, they are often created in pairs or small groups – harmoniously belonging to each other yet at the same time each one a free, independent vessel. Simplicity of form contrasts with intricately worked surfaces so dynamic that they evoke a vibrant life all their own.*

Wenz often works with white stoneware and porcelain but also experiments with new clays. A favorite step in the process is at the leather-hard stage when she carefully scrapes the surface with metal ribs, exposing the fine patterns and colors and refining the surface and contours. She fires her work twice, grinding the surfaces after the bisque and then again, grinding and polishing to a smooth soft finish after the final firing at 1160°C (2120°F).

In addition to her colored clay work Dorothee Wenz creates ceramic figurative sculptures of charm and wit. They appear as hybrid individuals with bodies of humans and heads of rabbits, donkeys, sheep, and fish. They are often seen in pairs or small groups, as are her vessels.

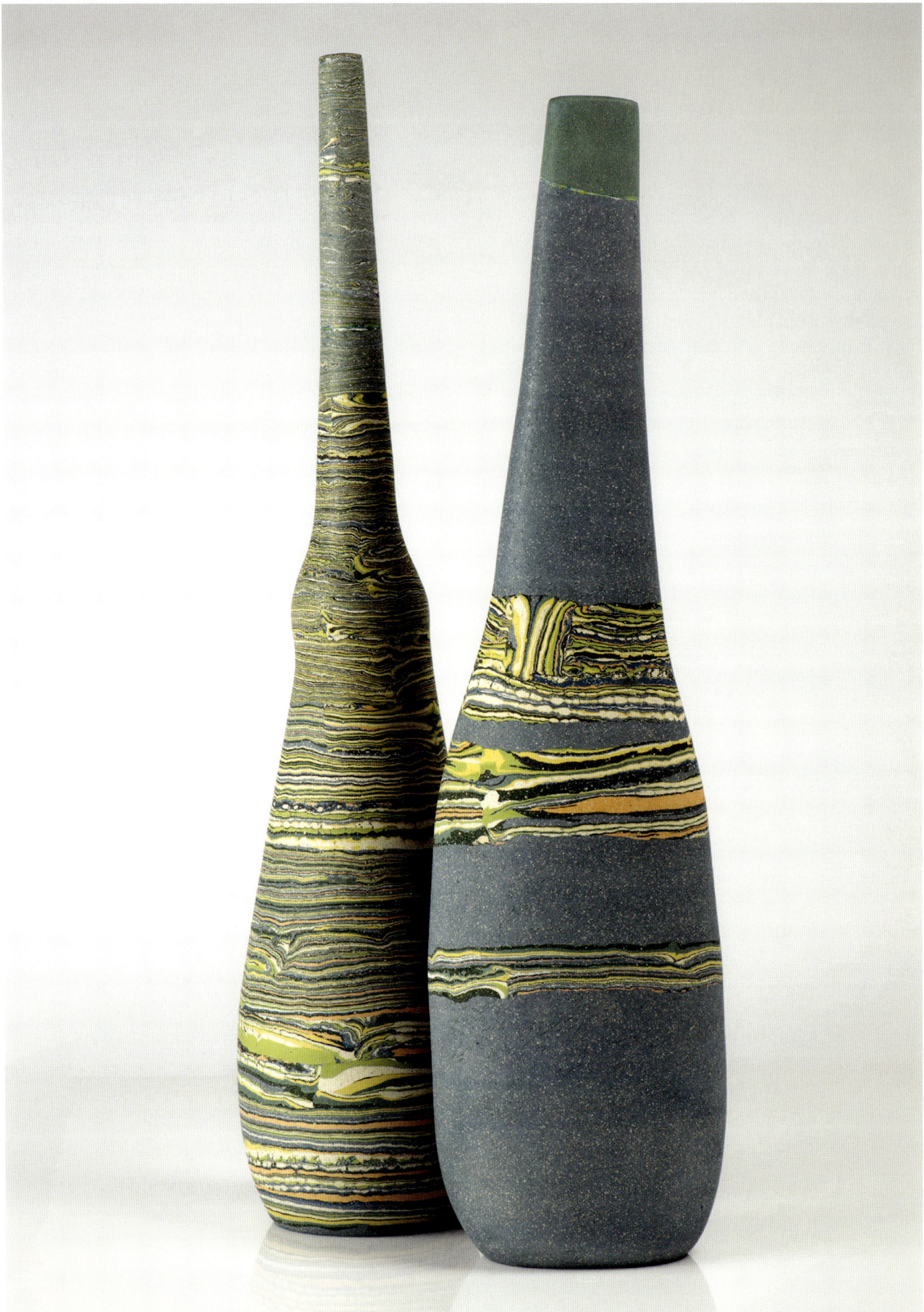

AWARDS AND EXHIBITIONS

Dorothee Wenz has had solo exhibitions in Hamelin, Leipzig, Staufen, Mainz, and Frankenberg, all in Germany; and Laren, the Netherlands. She has been included in numerous group exhibitions in Germany, the Netherlands, France, South Korea, Switzerland, Estonia, Australia, Japan, and New Zealand. She has also received many awards including the State Prize for Arts and Crafts 2022, Wilhelm Hack Museum, Ludwigshafen, 2022; the Perron Art Prize, Kunsthaus, Frankenthal, 2020, 2017; and the State Prize for Arts and Crafts Rhineland-Palatinate 2013. Her work is included in the collections of the Grassi Museum, Leipzig; the Hetjens Museum, Düsseldorf; the Bernhard Foundation, Berlin; and the Keramion Museum, Frechen, all in Germany.

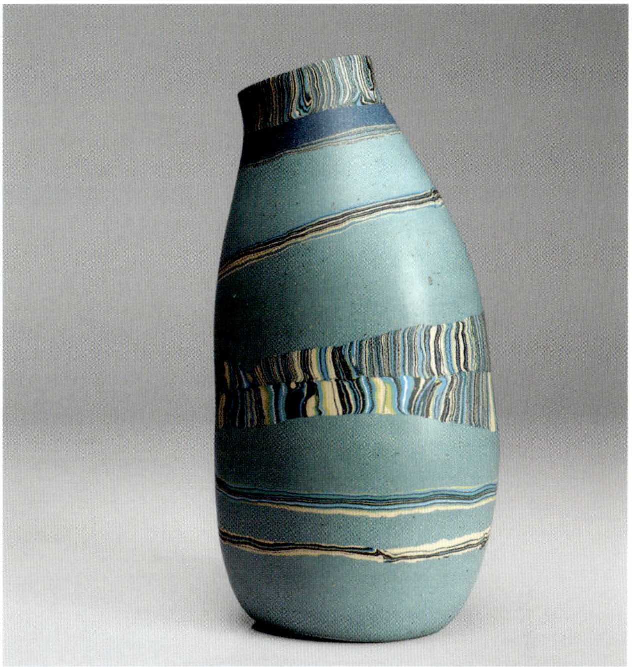

Vessel, 2018. Height: 19 inches, diameter: 9 inches (48 x 23 cm)

Colored clay and porcelain, hand-built, sanded, and polished.

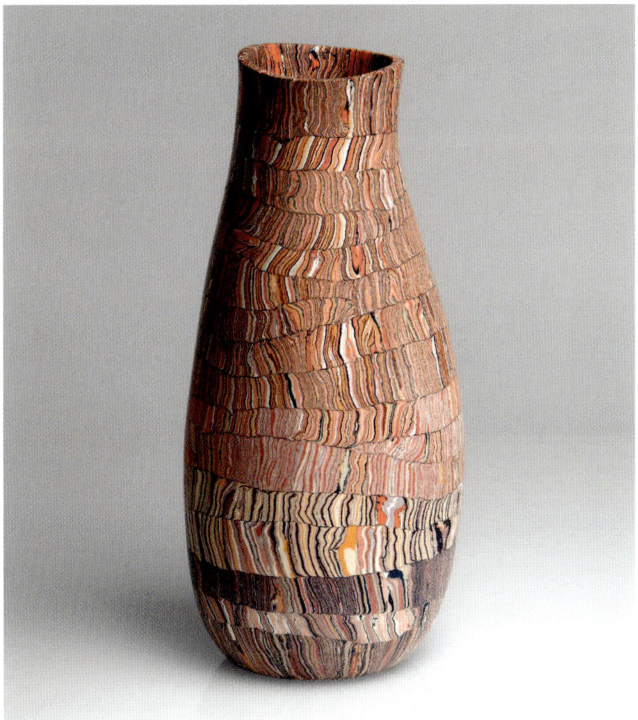

Vessel, 2020. Height: 15 inches, diameter: 6⅓ inches (38 x 16 cm)

Colored porcelain, hand-built, sanded, and polished.

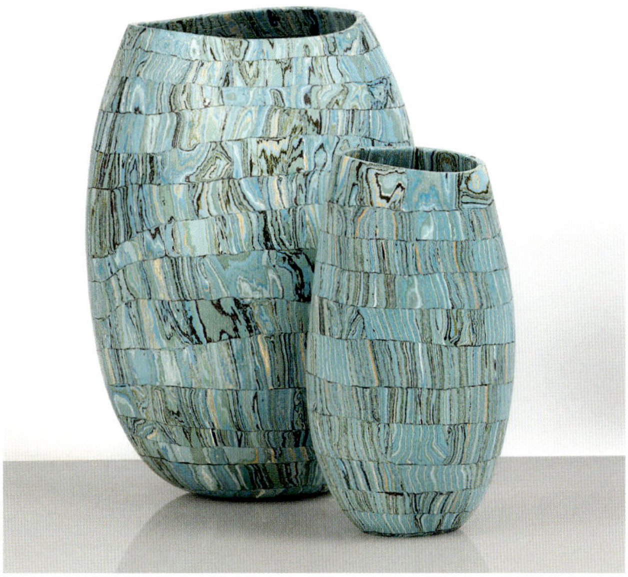

Two vessels, 2022. Left: height: 14½ inches, diameter: 10⅕ inches (37 x 26 cm); right: height: 11 inches, diameter: 6¾ inches (28 x 17 cm)

Colored clay and porcelain, hand-built, sanded, and polished.

◂ **Two vessel shapes, 2019. Left: height: 29 inches, diameter: 6¾ inches (74 x 17 cm); right: height: 28 inches, diameter: 7⅘ inches (71 x 20 cm)**

Colored clay and porcelain, hand-built, sanded, and polished. *These images all © Dorothee Wenz*

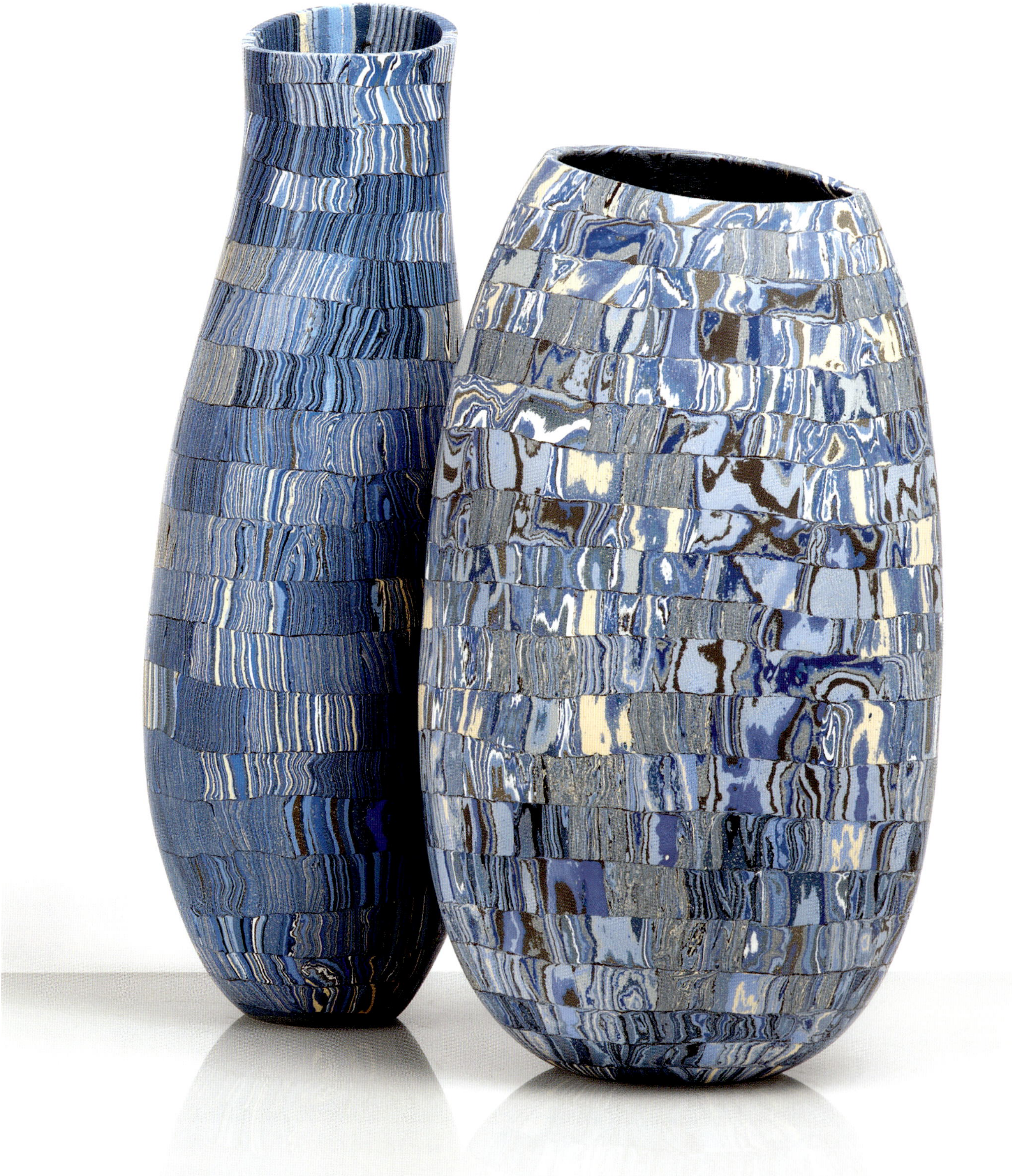

Two vessels, 2022. Left: height: 22 inches, diameter: 7⅛ inches (56 x 18 cm); right: height: 19⅕ inches, diameter: 10⅗ inches (49 x 27 cm)

Colored clay and porcelain, hand-built, sanded, and polished.

Lotte Westphael, Denmark

The breathtaking vessels of Lotte Westphael demonstrate a new and unique method of working with colored clay. Her porcelain cylinders are neither nerikomi nor neriage but, like a few others in this book, present a fascinating technique of artistic originality. Influenced by her Nordic roots and inspired by textiles and weaving, especially the work of Anni Albers and the paintings of Agnes Martin, Westphael initially works in two dimensions. She arranges small rectangles of colored porcelain, only 3 mm thick (approx. ⅛ inch), in stripes and grids which then miraculously become transformed into powerful cylindrical forms.

ARTIST'S STATEMENT

I call my vessels Syncopes, a musical expression referring to a disturbance or interruption of the regular flow of rhythm. This concept of syncopation gives me artistic freedom for the irregularity that comes from the technique and the material . . .

I am aware of the balance between perfection and imperfection. I work with geometric patterns – vertical and horizontal lines. There is a narrative of time in my pieces – both the time I have spent making them, but also the time to discover all the small offsets – off beats! One can be amazed by perfection but with small irregularities, my pieces are human with an invitation for a dialogue with the viewer.

Beauty is absolutely essential for me in my life and in my works, and I have a constant awareness of patterns, contrasts, lines, and colors. With respect to beauty, I would like to quote Agnes Martin, who's a great inspiration for me: "Beauty is the mystery of life. It is not in my eyes, it is in my mind."

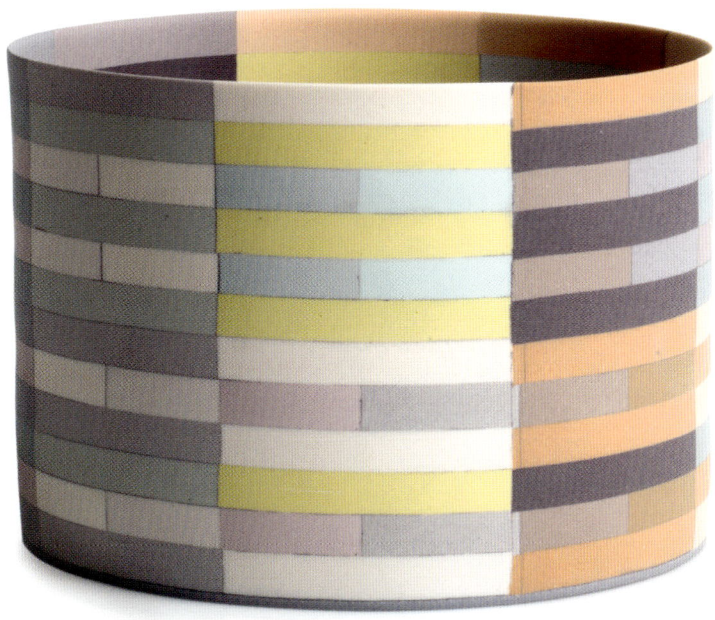

Color-Up Syncope, 2020. Height: 5½ inches, diameter: 7⅛ inches (14 x 18 cm). *Image © Lotte Westphael; photo by Erik Balle*

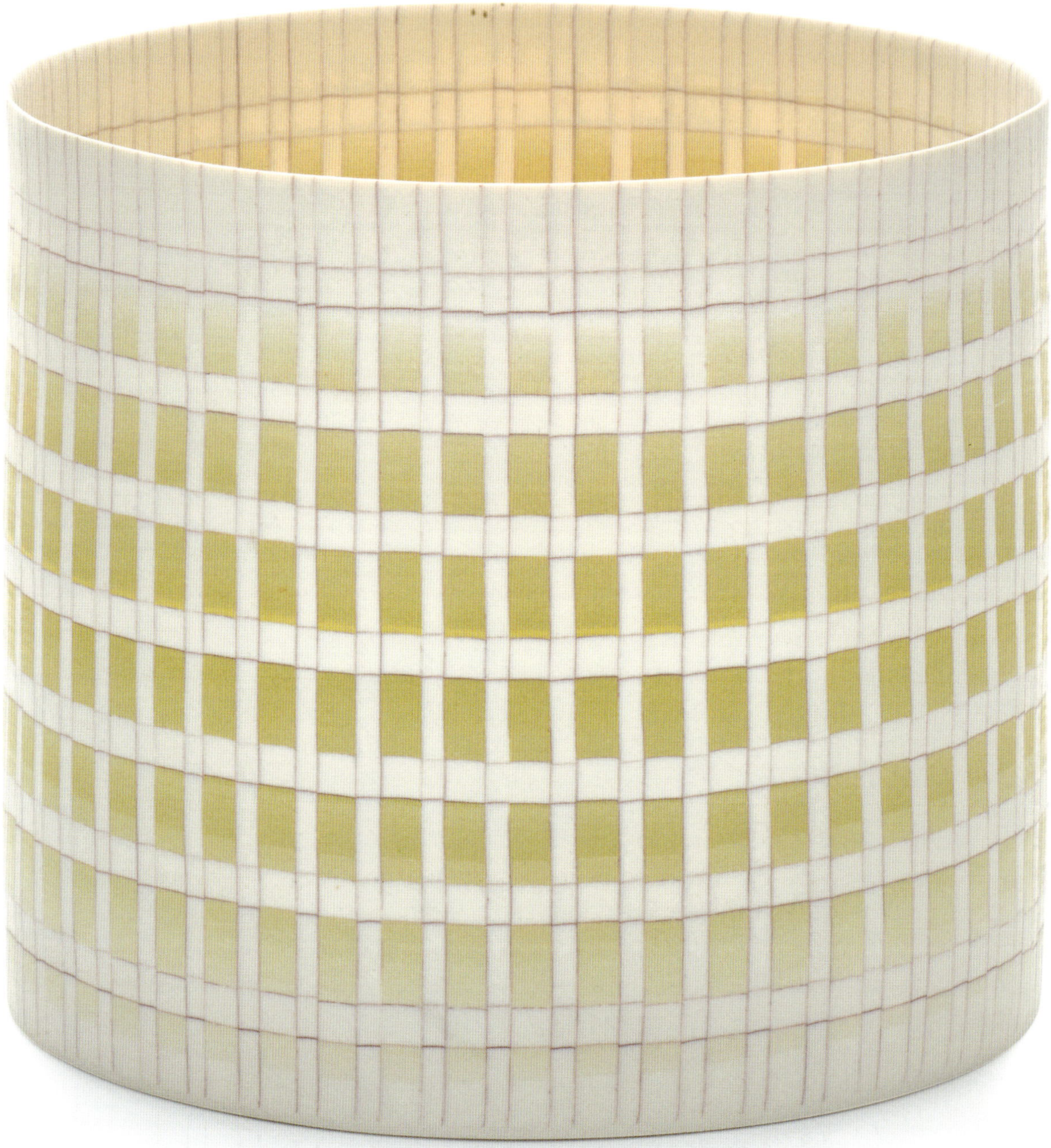

At an early age Westphael had a life-changing aesthetic experience.

My first aesthetic experience of ceramics was with Alev Siesbye's bowls, which I saw in an exhibition when I was 16 or 17 years old. There was a simple white room where her ceramics were displayed. It looked as if all the bowls were floating on their tiny, tiny feet. I remember the feeling of my mind opening for this beauty, and I remember the simplicity. It was magical.

Westphael began her education in Japan where she apprenticed for Kazu Nagayoshi at his studio in Kyushu (1987) as well as time with Donna Gilliss (1990) and Mamoru Teramoto (1990). She continued her studies at the Design School Kolding in Kolding, Denmark, a five-year Master's Degree program in ceramics and glass. She now works at her own studio in the countryside in Denmark, where she creates one-of-a-kind pieces for exhibitions.

As a very young girl Westphael had a desire to study mathematics, which reveals something about how she thinks:

I find that I think in systems, which is reflected in my approach. In my series I always have something fixed, and something that I change; a constant and a variable... It is in my DNA to systematize and try different variations within the same theme. Therefore, I consider every new vessel as an answer to a question the previous piece has raised. If all my works were placed chronologically in a row, there would be a clear reference to this approach. I make sketches, but when transferring them into a vessel, I always make slight changes (for example, in the rhythm of how the templates are assembled). Making a vessel involves a lot of advance planning. The colors in the unfired porcelain are very different from the finished colors. It is especially difficult to see the difference in the pastels, and it is important to have a kind of master plan. I have small paper tags with the numbers of the colors written on them and keep them lying on the clay through the whole process, otherwise I lose track... One can say that my vessels are a result of detailed planning and subsequent improvisation. When I am in the middle of working with the porcelain, I get inspired and encouraged by the process to make something other than what I had planned.

The technical challenges Westphael has set for herself require extremely careful planning and execution. Creating just one or two pieces takes up to a month, and another month is required for gradual drying. Up to 25–30 percent of the work is lost due to cracks. For those that remain intact there is a very gradual firing process in one of her electric kilns whereby the heat is slowly increased to 1250°C (2282°F) and then, again, slowly reduced to prevent thermal shock as it cools.

◄ ***Yellow Grid Gradient*, 2020. Height: 8¼ inches, diameter 8⁹⁄₁₀ inches (21 x 22.5 cm).**
Image © Lotte Westphael; photo by Sylvain Deleu

AWARDS AND EXHIBITIONS

Lotte Westphael is represented by Galerie Maria Wettergren, Paris, and has exhibited her work at the 5th Triennale of Kogei in Kanazawa, Japan, 2022; Design Miami-Basel, Switzerland, 2022; Ceramic Art Andenne, Belgium, 2022; TEFAF New York, 2022, 2023; and numerous other exhibitions in France, Belgium, the USA, South Korea, Switzerland, Denmark, Sweden, Germany, Romania, Hungary, and Russia. Her work is included in the public collection of Designmuseum Denmark, and several private collections. She is a member of the Danish Crafts and Design Association and in 2016 she was awarded with the Hetsch Medal of 1879 for unique artistry and craftsmanship.

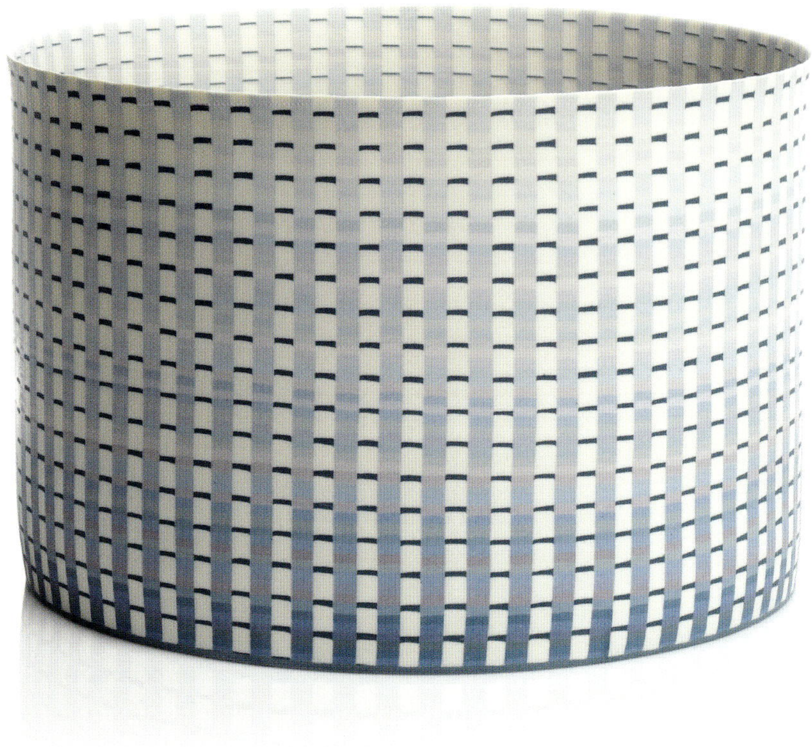

◀ ***Multilayered Polyrhythm*, 2021.** **Height: 8⅔ inches, diameter: 13 inches (22 x 33 cm).** *Image: © Lotte Westphael; photo by Erik Balle*

▼ (left) ***Pink Grid*, 2021. Height: 9⅞ inches, diameter: 10⅗ inches (25 x 27 cm).** *Image: © Lotte Westphael; photo by Erik Balle*

▼ (right) ***Untitled – homage to Agnes Martin*, 2021. Height: 5½ inches, diameter: 7⅛ inches (14 x 18 cm).** *Image: © Lotte Westphael; photo by Erik Balle*

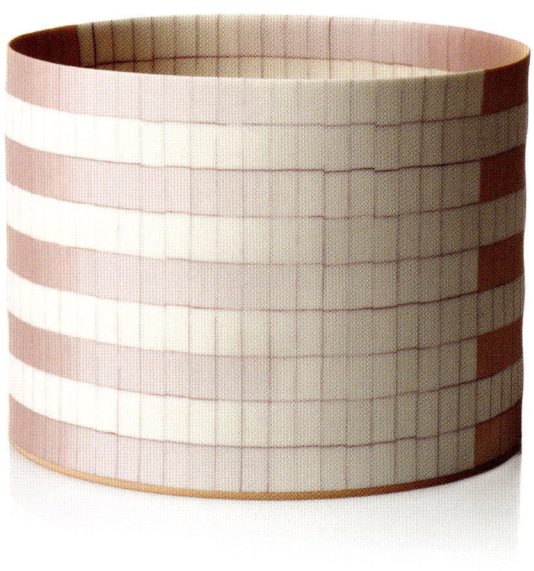

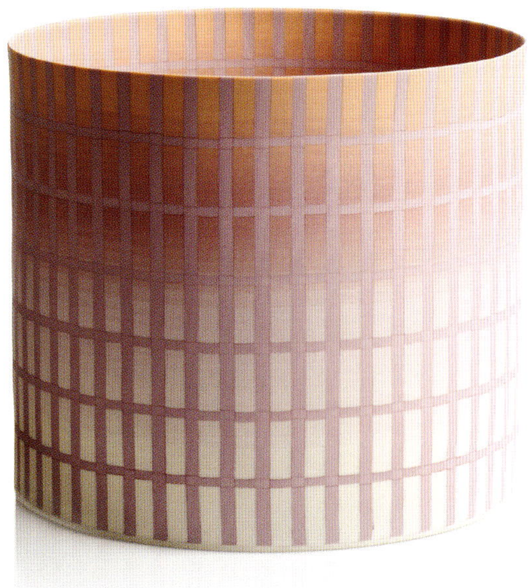

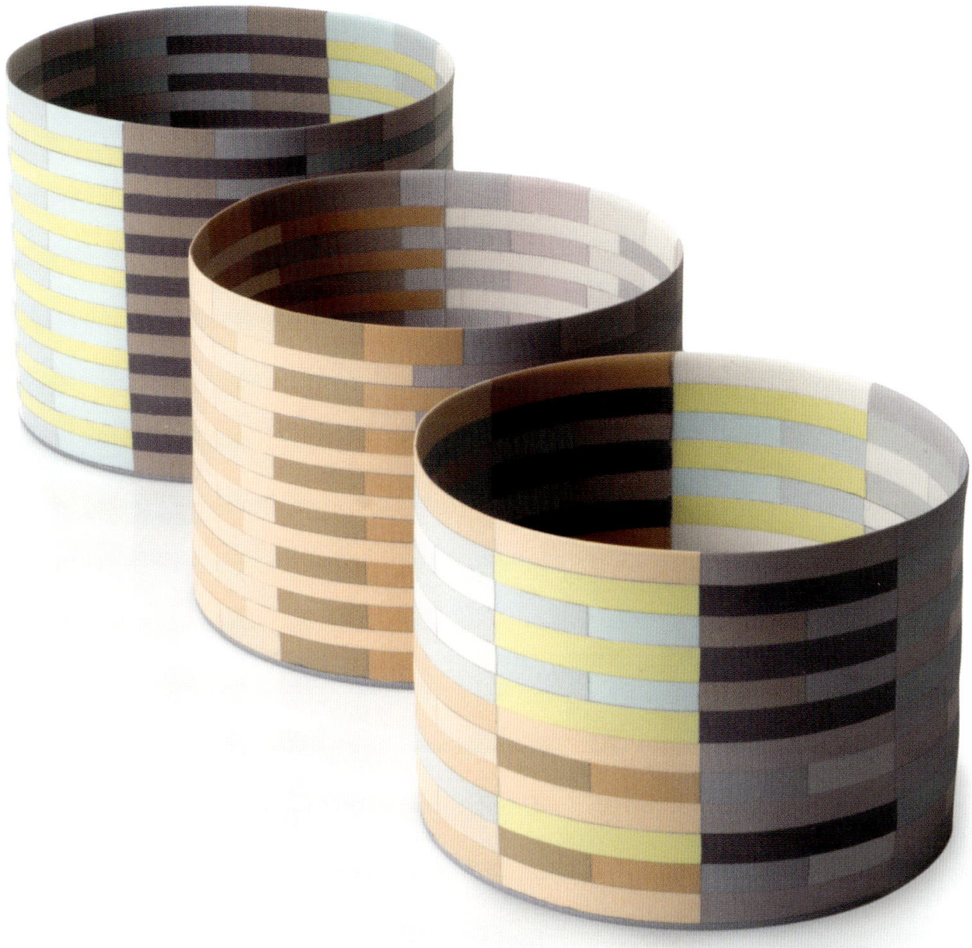

▲ Vessels from rear to front – *Lilac-Yellow Syncope*, 2022. Height: 5⁹⁄₁₀ inches, diameter: 7⅛ inches (15 x 18 cm); *Lilac-Rust Syncope*, 2022. Height: 5½ inches, diameter: 7⅛ inches (14 x 18 cm); *Color-Up Syncope*, 2022. Height: 5½ inches, diameter: 7⅛ inches (14 x 18 cm). *Image: © Lotte Westphael; photo by Erik Balle*

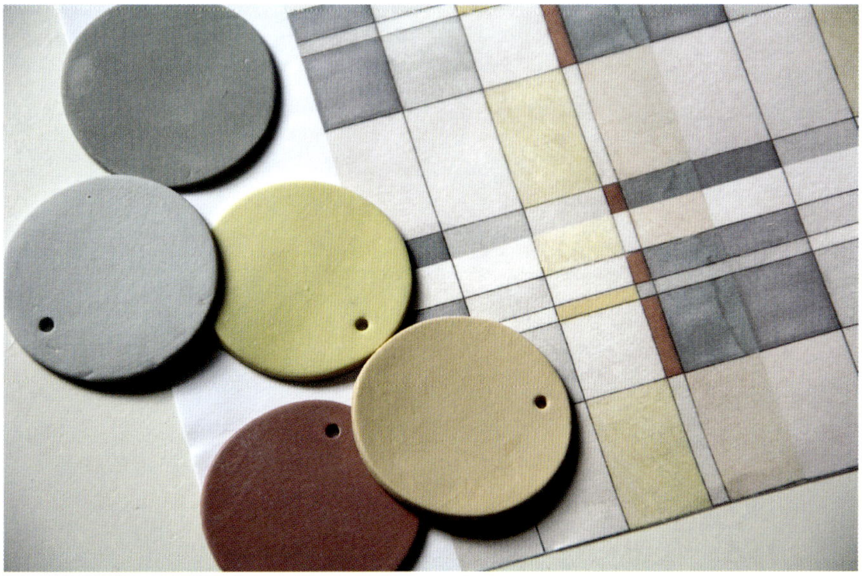

▶ Sketch and color tests. *Image: © Lotte Westphael; photo by Erik Balle*

Henk Wolvers, the Netherlands

Henk Wolvers is a man of many talents, especially with his chosen material, porcelain. Avoiding common pottery-making tools like potter's wheels, slip-casting molds, and even slab rollers, he keeps close to his materials, working with simple hand tools. Wolvers is a consummate record keeper, filling notebooks with firing schedules and test results from the wide variety of porcelain clays he has used. Despite this seemingly scientific side he is also a great experimenter, always looking for new directions or new techniques for working with this ancient Chinese invention. For all these reasons Wolvers has produced sizable collections of work ranging from functional objects to artistic vessels to wall-hung sculptural work and installations. The current selections represent examples of just one of these collections, work with colored porcelain. In the context of this book, Wolvers' creations occupy a space overlapping nerikomi and exploring unnamed areas of experimental colored clay work.

From an early age Wolvers and his three brothers were influenced by their father, an amateur painter. They would travel on holiday to various parts of the country in order to draw, be it a fishing port, river landscape, or heathland scene. His formal education took place at the Academy of Arts (AKI) in Enschede and at the Royal Academy of Art and Design in 's-Hertogenbosch, where he graduated in 1979. In 1994 Wolvers traveled to Japan for a residency in the city of Tokoname, where he worked at the INAX ceramics manufacturing company. The results of his work there netted him the 1994 INAX Design Prize, received by three worldwide artists per year. Since then he has returned to Japan several times. This was followed by a residency at the European Ceramic Work Centre (EKWC), 's-Hertogenbosch, in both 2005 and 2010 and another residency in Jingdezhen, China, in 2012.

Wolvers established his first studio in 1980 and has worked in the city of 's-Hertogenbosch since 1984. A general theme seen in many of his creations, whether with colored clay or not, is his desire to produce the design or pattern first, usually on a large slab on his work table, from which the ultimate three-dimensional form is made. This is the reverse of what many potters do – create the form first and then apply color and/or decoration. For the nerikomi world, Wolvers' approach is almost expected. Many of his works are in the general form of a loose cylinder with a flat base. The naturally undulating edge of a thin rolled slab is often used as the rim of the vessel, giving the piece a casual, flowing feeling. A similar looseness is also seen in his handling of strips of colored clay. While some show examples of slices taken from a patterned loaf – thin waving parallel lines of color embedded in the white porcelain – many are examples of layering, stretching, and rolling the clays which result in organic designs at times reminiscent of underwater flora waving in the ocean currents.

ARTIST'S STATEMENT

I have started to become increasingly driven by a desire for transparency. The fact that porcelain is translucent, that it lets light through, has always been one of the things that gave it its allure and made it so coveted. I want to take that even further, so I started looking for ways to increase and push the boundaries of that transparency. My porcelain is already extremely thin and delicate, but I wanted to take it a step further; so that you can really see through it, and not only see the outside but also the inside of the object, or at least experience it.

becomes one continuous development of invention, creativity, and achievement.

Given the relatively simple definitions of the Japanese terms 'nerikomi' and 'neriage,' the artists in this collection have demonstrated that these two basic concepts can manifest in wildly diverse variations, each artist following their own muse and recreating these techniques with their own distinctive expression. In addition to these two categories I have had the great fortune to discover artists who are opening up new frontiers in the colored clay world.

The work of Living National Treasure Matsui Kōsei (1927–2003) has provided a masterful and visually exciting range of new techniques and styles, all incorporating colored clay. His work has been a direct inspiration to many in this book and for some, a challenge to experiment and discover new ideas and methods.

While further new ways of working colored clay are emerging all the time, it is clear that the heritage of older techniques is kept alive by a number of modern ceramicists. Traditional neriage is best exemplified by Tsuneharu Tanaka's work as well as some of Lorraine Shemesh's, while the Japanese master Ogata Kamio has taken it in his own singular direction through carving textural surfaces. Traditional nerikomi, on the other hand, is seen in some of Dorothy Feibleman's amazingly detailed early work, and it is also seen in Narumi Ii's highly complex and colorful vessels.

Diverging further from these traditions, we have examined the highly intricate yet contemporary patterns and colors of David Pottinger, the more open and naturally inspired designs of Anne Mossman, and the casual mix of loose geometric patterns of Barbara Gittings. Sculptural forms created with colored clay are breaking into new territory with the organic flowing forms of Janny Baek, the intriguing plant and animal hybrids of Aya Murata, and the delicate and translucent sculptural vessels of Françoise Joris. Curtis Benzle often works with hybrid techniques, enhancing his basic nerikomi designs with other methods to create rich multi-layered work which is further enriched by its beautiful translucency. Robert Hessler has also opened new territory in his clever stacking of nerikomi logs to form a cylinder which is then thrown on the wheel, combining both nerikomi and neriage.

We have seen how Jongjin Park has developed an entirely new way of working with colored clay, and he continues to explore numerous design strategies all within his remarkable technique. Lotte Westphael's highly controlled cylinders have shown another unique way of utilizing stained porcelain in often subtle gradations of color, while Mieke Everaet's sculptural vessels have demonstrated a novel approach that expands on mosaic techniques. The multicolor twisted coils of both Janny Baek and Hans Munck Andersen, while not traditional nerikomi, have exhibited yet another strategy of melding color, design, and form into one basic operation. Like other self-taught colored clay artists, we have seen how Ben Davies explores alternate stacking and slicing strategies that are rare or perhaps brand new in the colored clay world.

Common to all this work is the lack of traditional glazes. Colored clay artists in general seem to focus more on controlled designs, or more properly, designs which are imposed on the clay but are then modified in often glorious ways by the plasticity of clay and the forming methods used while the clay is still soft – a true collaboration between artist and medium. The variety of this work and the exciting contemporary innovations all bode well for the future of the art of colored clay.

ACKNOWLEDGMENTS

I wish to thank many who have helped me with this book but foremost is Joan Mirviss and her Associate Director Bonnie Lee, of Joan B Mirviss Gallery. They provided access to information and images from Matsui Kōsei, Ogata Kamio, and Kondō Takahiro. In addition, they established a connection with an important Japanese museum, as well as offering me access to several books on Japanese ceramics. The Japanese Masters chapter would have been impossible without their help. Additionally I wish to thank Nana Onishi of Onishi Gallery for her help in obtaining images and information about Itō Sekisui V. Howard Clegg of The Stratford Gallery provided invaluable help in obtaining images from Nishi Koichi and information about his work and Kay Enokido was very helpful in obtaining additional information about this artist. My thanks to both. I also would like to thank Carol and Jeffrey Horvitz for access to their extensive Japanese ceramics collection and the use of their important collection of books about Japanese ceramics.

Early in my research phase I was given important advice from Dorothy Feibleman who has a wealth of knowledge about nerikomi and the world of ceramics in Japan. She also provided me with the images of Tokoname pottery seen in the chapter "A Colorful History." Good organizational advice was also shared by Anne Mossman early on.

A great deal of gratitude goes to all the worldwide artists included in this book. They shared images of their work as well as biographical information and worked with me closely over many months to get it all right.

Thanks to my editor Russell Butcher for inviting me to write this book and for his guidance along the way. Thanks also to Laura Shaw for creating such a beautiful layout and Anne Halliday for her meticulous copy editing and numerous improvements to my text.

Lastly, I wish to thank my wife Stephanie Hoadley for her support and advice throughout the project. She has also been my in-house editor and aided me in choosing artists for the book.

BIBLIOGRAPHY

Electronic sources

Ayres, Carly, "Beautiful Ceramics Made From...Paper Towels? Jongjin Park's unusual technique offers a fresh take on the medium," available at http://www.core77.com/posts/54024/Beautiful-Ceramics-Made-FromPaper-Towels (accessed November 4, 2022).

Chappell, Troy Dawson, "An English Pottery Heritage: A Survey of Earthenware and Stoneware 1630–1800," Northern Ceramic Society, available at https://www.northernceramicsociety.org/pottery-heritage/ (accessed June 15, 2022).

Dewald, Gabi, "The Porcelain of Maria ten Kortenaar: boogie-woogie feeling," available at https://www.thefreelibrary.com/The+porcelain+of+Maria+ten+Kortenaar%3A+boogie-woogie+feeling.-a0216897108 (accessed November 2, 2022).

D'Souza, Sara, "What Is Agateware and How Is It Created?" available at https://www.thesprucecrafts.com/how-to-create-agateware-ceramics-4141725 (accessed June 15, 2022).

Editors, history.com, "Silk Road," available at https://www.history.com/topics/ancient-middle-east/silk-road (accessed June 25, 2022).

Erickson, Michelle and Hunter, Robert, "Swirls and Whirls: English Agateware Technology," Ceramics in America 2003, available at https://www.chipstone.org/article.php/78/Ceramics-in-America-2003/?s=michelle%20erickson (accessed June 15, 2022).

Gallager, David A.,"English Agateware," available at https://doyle.com/specialists/david-gallager/stories/english-agateware (accessed June 15, 2022).

Hîrdo, Vasi, "David Pottinger," available at https://www.ceramicsnow.org/artists/david-pottinger/ (accessed September 9, 2022).

Hitomi, Kitamura, "The Legacies in the Colors of Japan," available at https://web-japan.org/niponica/niponica30/en/feature/feature04.html (accessed November 28, 2022).

Keys, Hillary, "Painstaking Patterns," available at https://en.tokyomk.com/post/painstaking-patterns (accessed December 20, 2022).

Lesser, Casey, "Cody Hoyt's Ceramic Vessels Break the Mold, and Reap the Rewards," available at https://www.artsy.net/article/editorial-cody-hoyts-ceramic-vessels-break-the-mold (accessed October 10, 2022).

Lotha, Gloria, "Agateware Pottery," *Encyclopedia Britannica*, available at https://www.britannica.com/art/agateware (accessed June 15, 2022).

Maas, Anna, "David Pottinger," available at https://www.skepsigallery.com.au/pottinger-david (accessed September 8, 2022).

Madlener, Adrian, "Jongjin Park: Porcelain Paper," *TLmagazine*, available at http://tlmagazine.com/jongjin-park-porcelain-paper/ (accessed November 4, 2022).

Miller, Richard, "Tea Canister," available at http://chipstone.dom5183.com/objects-1/info?query=Portfolios%3D%20%2238%22&sort=0&page=363 (accessed July 6, 2022).

Mirviss, Joan, "Kondō Takahiro," available at https://www.mirviss.com/artists/kondo-takahiro (accessed January 4, 2023).

Rodgers, Bill, "Kondō Takahiro: 'Waves of Ink: Painting in Porcelain' in NYC," available at https://cfileonline.org/exhibition-kondo-takahiro-contemporary-ceramic-art/ (accessed January 4, 2023).

Roos, Dave, "The Silk Road: 8 Goods Traded Along the Ancient Network," available at https://www.history.com/news/silk-road-trade-goods (accessed June 25, 2022).

Shield, Lucille, "Artist Study 2-1: Aya Murata, 11.18.16," available at https://lucilleshield.wordpress.com/2016/11/16/artist-study-4-aya-murata-11-18-16/ (accessed December 31, 2022).

Stewart, Courtney A., "Tracing the Development of Ceramics along the Silk Road," available at https://www.metmuseum.org/blogs/ruminations/2016/ceramics-along-the-silk-road, (accessed June 25, 2022).

Torres, Francesca, "Nerikomi Pottery: Japan Art of Ceramics Creation," available at https://craftshero.com/nerikomi-pottery/ (accessed July 6, 2022).

Truong, Alain, "Exhibition of Forty New Works by Kondō Takahiro Opens at Joan B Mirviss Ltd," available at http://www.alaintruong.com/archives/2016/09/13/34317106.html (accessed January 4, 2023).

Tung, Wu, "Beyond the Exhibit – EARTH TRANSFORMED: Chinese Ceramics in the Museum of Fine Arts, Boston," available at http://www.antiquesandfineart.com/articles/article.cfm?request=342 (accessed June 25, 2022).

Vreugdenhil, Zeniya, "Structured Rhythms: An Interview with Lotte Westphael," *TLmagazine*, available at https://tlmagazine.com/structured-rhythms-an-interview-with-lotte-westphael/ (accessed November 17, 2022).

Walthier, J.T., "Francoise Joris–Artist Profile," available at https://jtwceramics.com/2020/12/23/francoise-joris-artist-profile/ (accessed October 28, 2022).

Westphael, Lotte, ISSUU, available at https://issuu.com/lottewestphael/docs/studio-talk-2021-online/s/14711524 (accessed November 17, 2022).

Wodelonde, Morgaina, "An Exploration of Twisted Placenta Ceramics," available at https://athenaeum.baronyofmadrone.net/exhibits/an-exploration-of-twisted-placenta-ceramics/ (accessed July 5, 2022).

Yamaguchi, Lena, "Discover Tokoname the Home of Pottery and Beckoning Cats," available at https://www.nagoyaisnotboring.com/tokoname-pottery-and-beckoning-cats/ (accessed August 27, 2022).

Yasuhiro, Aoki, "A Master in Red Clay: Ceramic Artist Itō Sekisui V," available at https://www.nippon.com/en/guide-to-japan/b08505/?pnum=1 (accessed November 28, 2022).

Yi, Ji Hyun, Lee, Jyunsoo, and Kim, Songei, "An analysis of the appearance characteristics of Korean ceramics per era through statistical analysis of metadata annotated with a visual element classification system of ceramics," Heritage Science, available at https://heritagesciencejournal.springeropen.com (accessed June 27, 2022).

Anonymous online publications

"3 Great Sado Museums," available at https://gowithguide.com/blog/3-great-sado-museums-1268 (accessed December 4, 2022).

"3,500 years ago in ancient Egypt, this basic material was treasure," available at https://www.inverse.com/culture/ancient-origins-of-glass (accessed June 25, 2022).

"A Master in Red Clay: Ceramic Artist Itō Sekisui V," available at https://www.nippon.com/en/guide-to-japan/b08505/?pnum=2 (accessed December 4, 2022).

"Agateware," available at https://apps.jefpat.maryland.gov/diagnostic/ColonialCeramics/Colonial%20Ware%20Descriptions/Agateware.html (accessed June 16, 2022).

"Arching Oaks Arts & Cultural Center – Pottery," available at https://www.archingoaks.org/pottery.html (accessed June 25, 2022).

"Aya Murata," available at https://en.yodgallery.com/copy-of-en-artist-murata-1 (accessed December 31, 2022).

"Ceramic Carriers of Culture: The Impact of the Silk Road," available at https://carriersofculture.wordpress.com/the-impact-of-the-silk-road/ (accessed June 25, 2022).

"Ceramic Pillow," available at https://www.khanacademy.org/humanities/art-asia/imperial-china/song-dynasty/a/ceramic-pillow (accessed July 6, 2022).

"Ceramics from Different Countries," available at https://vk.com/wall-46365682_23787?lang=en (accessed November 22, 2022).

"David Pottinger," available at https://scottish-gallery.co.uk/artists/david-pottinger/ (accessed at September 8, 2022).

"Did You Know?: The Silk Roads Glass Trade in China and South East Asia," available at https://en.unesco.org/silkroad/content/did-you-know-silk-roads-glass-trade-china-and-south-east-asia (accessed June 25, 2022).

"Dorothee Wenz – Ceramicist – Schwabenheim, Germany," available at https://www.homofaber.com/en/discover/artisans-dorothee-wenz (accessed November 15, 2022).

"Dorothee Wenz Crazy Ceramics," available at https://www.craft2eu.net/en/artists/dorothee-wenz (accessed November 15, 2022).

"Earth & Alchemy Exhibition and Special Events," available at https://blogs.massart.edu/earthandalchemy/artists/ogata-kamio-japanese/ (accessed December 20, 2022).

"Egyptian glass – Glass making in Egypt," available at http://www.historyofglass.com/glass-invention/egyptian-glass/ (accessed June 25, 2022).

"Françoise Joris – Porcelain Maker," available at https://www.homofaber.com/en/discover/artisans-francoise-joris (accessed October 28, 2022).

"Glass and Faience in Ancient Egypt," available at https://factsanddetails.com/world/cat56/sub365/entry-6138.html#chapter-15 (accessed June 25, 2022).

"Kondō Takahiro," available at https://www.tokyo-gallery.com/en/artists/%E8%BF%91%E8%97%A4%E9%AB%98%E5%BC%98.html (accessed January 4, 2023).

"Living National Treasure Matsui Kōsei Shoretsu Neriage Vase," available at https://www.trocadero.com/stores/QualityChadoguJP/items/1355239/Living-National- Treasure-Matsui-Kosei-Shoretsu-Neriage (accessed November 22, 2022).

"Maria ten Kortenaar," available at https://www.kelliemillerarts.com/artistslist/maria-ten-kortenaar-artist-bio (accessed November 2, 2022).

"Matsui Kōsei," available at https://japanese-ceramics.com/matsui-kosei-%E6%9D%BE%E4%BA%95%E5%BA%B7%E6%88%90/ (accessed November 22, 2022).

"Matsui Kōsei," available at https://www.mirviss.com/artists/matsui-kosei (accessed November 22, 2022).

"Matsui Kōsei," available at https://www.shibuibkk.com/ceramics/matsui-kosei-1 (accessed November 23, 2022).

"Matsui Kōsei, Japan," available at https://loveforclay.tumblr.com/post/54250794197/matsui-k%C3%4sei-japan (accessed November 22, 2022).

"Matsui Kōsei Memorial Exhibition," available at https://corp.shiseido.com/art-house/en/exhibit/backward/2003.html (accessed November 22, 2022).

"Neriage Tsubo by LNT Kōsei Matsui," available at https://www.shirakuragallery.com/portfolio/neriage-tsubo-by-lnt-kosei-matsui?v=597d272a0a74 (accessed November 22, 2022).

"Nerikomi," available at https://en.wikipedia.org/wiki/Nerikomi (accessed August 27, 2022).

"Nerikomi Pottery: Japan Art of Ceramics Creation," available at https://craftshero.com/nerikomi-pottery/ (accessed August 27, 2022).

"Pottery Marbled Dish," available at http://www.benjanssens.com/portfolio/pottery-marbled-dish-2/ (accessed June 24, 2022).

"Puls Ceramics–Solo Show," available at http://www.pulsceramics.com/exhibitions/jongjin-park-2017/ (accessed November 4, 2022).

"Red Clay from the Gold Mines of Sado Island," available at https://www.teaforum.org/viewtopic.php?t=599 (accessed December 4, 2022).

"Special Gallery: Matsui Kōsei and the Art of Neriage," available at https://www.moaart.or.jp/en/?event=matsuikosei (accessed November 22, 2022).

"Stoneware on the Silk Roads: Ceramics from the Changsha Kilns," available at https://lammuseum.wfu.edu/exhibits/virtual/stoneware-on-the-silk-roads-ceramics-from-the-changsha-kilns/ (accessed June 25, 2022).

"Super Neriage Kamio Ogata," available at https://ogatak-m.jimdo.com/english/profile/ (accessed December 20, 2022).

"The Goryeo Dynasty (918–1392)," available at https://www.khanacademy.org/humanities/art-asia/korea-japan/goryeo-dynasty/a/the-goryeo-dynasty-9181392 (accessed July 5, 2022).

"The Silk Roads," available at https://education.nationalgeographic.org/resource/silk-roads (accessed June 25, 2022).

"Three Giants of the North: Kamoda Shōji, Matsui Kōsei, Wada Morihiro," available at https://www.mirviss.com/exhibitions/three-giants-of-the-north-kamoda-shoji-matsui-kosei-wada-morihiro (accessed November 22, 2022).

"Tokoname Ware Guide: Japanese Ceramics (Pottery)," available at https://en.thebecos.com/blogs/column/tokoname-ware-guide-japanese-ceramics-pottery/ (accessed August 27, 2022).

"Treasure in Matsui's Technique: Pattern with neriage technique that is thrown on a potter's wheel," available at https://www.us.emb-japan.go.jp/jicc/exhibits/crafted-treasures-matsui-kosei.html (accessed November 22, 2022).

"Works of Matsui Kōsei," available at https://kasamashiko.style/en/point/detail/40/ (accessed November 22, 2022).

Books and publications

Baek, Janny, "Chasing New Forms in Art and Life," Studio Talk, *Ceramics Monthly*, May 2022.

Baekeland, Frederick, translator, *Modern Japanese Ceramics in American Collections*, Japan Society, Inc., New York, NY, 1993.

Batrouney, Colin, "David Pottinger," *Ceramics Now*, September 2021.

Benzle, Curtis, "Kintsugi and Object History," *Pottery Making*, November/December 2016.

Bincsik, Monika, essay included in the exhibition catalog "Itō Sekisui V: Red Soil," Onishi Gallery, New York, NY, 2017.

Burkett, Richard, *Masters: Porcelain: Major Works by Leading Ceramists*, Union Square Co., New York, NY, 2008.

Chief Curator, Crafts Gallery, The National Museum of Modern Art, Tokyo, exhibition publication *The Works of Kōsei Matsui 1985–1990*, Kodansha, Japan, 1990.

Earl, Joe, Gotleib, Rachel, Welch, Matthew, and Adamson, Glen, exhibition publication *Transcendent Clay*, Lowe Art Museum, Miami, 2023.

Gittings, Barbara, "Masterclass with Barbara Gittings," *Ceramics Review*, May/June 2020.

Hansen, Valerie, *The Silk Road: A New History,* Oxford University Press, New York, NY, 2012.

Hasebe, Mitsuhiko, *Toh-Vol. 89: Kōsei Matsui*, Kyoto Soon Co. Ltd., Kyoto, 1993.

Hessler, Robert and Vida, Martha B., "Crystalline Glaze From the 19th Century Through Today: A Brief History and a Conversation with Studio Potter, Robert Hessler," *Journal to the American Art Pottery Association*, Fall 2020, Vol. 36, No. 4.

Ikutaro, Itoh, *Korean Ceramics,* Metropolitan Museum of Art, New York, NY, 2000.

Kemske, Bonnie, *Kintsugi: The Poetic Mend*, Herbert Press, London, 2021.

Kondō, Takahiro, *Kondo Takahiro: Vessel-Body-Void*, Mitsumura Suiko Shoin Publishing Co., Ltd., Kyoto, 2022.

McManus, Austin, "Cody Hoyt: The Long View," *Juxtapoz – Art & Culture*, May 2017.

Miller, Brian E., "Caressing Light," *Ceramic Review*, 226, July/August 2007.

Mossman, Anne, "Nerikomi: A Focus on Coloured Clay, the Process and Artists," *Journal of Australian Ceramics*, November 2018.

North, Alice, North, Halsey, and Cort, Louise Allison, *Listening to Clay: Conversations with Contemporary Japanese Ceramic Artists*, Monacelli, A. Phaidon Company, New York, NY, 2022.

Pak, Youngsook and Whitfield, Roderick, *Book of Korean Art, Earthenware and Celadon,* Yekyong Publishing Co., Seoul, South Korea, 2002.

Piepenburg, Robert, "A Visit with Beauty," *Ceramics Monthly*, March 2018.

Sato, Masahiko, *Chinese Ceramics: A Short History,* Weatherhill/Heibonsha, New York, NY, 1981.

Saunders, Tim, "Barbara Gittings – Profile," *New Ceramics*, July/August 2021.

Schoenmann, Evelyne, "In Studio with Angela Burkhardt-Guallini," *New Ceramics*, May/June 2020.

Spielmann, Prof. Dr. Heinz, *Gerd Hiort Petersen & Hans Munck Andersen*, Tidendes Tryk, Ronne, Denmark, 1999.

Suidman-Leeuwendal, leneke, "Henk Wolvers' Powerful Delicacy of Porcelain," *Ceramics: Art and Perception* No. 60, 2005.

Taylor, Louisa, *Ceramics Masterclass*, Thames & Hudson, London, 2020.

unknown, "Form Follows Function," *Ceramic Review*, September/October 2014.

unknown, "Structurally Sound," *American Craft Magazine*, December/January 2015.

Vink, Netty, "Henk Wolvers," *New Ceramics*, November/December 2018.

Warren, Larissa, "Smooth Lines," *Pottery Making Illustrated*, September/October 2019.

Warren, Larissa, "Wild Women, Wild Clay-Women Potters of Tamborine Mountain 1940–1980," *Journal of Australian Ceramics*, Vol. 59 No. 3, November 2020.

Watson, William, *Tang and Liao Ceramics*, Rizzoli International Publications, Inc., New York, NY, 1984.

Yates, Christopher A., "Curtis Benzle and Translucent Porcelain," published in Tarkasis, Kosta, *Clay in Art: International Yearbook*, Clayart International, Greece, 1987.

GLOSSARY

Agateware: Style of pottery in England starting in the seventeenth century, made from combining different colored clays to make patterns or designs in the clay itself, usually finished with a transparent glaze.

Bisque firing: A low-temperature firing that makes the clay permanent yet still porous so it can accept a glaze. This is followed by a high-temperature firing to melt the glaze and harden the clay body.

Cone: A small cone or wedge-shaped form made from ceramic materials designed to bend at specific temperatures. Cones are given specific numbers corresponding to certain temperatures. They can automatically turn off a kiln when it reaches a certain temperature or be seen through spy holes to judge when to manually end the firing.

Core method: Ancient Egyptian glass-making technique for making hollow forms before the development of glass blowing.

Jiaotai (mixed clay): Chinese term for pottery made from colored clay bodies.

Kintsugi: Japanese technique of mending cracked pottery with resin and gold.

Leather-hard: Semi-hard state of pottery as it slowly dries, soft enough to easily carve.

Mason stain: An American brand of colored stains used in the ceramics industry and by studio potters.

Mishima: A Japanese pottery technique of filling carved lines or impressed patterns in clay with a contrasting colored-clay slip.

Mumyoi-yaki: A type of pottery made from mumyoi, a reddish-brown ferric oxide clay extracted from gold mines native to Sado Island, Japan.

Neriage: A Japanese pottery technique involving colored clays that are gently mixed together and thrown on the wheel – from the root words "neri" (to mix or knead) and "age" (to pull up).

Nerikomi: A Japanese pottery technique involving colored clays that are stacked or joined, often in elaborately planned designs, to form a block or loaf of soft clay which is then sliced. Individual slices are joined in a variety of ways to form pottery or sculpture.

Paperclay: Sometimes referred to as fiberclay, paperclay is any clay body to which processed cellulose fiber (paper being the most common) has been added.

Pottery: A broad term used to describe vessels made from clay and fired to a temperature high enough to make the clay hard and permanent. This term includes the categories of earthenware, porcelain, and stoneware.

Shellac: A common wood, finishing material made from naturally occurring resin.

Silver mist (gintekisai) glaze: A specialized glaze-like coating, conceived and formulated by Kondō Takahiro, which when fired resembles water droplets.

Slip: Fluid clay the consistency of cream made from clay and water, sometimes colored with ceramic stains.

Sometsuke: The Japanese tradition of underglaze cobalt blue decoration, usually added to porcelain ceramic forms and painted with calligraphy brushes.

THOMAS HOADLEY

Resumé and Publications

EDUCATION

1970 Skowhegan School of Painting and Sculpture, Skowhegan, Maine

1971 Amherst College, Amherst, Massachusetts, BA cum laude

1974 Apprentice to Malcolm Wright, Marlboro, Vermont

1977 Illinois State University, Normal, Illinois, MS in Ceramics

NATIONAL EXHIBITIONS

1978 "Young Americans: Clay/Glass" Tuscon Museum of Art, Tuscon, Arizona. Museum of Contemporary Crafts, New York, NY

1980 "Westwood Clay National" Otis Art Institute, Los Angeles, CA

1980 "American Porcelain" Renwick Gallery, Washington, D.C.

1982 "Scripps College 38th Ceramic Annual" Scripps College, Claremont, CA

1982 "Westwood Clay National 1982" Downey Museum of Art, Downey, CA

1983 "Masterworks – Glass, Wood, Clay, Fiber" Jesse Besser Museum, Alpena, MI

1985 "Ceramic Invitational" The Lawrence Gallery, Portland, OR

1986 "Designed and Made for Use" American Craft Museum, New York, NY

1986 "International Ceramics Festival Mino" Tajimi City, Japan

1990 "Art That Works: Decorative Arts of the Eighties, Crafted in America" Mint Museum of Art, Charlotte, NC, and 13 other cities

1992 "American Crafts: The Nation's Collection" Renwick Gallery of the National Museum of American Art, Washington, DC

1993 White House Collection of American Crafts, Washington, DC

1995, 1996 "The White House Collection of American Crafts" National Museum of American Art, Washington, DC: also Rochester, NY; New York, NY; Springfield, MA; Los Angeles, CA; other cities

1995 "International Ceramics Festival Mino" Tajimi City, Japan

1998 The Fletcher Challenge, Auckland, New Zealand

1998 "Re:FORM," Fuller Museum of Art, Brockton, MA

2000, 2004 "Armory National Ceramics Invitational" Armory Art Center, West Palm Beach, FL

2003 "Vessels, Boxes and Baskets" Function + Art, Chicago, IL

2003 "21st Century Ceramics in the United States and Canada" Columbus College of Art and Design, Columbus, OH

2003 "The Power of Design and Art" The Museum of Modern Ceramic Art, Gifu, Japan

2004 "San Angelo National Ceramic Competition" San Angelo Museum of Fine Arts, San Angelo, TX

2004 "Creations in Clay: Contemporary New England Ceramics" Currier Museum of Art, Manchester, NH

2005 "Ceramics Invitational 2005," Texas A&M University, Commerce, TX

2011 "International Ceramics Festival '11, Mino" Tajimi City, Japan

SOLO AND GROUP SHOWS (SELECTED LIST)

1981, 1983 The Jackie Chalkley Gallery, Washington, DC – solo show

1981 "Porcelain Invitational" Holstein Galleries, Palm Beach, FL

1982 "Clay Invitational 1982" DBR Gallery, Cleveland, OH

1982, 1983, 1984 "The Clay Pot" Brooklyn, NY – solo show

1983, 1985 Venture Gallery, Lathrup Village, MI – solo show

1984 Elaine Potter Gallery, San Francisco, CA – solo show

1984 Swan Gallery, Philadelphia, PA – solo show

1985 "Malcolm Wright and Apprentices" Benchmarks Gallery, Washington, DC

1987 Lawrence Gallery, Portland, OR – solo show

1988, 1991, 1993 Mendelson Gallery, Washington Depot, CT – solo show/group clay show

1995 Joanne Rapp Gallery, Scottsdale, AZ – feature

1999 Gallery W.D.O., Charlotte, NC – three-person show

1999 Zazen Gallery, Belleville, WI – group show

2001 Running Ridge Gallery, Santa Fe, NM – solo show

2003 Gallery W.D.O., Charlotte, NC – solo show

2004 "The Fine Art of Ceramics" Guadalupe Arts Center, Austin, TX

2004, 2005 Haddad Lascano Gallery, Great Barrington, MA – group show, paintings

2004 "Coloured Porcelain" St. Joseph Galerie, Leeuwarden, Netherlands – group show

2005 "100 X 2, 10 Years St. Joseph Galerie" St. Joseph Galerie, Leeuwarden, Netherlands

2005 "Naked Clay" Rufford Craft Centre, Newark, Nottinghamshire, UK
2005 Sherrie Gallerie, Columbus, OH – solo show
2006 Lascano Gallery, Great Barrington, MA – group show, paintings
2007 Kasten Fine Art, Great Barrington, MA – group show, paintings
2007 Carrie Haddad Gallery, Hudson, NY – group show, paintings
2009 "Patchwork" Armstrong's Gallery, Pomona, CA – group show
2009 "Out of America" The Gallery at Bevere, Worcester, UK – three-person show
2009, 2010, 2011 Hoadley Gallery, Lenox, MA – two-person show, paintings
2011 "New England Collective II" Galatea Fine Art, Boston, MA
2012 "PSA National Ceramics Invitational" Paducah School of Art, Paducah, KY
2017 "Three Masters in Porcelain" Sherrie Gallerie, Colombus, OH

COLLECTIONS

American Museum of Ceramic Art, Pomona, CA
Crocker Art Museum, Sacramento, CA
Delaware Art Museum, Wilmington, DE
Henry Ford College, Dearborn, MI
Los Angeles County Museum of Art, Los Angeles, CA
MCI Collection, Washington, DC
Mint Museum, Charlotte, NC
Mobile Museum of Art, Mobile, AL
Museum of Fine Arts, Boston, MA
Museum of Modern Ceramic Art, Gifu, Japan
National Museum of American Art, Renwick Gallery, Washington, DC
Philadelphia Museum of Art, Philadelphia, PA
Racine Art Museum, Racine, WI
San Angelo Museum of Fine Arts, San Angelo, TX
University of Iowa Museum of Art, Iowa City, IA
William J. Clinton Presidential Library and Museum, Little Rock, AR
Yale University, New Haven, CT

AWARDS AND HONORS

1981 Massachusetts Artists Fellowship Finalist
1981 Philadelphia Craft Show – Judges' Award, Craftsmen's Award
1982 Philadelphia Craft Show – Judges' Award
1985 Massachusetts Artists Fellowship, The Artists Foundation, Boston, MA
1985 American Craft Museum Design Award –"Designed and Made for Use" competition
1986 International Ceramics Festival '86, Mino, Japan – Honorable Mention
1988 Philadelphia Craft Show – Franklin Mint Ceramics Award
1989 The Washington Craft Show – Award of Excellence
1989 Massachusetts Artists Fellowship Finalist
1990, 1992 National Endowment for the Arts – Visual Artists Fellowship
1993 Porcelain Bowl chosen by Joan Mondale to be shown in the US Ambassador's residence in Tokyo
1993 Large Porcelain Bowl chosen to be part of the White House Collection of American Crafts, now in the collection of the William J. Clinton Presidential Library and Museum in Little Rock, AR
1995 International Ceramics Festival '95, Mino, Tajima City, Japan – Bronze Award
2011 International Ceramics Festival '11, Mino, Tajima City, Japan – Special Judge's Award

PUBLICATIONS FEATURING THOMAS HOADLEY'S WORK

American Ceramics, 1 Apr 1982.
American Craft, Aug/Sept 1984.
Axel, J. and McCready, K. *Porcelain, Traditions and New Visions*. New York: Watson Guptil, 1981.
Birkes, T. *The Complete Potter's Companion*. London: Conran Octopus, 1993.
Ceramics Monthly, Sept 1983.
Daniels, D. L. and Lynn, M. D. *The Vase and Beyond: The Sidney Swidler Collection of the Contemporary Vessel*. Sacramento: Crocker Art Museum, 2011.
Deitz, P. "American Pottery," *The New York Times Magazine, Home Design*, 13 Apr 1986.
Frith, D. E. *Moldmaking for Ceramics*. Boston: Chilton Book Co., 1985.
Herman, L. E. *American Porcelain, New Expressions in an Ancient Art*. Portland: Timber Press, 1981.
Herman, L. E. *Art That Works*. Seattle: University of Washington Press, 1990.
Hoadley, T. "Pattern and Color," *The Studio Potter*, Dec 1994.
Hopper, R. *Making Marks: Discovering the Ceramic Surface*. Iola: Krause Publications, 2004.
Hunt, B. *21st Century Ceramics*. Westerville: The American Ceramic Society, 2003.

Lane, P. *Ceramic Form, Design and Decoration*. New York: Rizzoli, 1988.

Lane, P. *Contemporary Porcelain*. Boston: Chilton Book Company, 1995.

Lane, P. *Contemporary Porcelain*. 2nd ed. Philadelphia: University of Pennsylvania Press, 2003.

Lynn, M. D. *Clay Today*. San Francisco: Chronicle Books, 1990.

Monroe, M. *The White House Collection of American Crafts*. New York: Abrams, 1995.

Museum of Modern Ceramic Art ed. *The Power of Design and Art*. Gifu: Museum of Modern Ceramic Art, 2002.

Ostermann, M. *The Ceramic Surface*. London: A&C Black, 2002.

Pearson, K. *American Crafts: A Sourcebook for the Home*. New York: Stewart, Tabori & Chang, 1983.

Perryman, J. *Naked Clay: Ceramics Without Glaze*. Philadelphia: University of Pennsylvania Press, 2004.

Shaw, K. "Abstractions recall earth and archaeology," *Berkshires Week*, 5 Aug 2010.

Waller, J. *Colour in Clay*. Ramsbury: The Crowood Press, 1998.

Whyman, C. *The Complete Potter: Porcelain*. Philadelphia: University of Pennsylvania Press, 1994.

Zakin, R. *Ceramics: Mastering the Craft*. 2nd ed. Iola: Krause Publications, 2001.

Zakin, R. *Electric Kiln Ceramics*. 3rd ed. Iola: Krause Publications, 2004.

INDEX

References to images are in *italics*.